BEGINNING STATISTICS

R. Lowell Wine

Hollins College

42679

Winthrop Publishers, Inc.
Cambridge, Massachusetts

Library of Congress Cataloging in Publication Data

Wine, Russell Lowell
  Beginning statistics.

  Bibliography: p. 400
  Includes index.
  1. Statistics. I. Title.
HA29.W536   519.5   75–33409
ISBN 0-87626-062-8

Cover design by David Ford

© *1976 by Winthrop Publishers, Inc.*
    17 Dunster Street, Cambridge, Massachusetts 02138

All rights reserved. No part of this book may be reproduced in any form or by any means without permission in writing from the publisher. Printed in the United States of America. 10 9 8 7 6 5 4 3 2 1

CONTENTS

PREFACE / vii

1 INTRODUCTION / 1

    1.1. The World of Statistics / 1
    1.2. What Is Statistics? / 2
    1.3. What Do Statisticians Do? / 5
    1.4. Opportunities and the Challenge of the Future / 9
    1.5. Important Concepts in Statistics / 11
         Bibliography / 12

2 FREQUENCY DISTRIBUTIONS / 13

    2.1. Grouped Frequency Distribution Tables / 13
    2.2. Construction of Tables for Numerical Data / 16
    2.3. Exercises 2.1–2.12 / 19
    2.4. Graphic Methods of Presenting Data / 22
    2.5. Population and Sample / 28
    2.6. Exercises 2.13–2.22 / 31

3 STATISTICAL MEASURES / 35

    3.1. Introduction / 35
    3.2. Measures of Location / 35
    3.3. Exercises 3.1–3.12 / 39
    3.4. Measures of Location—Grouping and Coding / 43
    3.5. Exercises 3.13–3.26 / 50
    3.6. Summation Properties / 52
    3.7. Exercises 3.27–3.41 / 55
    3.8. Measures of Variation / 57

3.9. Some Applications of the Standard Deviation / 59
3.10. Exercises 3.43–3.52 / 62
3.11. Measures of Variation—Grouping and Coding / 65
3.12. Approximations in Calculations / 68
3.13. Exercises 3.53–3.64 / 73

4 DISCRETE PROBABILITY / 76

4.1. Introduction / 76
4.2. Sample Space / 76
4.3. Exercises 4.1–4.6 / 81
4.4. Events / 81
4.5. Combinations of Events / 84
4.6. Venn Diagrams / 87
4.7. Exercises 4.7–4.20 / 90
4.8. Probability: Finite Number of Equally Likely Outcomes / 93
4.9. Probability of Two or More Events: Equally Likely Outcomes / 95
4.10. Exercises 4.21–4.44 / 103
4.11. Conditional Probability: Equally Likely Outcomes / 106
4.12. Three Definitions of Probability / 111
4.13. Summary of Probability / 118
4.14. Exercises 4.45–4.82 / 119

5 MODEL DISTRIBUTIONS / 125

5.1. Random Variables and Probability Functions / 125
5.2. Formulas for Counting / 130
5.3. Exercises 5.1–5.26 / 133
5.4. Binomial Distributions / 136
5.5. The Mean and the Standard Deviation of Probability Models / 143
5.6. Exercises 5.27–5.66 / 147
5.7. Continuous Distributions of One Variable / 154
5.8. Continuous Uniform Distributions / 157
5.9. Normal Distributions / 158
5.10. Exercises 5.67–5.88 / 165
5.11. Normal Approximations for Binomial Distributions / 168
5.12. Exercises 5.89–5.94 / 173

6 SAMPLING DISTRIBUTIONS / 174

6.1. Introduction / 174
6.2. Sampling Distributions of the Mean / 175
6.3. Central Limit Theorem / 184
6.4. Exercises 6.1–6.22 / 187

Contents

    6.5. Expected Value / 190
    6.6. Sampling Distributions of the Variance and of the Standard Deviation / 191
    6.7. Sampling Distribution of the Difference in Sample Means / 194
    6.8. Exercises 6.23–6.34 / 197
    6.9. Random Numbers and Simulated Experiments / 199
    6.10. Sample Plans / 206
    6.11. Exercises 6.35–6.51 / 209

7 ESTIMATION AND TESTS OF HYPOTHESES / 212

    7.1. Introduction / 212
    7.2. Estimation / 213
    7.3. Properties of Point Estimation / 215
    7.4. Exercises 7.1–7.15 / 220
    7.5. Relationships of Estimates to Sample Size, Error, and Risk / 225
    7.6. Confidence Interval / 231
    7.7. Exercises 7.16–7.31 / 235
    7.8. Tests of Hypotheses / 236
    7.9. Tests of Hypotheses: Two Kinds of Errors / 243
    7.10. Tests of Hypotheses: Summary / 250
    7.11. Tests of Hypotheses: Differences Between Two Means / 252
    7.12. Exercises 7.32–7.51 / 254

8 CHI-SQUARE DISTRIBUTIONS AND THEIR APPLICATIONS / 257

    8.1. Chi-Square Distributions / 257
    8.2. Inferences About Variances and Standard Deviations / 259
    8.3. Exercises 8.1–8.12 / 264
    8.4. Degrees of Freedom / 266
    8.5. Distribution of z^2 / 267
    8.6. Goodness of Fit / 272
    8.7. Exercises 8.13–8.29 / 278
    8.8. Contingency Tables / 280
    8.9. Exercises 8.30–8.44 / 290

9 DISTRIBUTION-FREE METHODS / 296

    9.1. Introduction / 296
    9.2. Tests of Two Correlated Samples / 297
    9.3. Exercises 9.1–9.16 / 305
    9.4. One-Sample Tests / 309
    9.5. Exercises 9.17–9.34 / 318

9.6. Tests for Two Independent Samples / 320
9.7. Summary and Comparisons / 332
9.8. Exercises 9.35–9.52 / 334

10 APPLICATIONS OF STUDENT t DISTRIBUTIONS / 339

10.1. Properties of Student t Distributions / 339
10.2. Applications of t Distributions: Single Sample / 342
10.3. Exercises 10.1–10.10 / 344
10.4. Applications of t Distributions: Two Samples / 345
10.5. Exercises 10.11–10.26 / 354

11 LINEAR REGRESSION / 357

11.1. Introduction / 357
11.2. Fitted Regression Line of y on x / 362
11.3. Distribution of Sample Regression Coefficient b / 372
11.4. Distribution of Adjusted Mean $\hat{\eta}$ / 375
11.5. Exercises 11.1–11.17 / 379

12 CORRELATION ANALYSIS / 383

12.1. The Pearson Coefficient of Correlation / 383
12.2. Inferences About the Correlation Coefficient / 388
12.3. Rank Correlation / 393
12.4. Other Measures of Correlation / 395
12.5. Exercises 12.1–12.18 / 396

ANNOTATED REFERENCES AND BIBLIOGRAPHY / 400

STATISTICAL TABLES / 403

I Poisson Probabilities / 403
II Cumulative Standard Normal Distribution / 405
III Random Digits / 406
IV Percentage Points of Chi-square Distributions / 409
V Percentage Points of Student t Distributions / 411
VI Confidence Belts for the Correlation Coefficient ρ / 412

ANSWERS TO EVEN-NUMBERED EXERCISES / 414

INDEX / 427

PREFACE

This book is designed for an elementary course in statistics for students of liberal arts, the social and physical sciences, and other areas where there is a need to understand some basic principles and methods of statistics. Usually the intuitive approach, accompanied by an illustration, is used to present each concept. In this way an understanding of the underlying assumptions and meanings is developed without introducing mathematics beyond some simple algebra.

A strong and sustained effort has been made to present statistics (1) as a lively and timely subject worthy of study for its own special use and appeal and (2) with such simplicity and clarity as to make the material almost self-instructive. Formulas are stated as abbreviated descriptions of concepts or procedures, but they are not derived. However, conditions under which formulas may be applied are usually stated and illustrated.

Each chapter except the first contains a copious supply of exercises which are applicable to many and widely different fields. Exercises are designed to give the readers practice in applying the material of the text, to extend both their practical and theoretical concepts of statistics, and to encourage them to look carefully at new concepts which are outlined only in the exercises. Many exercises are directly related to the material of sections of the chapter, but some exercises introduce new concepts and are designated by ^{E}. Exercises that should be done with the aid of a desk calculator or computer are designated by ^{C} and those which are difficult are designated by *.

A broad picture of what statistics is, what statisticians do, and what methods are used is presented in Chapter 1. Basic ways of describing data are discussed in Chapters 2 and 3. Chapters 4 and 5 introduce some concepts of probability and a few probability distributions. The remaining seven chapters deal with the principles and methods of statistical inference which, in the opinion of the author, are central to an understanding of statistics. Chapters 3, 5, 6, and 7 are necessary for an understanding of the last five chapters. However, Chapters 8, 9, and 10 are largely independent of each other.

In colleges and universities this book may be used for a one-semester or a two-quarter course. Almost all of the material may be covered in a two-quarter course. For a one-semester course there is enough material to satisfy many different interests and needs. The author, in a one-semester course, takes one class period on Chapter 1, usually abbreviates Chapter 2, takes about three periods on selected parts of Chapter 4 (such as concept of sample space, probability of events for equally likely outcomes, much of Section 4.12, and Section 4.13), covers most of Chapters 3, 5, 6, 7, and two other chapters.

Many persons have contributed to the development of this book. I wish to acknowledge my gratitude to the several classes of high school and college students who have read and made helpful suggestions on parts of the manuscript; to each reviewer who has made comments and criticisms; and to other persons who have published data which are reproduced and acknowledged at the appropriate places in the text. Also, I am indebted to the College Science Improvement Program of the National Science Foundation and to the Hollins College Travel and Research Committee for funds used in developing and typing the manuscript. Further, my appreciation goes to Sue Ross and Mrs. Lewis Whitescarver for their efficient and tireless assistance in typing the manuscript. My greatest personal indebtedness is to my wife, Ruth Wine, without whose patient understanding the book could not have been written.

I am indebted to the *Biometrika* trustees for permission to reproduce parts of Tables VIII and XV from their *Biometrika Tables for Statisticians*. I am grateful to the Literary Executor of the late Sir Ronald A. Fisher, F.R.S., to Dr. Frank Yates, F.R.S., and to Longman Group Ltd., London, for permission to reprint part of Table III from their book *Statistical Tables for Biological, Agricultural and Medical Research* (6th ed., 1974). Also, I wish to express my appreciation for permission to reproduce Sections 1.1–1.5, Chapter 1, from *Careers in Statistics*; Table 9.3 from a publication by the American Cyanamid Company; Table 9.5 from the *Journal of the American Statistical Association*; Table 9.6 from the *Annals of Mathematical Statistics*; Table 9.9 from *Biometrics*; and Tables I and III from the *Handbook of Tables for Probability and Statistics*.

To my parents

1
INTRODUCTION

In this chapter an attempt is made to give a broad picture of what statistics is, what statisticians do, and what methods are used. Sections 1.1, 1.2, and 1.5 are quotations from Sections I, II, and V of *Careers in Statistics,** second edition (1966) and Sections 1.3 and 1.4 are quotations from pages 6 through 13 of *Careers in Statistics** (1973). Each of these booklets was prepared by several noted statisticians under the auspices of the Committee of Presidents of Statistical Societies—the American Statistical Association, the Institute of Mathematical Statistics, and the Biometrics Society (Eastern and Western North American Regions)—and published by the American Statistical Association, 806 15th Street, N.W., Washington, D.C.

1.1. THE WORLD OF STATISTICS

"The average person thinks of 'statistics' as the columns of figures on sports pages or in the business section of newspapers, and illustrated with zigzag graphs, stacks of dollars, and rows of people.

"This may have been a good picture years ago. But the statistician of today is not one who compiles tables of figures and illustrates them graphically. He works on scientific and professional tasks that are exciting and complicated. His job is to help design experiments, interpret the data from the observations that are made, and to set forth the results so as to make it easier to reach reasonable decisions for action.

"The ideas and instruments of modern statistical methods are useful for investigating many kinds of problems. Here are some examples of recent questions in which statistical methods helped to find the answers.

"How does one test the effectiveness of a vaccine for polio? How can one tell the difference between an atomic bomb explosion and a small earthquake from a distance of several thousand miles? How may you distinguish between the natural noise in a Geiger counter and regular cosmic radiation? How may

*Reproduced by permission.

one evaluate, in a short space of time, the inventory of materials in process in a large manufacturing concern, with sufficient precision for the corporation's financial statement? How can one tell whether the recent change in the Consumer Price Index is a seasonal fluctuation or a small random deviation? How much, if any, does smoking cigarettes increase your chances of having lung cancer? What is the effect of overweight on longevity of life?

"Modern statistical methods help to answer such questions. And practically every day the statistician encounters entirely new problems, or new variations of old ones, which require the development of new statistical methods or the adaptation of known methods.

"How many rounds of ammunition should one fire in order to test a rifle? How may one interview several hundred or a thousand people and closely estimate the number of persons in the nation who are working on farms, or the number having a specified illness, or the number who currently prefer one candidate for President rather than another? What is the condition of the crop of wheat this year? How many mass-produced light bulbs in a large lot are defective? Statistical methods deal with such questions.

"Skills, funds, instruments, and other facilities are always limited in any research endeavor, whether it be in industry, a government statistical agency, an agricultural experiment station, or an institute for medical research. The problem is how to extract the greatest amount of information within the limits of available resources. The work of the statistician is to provide guidance in laying out a plan of investigation that will allocate manpower, skills, and other resources into the various areas, materials, or subjects to be studied, so as to achieve the required precision and answers at lowest cost.

"Originally the word 'statistics' was derived from the word 'state,' and it is still true that governments find increasingly important uses for counting or measuring all kinds of things and activities. The government needs to keep track of the population, births, deaths, the employed, the unemployed, business establishments, cost of living, and many other characteristics of our society. Despite the fact that we have had many years of experience in collecting such data, there continue to be complications and opportunities for scientific contributions in collecting, processing, and interpreting such data efficiently. Today many kinds of activities are concerned with statistics, and many kinds of occupations involve the use of statistical methods."

1.2. WHAT IS STATISTICS?

"A current official classification document of the Civil Service Commission says that 'statistics means the science of the collection, classification, and measured evaluation of facts as a basis for inference.' This is elaborated as 'a body of techniques for acquiring accurate knowledge from incomplete information, a scientific system for the collection ... organization, analysis, interpretation, and presentation of information which can be stated in numerical form.'

"Over the past fifty years statisticians have developed a science out of their profession. They have been able to provide a logical basis to a variety of procedures for assisting in making decisions when less than the total information about a problem is available.

"Why should anyone tolerate incomplete or anything less than total information? It is sometimes *desirable*, for various reasons, to settle for less, and at other times it is a harsh *necessity*. Take, for instance, a manufacturer of footballs. He may use statistical methods to develop a better football, and also to determine the right amount of inspection of his product to keep the outgoing quality high. For the latter, he would like to have assurance of the quality of his product and yet spend as little as possible on inspecting the footballs that he makes. Doctors must have ways of determining the effectiveness of a new drug without waiting a period of many years for it to be tried on every single person who might have the disease, as well as assessing its potentially harmful effects.

"Moreover, a manufacturer of television picture tubes cannot obtain exact information on the length of life of every one of his tubes unless he is willing to end up with no tubes to sell. One cannot fire every bullet to decide how many will be duds. Methods that do not require the examination of every item are essential if these manufacturers are going to have a product to sell. The manufacturer may be willing to utilize a fraction of his product to estimate the proportion of defectives, and this is where statistical methods provide guidance.

"In practical situations a decision *has* to be made and *will* be made in regard to the quality of the footballs, the effectiveness of the drug, and the life of the television tube. One contribution of statistics is to provide a procedure that helps to minimize the net economic loss from the risks that confront the decisions to be made. One of the earliest uses of statistics was in connection with the development of the actuarial basis of insurance, and statistical training is still an important part of the actuary's training. Although actuarial science begins with the concept of risk, this concept in a much broader sense permeates also practically every decision that one must make in science, in business, or in government, and has had considerable attention paid to it by statisticians. For example, in the design of a survey, one could easily design the sample to be so complex and large that it delivers more precision than is necessary. It is then too costly, as well as being subject to special errors that accompany large-scale operations. On the other hand, one could as easily design a survey to deliver so little precision that it settles nothing. The cost of the survey is then a dead loss. The statistician, through his knowledge of the theory of probability, saves money by designing samples so that the investigator gets nearly the precision he requires, and is able to compute afterward the precision achieved, not as opinion, but as a mathematical consequence.

"There is also the matter of economy of time, effort, and money in making observations. Some observations, like reading a clinical thermometer, may be quick and inexpensive to make but others may take years and cost thousands of dollars each. Agricultural experiments, for instance, may take years to complete. Whether it be a laboratory experiment or a survey of people, some arrangements of the investigation will be better than others for getting information—they will yield more precise results for a given number of observations. Or looking at the matter the other way around, certain designs will allow a required level of precision with fewer observations. The proper *design of experiments and surveys* is a field to which modern statistics makes many contributions.

"In almost every scientific investigation, the observations that are made give only a portion of all possible information on the point at issue. It is essential that this portion, referred to as a 'sample' be selected in such a way as to give useful information about the whole. It is equally essential to use the theory of sampling for drawing proper conclusions from the data. Dependable methods of sampling and of drawing inferences are important features of the statistical process.

"An equally important aspect of statistics is the tabulation of samples of complete aggregates and the development of appropriate and useful classes and subclassifications. These tasks are perhaps best represented by the government's complete Censuses of Population, Housing, Manufactures, and Agriculture, but are paralleled in industry by complete compilations of sales, production, inventories, payrolls and other internal data. Statistical tables so produced are useful tools for management, but they also are statistics which may be compared with similar data, or analyzed and projected as time series.

"The field of statistics also encompasses other aspects of research and the management of operations. Thus, mathematically-oriented statisticians may spend their time on the invention of important mathematical descriptions of physical and social phenomena. Most of us are familiar with the fact that mathematics provides a theoretical framework for many of the sciences. In fact, you probably know some of the mathematical laws governing electricity, light, magnetism, energy, projectiles, and satellites. It is not so well known, however, that other mathematical theories are currently being developed in nearly every field of endeavor by persons who are trained in statistical theory and probability. To mention a few:

> in biology, for theories of mutation and inheritance;
>
> in psychology, for theories of learning and behavior;
>
> in sociology, for theories explaining migration;
>
> in public health, for theories of epidemics;
>
> in insurance, for theories of the distributions of claims;
>
> in public works, for statistical models to assist in the building of roads and dams.

"Statisticians play an important role in every type of organization, working either independently or in close cooperation with specialists in other fields. Despite the wide divergence of the fields in which they work and the problems which they handle, all statisticians are linked by a common desire to perceive the statistical regularity that exists in situations involving variability and uncertainty. The next section will attempt to present a cross section of them."

1.3. WHAT DO STATISTICIANS DO?

"Statistics is a changing field with new methods being generated constantly. The user of statistical techniques regularly has more and better tools at hand to help solve his problems.

"There seems to be no limit to the areas of challenge to the statistician. Let's look at some specific examples of statistical problems, in widely ranging fields."

MODIFYING THE WEATHER. "In the not too distant future man may be able to control the weather. In fact, by seeding clouds with chemicals, scientists are currently attempting to dissipate potential hurricanes before they reach disaster proportions.

"Recently, a large contingent of U.S. Naval vessels and Air Force planes gathered in the Caribbean, under the scientific direction of a meteorologist, to test the effects of seeding clouds with silver iodide. Statistical methods were crucial in determining the proper design of the experiment. Earlier trials carried out on a smaller scale had been sharply criticized because the meteorological scientist may have subconsciously picked those clouds for seeding that were likely to grow and precipitate anyway. To counteract this criticism, a statistician with the National Bureau of Standards constructed a randomized design: first a cloud would be chosen for observation; then a sealed envelope would be opened to learn whether that particular cloud was to be seeded or not. The design was arranged so that about two out of every three clouds were seeded. The physics of precipitation is complicated and, as yet, not well understood. The randomized experiment, however, played an important role in insuring the objectivity and accuracy of these meteorological tests.

"In addition to its value in designing the experiment itself, statistics was vital to the analysis of the data gathered. A Weather Bureau statistician supervised the analysis, helping to reach the conclusion that seeded clouds do grow larger, and these larger clouds presumably produce increased rainfall.

"Now political scientists, working with statisticians, will have to determine the circumstances under which increased rainfall is really desired by the community. Farmers may want it during a particular time of the year while vacationers and resort owners may not."

THE ENERGY CRISIS—NATURAL GAS AND OIL WELLS. "America is facing a shortage of energy. Both Canada and the United States see their domestic oil resources dwindling at a time when demand is increasing, especially for non-polluting sources of energy. Here are two examples of how statisticians are helping in the search for sources of energy.

"Natural gas is relatively pollution-free, and therefore a vital resource. But how much is there, and where is it? To discover the extent of our actual energy reserves the United States National Gas Survey carried out a statistical sampling experiment in which gas fields in every gas producing region in the U.S. were carefully examined. This was an immense task. Each field had to be examined by a team composed of a reservoir engineer and a geologist. There are almost 10,000 economically producible gas fields in the continental U.S. If every field had to be examined it would have taken roughly 900 team years. That's 9 years for 100 teams of 2 individuals! Accordingly, the project's statisticians designed a sampling experiment balancing time and cost against accuracy of measurement of the characteristics of the totality of U.S. gas fields. By judiciously selecting the sample and the means of analyzing it, they helped to develop procedures for valid gas reserve estimation.

"While most *oil fields* are small, the bulk of the world's oil is concentrated in a few very large fields. Before large amounts of money are spent exploring for petroleum in a geological basin, it is essential to describe the relative frequency with which such large fields appear in the basin.

"The Alberta (Canada) Oil and Gas Commission sponsored an ambitious program to gather and analyze data from about 24,000 exploratory wells drilled in Alberta. Data from these wells were systematically coded and put onto magnetic computer tape. After establishing an operational definition of an oil field, geologists and statisticians working together analyzed the data and came up with a mathematical function (a 'model') that closely approximates the relative frequencies of sizes of fields. This model yielded a plausible answer to the question: Are giant oil fields geologically unique? Since the frequencies of both large and small fields were well approximated by the same model, the investigators concluded that all oil fields could be considered as common members of a single population. Hence, there is no reason to consider large fields as 'freakish.'

"Such studies provide important clues in the search for oil and gas. Many large oil and gas companies are presently undertaking similar studies, employing full-time statisticians to help carry them out."

TESTING THE POLIO VACCINE. "Many readers may be too young to know the panic felt by parents each summer when the poliomyelitis season arrived with its threat to strike children with death or paralysis. In fact, relatively few children were its victims, but the fear was there.

"Considerable hope and excitement were evident therefore when, in 1954, over one million public school children were chosen to participate in the largest *controlled* public health experiment in the history of the United States. The study was intended to determine the effectiveness of a vaccine—the Salk vaccine—as a protection against polio.

"A key part of the study involved 400,000 children, half of them to receive the vaccine, and the other a harmless salt solution. The selection was random; each child could have received either solution. Such extensive testing was required because the number of children visibly affected by polio were few and because its incidence varied widely from year to year. Only a very large sample would reliably reveal whether the vaccine was effective.

"Fewer than 200 cases of polio occurred among the 400,000 children. The rate of paralysis, however, was three times greater among the children who received the salt solution than among those who received the vaccine. This showed the effectiveness of the vaccine. Statisticians had pressed for a controlled experiment of substantial proportions, and had successfully applied various statistical methods in evaluating the possible effects of chance fluctuations on the relatively small numbers of polio cases and deaths."

WOMEN JURORS. "In a well-publicized court case in Massachusetts, Dr. Spock, the author of a famous book on child care, and several others were initially convicted of conspiracy in connection with advocating draft evasion during the Vietnam War. During the trial the defense argued that the absence of women on the jury hurt the defendants. Might this absence have occurred by chance?

"A statistician with legal training undertook a special investigation for the defense. He examined jury lists over a two and one-half year period in the court district, and found some surprising evidence. The judge who tried the conspiracy case regularly had a lower proportion of women on his jury lists than the other six judges in the district. A variety of statistical tests showed that it was close to impossible for such differences to have occurred by chance.

"The conviction was eventually set aside on other grounds. Yet, the statistician's investigation did have some impact. Individuals are now assigned at random to the jury lists in all federal courts."

THE SST: FLYING INTO DANGER? "A Supersonic Transport (SST) has been built by Great Britain and France. Should the United States build a fleet of these? Does the SST present a health hazard? The SST will fly so high as to disturb the ozone blanket over the earth, permitting more ultraviolet radiation to reach the earth's surface and the people on it. Skin cancer (including a fatal form, melanoma) is said to be related to ultraviolet radiation. (Doctors have long suspected that there is more skin cancer in people living closer to the equator. The closer to the equator, the more intense the ultraviolet radiation.)

"Statisticians at the National Cancer Institute and several other government agencies were called upon to:

a. find out if the relationship between ultraviolet radiation and skin cancer was really true, and

b. find out how much change in skin cancer would be likely to come about with each unit change in ultraviolet intensity, and relate this to ozone and the size of the SST fleets.

They discovered that ultraviolet measurements around the world were not very good, nor were there really good field data on the relationships of changes in ultraviolet radiation to changes in ozone. The ozone measuring instruments were expensive, and so a plan had to be developed for proper nationwide spacing of the few ozone instruments, to coincide with the placing of ultraviolet radiation measuring instruments, and with skin cancer reporting districts. Through cooperation among the National Oceanic and Atmospheric Administration (the meteorologists), the Department of Transportation (people interested in the SST) and the National Cancer Institute (people studying skin cancer) a measurement program has been set up, and some answers are on their way. So far one conclusion has been reached by the statisticians examining the data: deaths from melanoma appear to be related to the latitude at which they occur."

SEATBELTS. "All automobiles sold in Canada and the United States are now made with seatbelts. Unquestionably the use of seatbelts reduces serious injury and death in motor vehicle accidents. But the use of seatbelts is disappointingly low, despite the obvious benefits. In 1971, the Ontario Department of Transport carried out a seat-belt education program in a set of elementary schools in Toronto to teach the children the benefits of seatbelts use not only for themselves, but for the whole family. Statisticians on the Department's research staff planned how the program would be carried out and also designed a study to evaluate the program's effectiveness.

"Effectiveness was measured by the increased use of belts by the parents since the children themselves were too young to drive. The statisticians also decided, after careful consideration, not to undertake a before-treatment survey but to use a control. Half of the children in the schools would get the training; the other half would not.

"They learned two basic facts from the survey: (1) the use of seatbelts could be increased through special education programs and (2) this increase could not be maintained. After six months seatbelt use dropped to where it was before the training.

"Follow-up evaluations are an important aspect of effective statistical application."

VANISHING WHALES. "In the 1950's scientists associated with the International Whaling Commission went on a hunt—not to kill whales, but to count them. The idea was to estimate the number of blue and fin whales remaining in the world. Samples of whales of each type were located, and small metal cylinders were then fired into the thick layer of fat which a whale has under its skin. The marked whales were free to roam and intermingle with the remaining unmarked whales. Whaling factory ships from around the world captured whales and reported to the Commission the numbers of marked and unmarked whales which they captured.

"This approach, sometimes called the *mark-recapture method*, is a simple and reliable measuring technique. Statisticians working for the Commission using this method in conjunction with a variety of other statistical techniques found that there may be as few as 1000 blue whales left in all of the Southern Hemisphere. As a result, the International Whaling Commission banned the taking of blue whales to prevent this species from becoming extinct.

"The same statistical methods have been applied to the sampling and counting of other animals, such as deer, wolves, and rabbits, and have been used by demographers—scientists involved with human population studies—in India to estimate the number of births and deaths in various regions of the country."

1.4. OPPORTUNITIES AND THE CHALLENGE OF THE FUTURE

"Some of the diverse fields in which statistical methodology has had extensive applications are:

Actuarial Science: determining premium rates for different insurance risks; designing pension plans for private and public groups; measuring effectiveness of loss prevention and loss control programs; ...

Agriculture and Fisheries: developing superior varieties of grain; increasing egg and milk production; assessing the effectiveness and potential dangers of pesticides; management and allocation of natural fishery resources; ...

Biology: exploring the interactions of species with their environment; creating theoretical models of the nervous system; studying genetical evolution; ...

Business: estimating the volume of retail sales; designing inventory control systems; producing auditing and accounting procedures; improving working conditions in industrial plants; assessing the market for new products; ...

Economics: measuring indicators such as volume of trade, size of labor force, and standard of living; analyzing consumer behavior; ...

Engineering: working out safer systems of flight control for airports; improving product design and testing product performance; determining reliability and maintainability; ...

Health and Medicine: developing and testing new drugs; delivering improved medical care; preventing, diagnosing, and treating disease; ...

Psychology: measuring learning ability, intelligence, and personality characteristics; creating psychological scales and other measurement tools; studying normal and abnormal behavior; ...

Quality Control: determining techniques for evaluation of quality through adequate sampling; in-process control; consumer surveys and experimental design in product development; . . .

Sociology: testing theories about social systems; designing and conducting sample surveys to study social attitudes; exploring cross-cultural differences; studying the growth of human populations; . . .

"In all of these areas, and many others, statisticians work closely with other scientists and researchers to develop new statistical techniques, adapt existing techniques, design experiments, and direct the analysis of surveys and retrospective studies.

"These aspects of statistical work will surely continue, and new areas of application are constantly opening up for the enterprising statistician. Anthropology, archeology, history, library science, law and public policy are among the many disciplines providing new avenues for statistical inquiry. Statisticians are presently expanding their role in the development of components of computer systems, and in the equally important evaluation of computer system performance.

"The federal governments of both Canada and the United States employ professional statisticians at various levels of responsibility and policy making. In the United States, statisticians play a vital role in the activities of the Department of Commerce (e.g., the National Bureau of Standards and the Bureau of the Census), Department of Defense, Department of Health, Education and Welfare (e.g., the National Institutes of Health), the Department of the Interior, the Environmental Protection Agency, and in the Office of Management and Budget, the Bureau of Labor Statistics and many other agencies. In Canada, statisticians employed by the federal government engage in similar activities, and many work for Statistics Canada, a special agency in charge of the collection and interpretation of economic and social data. At state, provincial, and local levels statistical experts help solve problems concerning the environment, urbanization, finance, transportation, and public health.

"Increasingly, the legal profession is turning to statisticians to help weigh evidence and assess reasonable doubt. As concerned citizens continue to press for equal rights and opportunities, statisticians will be needed to help formulate and then measure criteria for such equality. Protecting consumer rights, preserving the environment and searching for and evaluating new sources of energy are other fertile fields for statisticians who want to make direct contributions to improving the quality of life.

"Universities employ many statisticians not only as faculty in Departments of Statistics, Mathematics, Biostatistics, Biomathematics, Management Science, and Operations Research, but also in various capacities with Departments of Economics, History, Biology, Genetics, Computer Science, Engineering, Psychology, Sociology, and Anthropology, and Schools of Business, Medicine, Education, Dentistry, and Public Health."

1.5. IMPORTANT CONCEPTS IN STATISTICS

"Every applied field deals with practical problems. Statistics, too, is concerned primarily with practical problems. We shall not discuss statistical methods in detail here, but shall illustrate certain concepts which come up again and again in the application of statistics to practical problems.

"A basic and valuable statistical idea is that one can say a good deal about a whole class of objects by studying a *sample* of them—if it is properly drawn from the larger class. One must make careful observations on the units included in the sample in order to make correct approximations as to the nature of the large group—called the *population*. When the sample is properly drawn and the observations are sufficiently accurate, statistical theory enables us not only to estimate, for example, the proportion of the population that has a special characteristic, but also to judge how accurate our estimate is likely to be.

"In discussing the sampling of a simple population, statisticians often picture a bag or box of marbles. Suppose there are 10,000 marbles in a bag, all alike except for color—and we cannot see them, but we know only that some are red and the rest are white. Suppose that we take 100 marbles at random from the bag, and find that we have seventy white and thirty red marbles. Can we conclude that the same proportion of red and white would hold if we could count all the marbles? If not, what can we reasonably conclude about the proportion of white marbles in the bag?

"A thoroughly conservative person might insist that our experiment has established only that at least seventy of the 10,000 marbles are white and at least thirty are red. But even our conservative person, having seen the result, would say it was more rational to bet that a complete count would show more white than red marbles rather than the other way around.

"The theory of statistics helps make more precise and useful our intuitive feeling that a sample gives us some kind of information about the universe or population from which it was drawn, if it was chosen in such a way that each member had an ascertainable chance of being included. If it is a random sample, each member has an ascertainable probability. Thus, the theory can answer such questions as, 'How large a sample should we draw from the bag if we want the odds to be at least two to one (or ten to one or one hundred to one) in favor of our sample showing a proportion of white marbles within two percentage points of what a complete count would show?'

"Application of our bag-of-marbles theory to real life becomes complicated, but this is the job of the statistician. Suppose that we wish to know which of several dietary formulas gave newborn babies the best chance of gaining weight and growing during the first year of life. It is not easy to draw a random sample from the population of newborn babies. If we were to set up an experiment in which pediatricians in a hospital gave one diet to one group of babies and a different diet to another group, we would have many decisions to make about the class of babies being treated. For example, if we dealt only with babies that had been available from this one hospital, we might have a different population of babies than the total population of *all* the babies in the United States. And we would wish to design the study so that the babies on each diet were as much alike as possible in all respects except diet. But this is unrealistic in our world and the statistician must know how to study variations and causes of variations so that he can try to get useful answers in spite of these differences. The decision about which diet is superior will have to be made, and with proper statistical guidance it is more likely to be correct.

"Determination of how best to design such studies and how best to analyze the results involves a thorough understanding of statistics as well as of the area of the research itself. The design requires an understanding of what is meant by population, sample, and probability, and how the methods of selection and analysis affect the conclusions which may be drawn. In addition to the concepts of sampling and the design of studies, proper analysis of results requires the use of varied statistical tools and techniques. For instance, in an attempt to understand our environment and the forces which shape it, statisticians frequently encounter the problem of time series analysis. A time series is a set of chronologically ordered observations on a quantitative characteristic of either an individual or some collective phenomenon taken at different points of time. Monthly or annual government and industry statistics of production, sales, employment, and income are all examples of time series. Through the use of specialized tools, statisticians and other scientists analyze the components which make up the sources of variation observed in time series. Techniques have been developed to depict the long-time underlying trend, seasonal, cyclical, and irregular fluctuations, as well as measuring the correlation of these variables to other factors.

"Many kinds of scientists and researchers use these statistical tools and concepts in many ways, and of course not all of these persons are statisticians. But the use of these ideas and of many processes growing out of them is the particular province of statistics, and the more complex problems require special training for the statistical problem-solver."

BIBLIOGRAPHY

A collection of 44 essays written in nontechnical language and covering a broad range of statistical applications may be found in *Statistics: A Guide to the Unknown*, edited by Judith Tanur and published in 1972 by Holden-Day, Inc., San Francisco.

2

FREQUENCY DISTRIBUTIONS

In this chapter our primary objective is to describe some typical tabular and graphic methods which display data in an organized way. Also, we introduce some terms basic to an understanding of statistics.

2.1. GROUPED FREQUENCY DISTRIBUTION TABLES

When a set of data contains only a few entries, a simple listing of the observations might be sufficient for most analyses. However, when there are many entries, the data are usually organized into groups called **classes** or **categories** and presented in a table which gives the frequency (or relative frequency or percentage) in each group. Such a frequency table gives a better overall view of the distribution of data and enables an investigator to rapidly assess important characteristics of the data.

The collection and organization of data may require a tremendous amount of time. Fortunately, in many cases such time-consuming work has already been done. For example, if we wanted information on incomes, we could look to publications by the federal government. There we can find many tables already prepared, and they are useful in answering a variety of questions. As an illustration, Table 2.1 shows the number of families in each of 10 classes of income level in 1966.

Table 2.1 is an example of a frequency table for **numerical** or **quantitative data**. The table shows income classes arranged in ascending order of magnitude and their corresponding frequencies. Such a table shows the **frequency distribution** (or, briefly, **distribution**) of a set of data. (Table 2.1 may also be termed a **percentage distribution** since it shows the percentages in the arranged income classes.)

Not all groups are numerical. Table 2.2 is an example of a frequency table for **categorical** or **qualitative** data. Such categories are listed alphabetically or in order of decreasing frequencies or in some other conventional way.

Table 2.1. Money Income by Families in 1966*

Income Level (in dollars)	Number of Families (in thousands)	Percent of Families
Under $1,000	1,100	2.3
1,000 to 1,999	2,600	5.4
2,000 to 2,999	3,200	6.6
3,000 to 3,999	3,300	6.8
4,000 to 4,999	3,400	7.1
5,000 to 5,999	4,100	8.4
6,000 to 6,999	4,500	9.4
7,000 to 9,999	11,800	24.4
10,000 to 14,999	9,900	20.4
15,000 and over	4,400	9.2
Total	48,300	100.0

*Department of Commerce, Bureau of the Census, Series P-60

Table 2.2. Deaths for Selected Causes, 1966*

Cause of Death	Number
Diseases of cardiovascular system	1,010,812
Malignant neoplasms (cancer)	303,736
Accidents (motor vehicle—53,041)	113,563
Influenza and pneumonia, excluding newborn	63,615
Diseases of early infancy	51,644
Diabetes	34,597
Cirrhosis of liver	26,692
Suicide	21,281
Tuberculosis	7,625
Syphilis and its sequelae	2,193
All other causes	227,391
All causes	1,863,149

*United States Department of Health, Education and Welfare

The two illustrations just presented display some properties that all distribution tables should have in common. In general,

> A **grouped frequency table** shows the table number, the title, column headings, and a minimum of two columns—the first has the classes arranged in some useful order, and a second has the corresponding frequencies or values computed from frequencies.

It is from such tables that we derive graphs, charts, other tables, and a great variety of numerical information discussed in the remainder of this chapter and in Chapter 3.

2.1. Grouped Frequency Distribution Tables

Certain information which can be found from the **raw data** (i.e., original unorganized data) cannot be obtained from grouped frequency distributions. For example, we cannot find the largest family income from Table 2.1, nor can we determine from Table 2.2 the number of people who died of lung cancer or the number of females who died from syphilis. However, many useful questions which relate to the total collection of data may be answered by use of such tables.

The first column in a frequency distribution table is for the **variable** (or variables) of a distribution. For example, in Table 2.1 "family income" is the variable of the distribution, and the income of a particular family, say $12,251, is a **value of the variable**. Thus, a class is a collection of observed values of the variable.

It is desirable that we introduce some more terminology before constructing a frequency table from raw data. To illustrate the meaning of these new terms we use Table 2.1 and assume the raw data were recorded to the nearest dollar. For example, the possible data values in the second class are $1,000, $1,001, $1,002, ..., $1,998, $1,999. The value $1,000 is the **lower class limit** and the value $1,999 is the **upper class limit** of the second class. Generally,

> The smallest and largest raw data values which can fall in a given class are called its **lower** and **upper class limits**, respectively.

Note that the first class does not have a lower limit and the tenth class does not have an upper limit. Thus, both these classes are said to be **open**.

Since the raw data were recorded to the nearest dollar, the **true values** in the second class could be no less than $999.5 nor no greater than $1,999.5. Thus, $999.5 is called the **lower class boundary** and $1,999.5 the **upper class boundary** of the second class. Note that $999.5 is also the upper class boundary of the first class and $1,999.5 is the lower class boundary of the third class. Generally,

> That numerical value which is halfway between the upper limit of one class and the lower limit of the following class is the **boundary** between the two classes and is called the upper boundary of the first class and the lower boundary of the following class.

Class boundaries should contain one more decimal place than the raw data so that each raw datum value falls in a unique class. (The reader should note that the interval of values between two consecutive boundary values contains infinitely many real numbers, but that only a finite number of these values are actual data values—the two extreme values being class limits.)

> The **class length** of a class is the numerical difference between the upper and lower boundaries of that class.

Using Table 2.1 we find the class lengths of the second and ninth classes to be $1,999.5 - 999.5 = 1,000.0$ and $14,999.5 - 9,999.5 = 5,000.0$, respectively. If all classes have the same length, then boundaries of the extreme classes are obtained by decreasing the smallest boundary value by the common class length and by increasing the largest boundary value by the common class length; otherwise, the **extreme boundaries** are assigned when the problem is specified *or* there are no extreme boundaries, as in Table 2.1.

2.2. CONSTRUCTION OF TABLES FOR NUMERICAL DATA

After the raw data have been collected there is still much to do before a suitable frequency table can be constructed. Some key steps in the preparation of a frequency distribution for grouped data are the

1. Selection of non-overlapping classes
2. Enumeration of data values that fall in each class
3. Construction of the table

We use the hypothetical standardized mathematics scores of 125 students shown in Table 2.3 to illustrate these three steps for numerical (quantitative) data. Note that the data of Table 2.3 is a jumble of numbers and conveys little information without analysis; a frequency distribution, however, displays specific information which can be visualized at a glance.

Table 2.3. Standardized Mathematics Test Scores for 125 Students

68	84	53	84	71	50	54	75	51	76
48	78	60	64	63	66	65	73	71	54
77	70	56	73	81	70	68	70	92	72
66	72	66	73	68	70	59	75	72	69
83	51	74	81	72	64	84	64	64	60
66	55	59	77	71	67	72	49	84	64
69	73	72	74	70	64	58	64	71	84
71	67	57	79	68	80	79	77	77	81
93	66	48	89	70	81	63	65	67	68
75	66	66	81	74	65	71	61	65	74
70	71	74	54	69	57	67	66	65	73
74	66	58	73	64	64	65	64	81	74
74	66	63	42	78					

First, we decide on the **number of classes** and the **set of values** which fall in each class. Each of these choices, to some extent, depends on the nature of the data and the objective of the study. (This is an area of statistics where experience is a good teacher.) When grouping is desirable, experience indicates that we should normally choose between 5 and 20 classes. These classes should be the most natural (or convenient) classes, and they should be selected so that the

2.2. Construction of Tables for Numerical Data

resulting frequency distribution is as smooth as possible. It is highly desirable that the class lengths be the same unless there are strong reasons to the contrary.

Turning to the 125 numbers in Table 2.3 we find that the largest and smallest scores are 93 and 42, respectively. Thus, the range of scores is $93 - 42 = 51$. For a common class length of 5 units we would need at least $51/5 = 10.2$ classes. Thus, we decide to use 11 classes each of length 5. (It might be just as desirable to use 6 classes each of length 10, but classes of lengths 4, 6, 7, 8, 9, 11 seem less desirable since we typically think in terms of intervals of five, ten, and so on.)

Next, we decide where to start the first class. We could let the class limits of our first class be either 38–42, 39–43, 40–44, 41–45, or 42–46. For our first class we choose 40–44 and use classes

$$40\text{–}44,\ 45\text{–}49,\ 50\text{–}54,\ \ldots,\ 90\text{–}94$$

since tallying the scores is easier (and possibly more natural) than with other classes of length 5. The resulting frequency distribution, including a tally count, is shown in Table 2.4.

Table 2.4. Frequency Distribution of Test Scores of Data of Table 2.3

Scores	Tally	Frequency
40–44	|	1
45–49	|||	3
50–54	ⵌ ||	7
55–59	ⵌ |||	8
60–64	ⵌ ⵌ ⵌ |	16
65–69	ⵌ ⵌ ⵌ ⵌ ⵌ |||	28
70–74	ⵌ ⵌ ⵌ ⵌ ⵌ ⵌ ⵌ	34
75–79	ⵌ ⵌ ||	12
80–84	ⵌ ⵌ |||	13
85–89	|	1
90–94	||	2
	Total	125

According to the definition, the boundary between the first two classes is 44.5. Since each class has length 5, the lower boundary of the first class is $44.5 - 5 = 39.5$. By successively adding 5 units we find the other boundaries to be 49.5, 54.5, ..., 89.5, 94.5. The reader should note that the two limits of a given class are 4 units apart, but that consecutive lower limits (or upper limits) are 5 units apart.

The mid-value of a class is called the **class mark** and is obtained by adding its two boundaries and dividing by two.

The class marks for Table 2.4 are 42.0, 47.0, 52.0, ..., 87.0, 92.0. For Table 2.1, the first and last classes do not have class marks; the marks of the other classes are 1,499.5, 2,499.5, 3,499.5, 4,499.5, 5,499.5, 6,499.5, 8,499.5, 12,499.5. (The reader should note that even though the class length is defined in terms of its boundaries other methods may be used to find class length if proper caution is applied. For example, the common class length for Table 2.4 may be obtained by taking the numerical difference of two consecutive class marks. For Table 2.1 the class lengths cannot all be obtained by this method.)

We have seen that a frequency distribution table should have two columns —one for the variable and one for corresponding frequencies. The "tally" part of Table 2.4 does not actually belong to the final presentation of the frequency distribution. For a quantitative variable, values of the variable are arranged in either increasing or decreasing order of magnitude, and, when the total frequency is large, class groupings should be made. Some useful rules of grouping are to

1. Select from 5 to 20 classes except in rare cases
2. Define classes so that each observation belongs to exactly one class
3. Select classes so that their lengths are equal unless there is good reason to the contrary

Classes may be designated in many ways—we have seen two. Class limits are often used, but class boundaries, class marks, and other values may be used to designate classes (see Exercise 2.5).

The frequency column of a frequency distribution table may also be expressed in several ways, depending on the purposes of the investigation. For example, if two or more similar tables are to be compared, **relative frequencies** or percentages should be used (see Exercises 2.5 and 2.6).

In many situations we need to know the number of observations which fall below (or above) a specified value. For example, we might use Table 2.4 to find the number of test scores below 80, but there is an easier way, as illustrated in Table 2.5.

> The "less than" **cumulative frequency** of a specific class is the total frequency of all data values less than the upper boundary of that class. A "less than" **cumulative frequency distribution table** shows each class boundary of a distribution and the corresponding "less than" cumulative frequency.

When frequencies are replaced by relative frequencies (or percentages) we call the resulting distribution a "less than" cumulative relative frequency distribution (or "less than" cumulative percentage distribution). Some people prefer to construct such distributions in terms of limits, since boundaries are not possible data values. In such cases the resulting distribution may be termed a "less than or equal to" cumulative distribution. Using Table 2.4 we obtain the "less than"

Table 2.5. "Less Than" Cumulative Distribution of 125 Mathematics Test Scores

Test Score Boundary	Cumulative Frequency of Scores Below the Boundary	Cumulative Percentage of Scores Below the Boundary
39.5	0	0.0
44.5	1	0.8
49.5	4	3.2
54.5	11	8.8
59.5	19	15.2
64.5	35	28.0
69.5	63	50.4
74.5	97	77.6
79.5	109	87.2
84.5	122	97.6
89.5	123	98.4
94.5	125	100.0

cumulative distribution shown in Table 2.5, expressed both in terms of frequencies and percentages.

From Table 2.5 it is easy to find what (or approximately what) percentage of the values falls below a specific value of the variable. Also, we can find that value (or approximate value) which has a specified proportion of scores below it. For example, 63 of the students have scores below 69.5, and 64.5 is the value below which 28 percent of the scores fall. Other applications of such tables, including "greater than" cumulative distributions, are discussed in later sections and introduced in the exercises.

In the construction of any table great care should be taken to make it accurate, easy to read, self-contained, and to the point. There should always be a table heading as well as column headings. If some or all of the data in a table are taken from another source, a reference should be indicated at the base of the table.

2.3. EXERCISES

2.1. For a distribution of measurements (to the nearest inch) the class boundaries are 29.5, 34.5, 39.5, 45.5, 53.5, and 63.5 inches. Find the class lengths, marks, and limits.

2.2. The class marks of a distribution of test scores (given in whole numbers) are 43, 56, 69, 82, and 95. Determine the class lengths, boundaries, and limits.

2.3. The maximum temperature in Richmond, Virginia, for 92 days in the summer of 1970 varied from 71 to 103 degrees fahrenheit. Construct a table with seven classes into which the 92 values might be grouped. Give class limits and class marks.

2.4. Use Table 2.4 to determine the number of scores
(a) greater than 79;
(b) less than 55;
(c) not greater than 64;
(d) not less than 80.
(e) Check each of the four above answers by using Table 2.5.

2.5. Table 2.6 shows the frequency distributions of grades in a specific course at Alpha College.

Table 2.6. Grades in a Course at Alpha College

Grade	Freshmen	Frequency for Sophomores	Juniors
A	2	4	4
B	13	20	17
C	12	58	30
D	8	17	9
F	5	1	0
Total	40	100	60

(a) Construct a percentage distribution table for each class standing (i.e., freshmen, sophomores, and juniors).
(b) Construct a percentage distribution table for the grade received in the course, ignoring the class standing of the students.
(c) Compare the grade distributions for the three classes of students.

2.6. The percent of families at each income level in the years 1955, 1960, and 1965 is shown in Table 2.7. Write a report in which you compare the distribution of incomes across the three years.

Table 2.7. Money Income by Families*
(Percent Distributions by Income Level)

Income Level	1955	1960	1965
Under $1,000	7.7	5.0	2.9
1,000 to 1,999	9.9	8.0	6.0
2,000 to 2,999	11.0	8.7	7.2
3,000 to 3,999	14.6	9.8	7.7
4,000 to 4,999	15.4	10.5	7.9
5,000 to 5,999	12.7	12.9	9.3
6,000 to 6,999	9.5	10.8	9.5
7,000 to 9,999	12.9	20.0	24.2
10,000 to 14,999	4.8	10.6	17.7
15,000 and over	1.4	3.7	7.6

*Department of Commerce, Bureau of Census, Series P-60

2.3. Exercises

2.7. On the first day of school a bus brought 40 children to a combined elementary-high school. Their ages were as follows:

13	11	6	5	8	17	14	13	6	6
7	8	15	16	9	10	10	9	7	8
15	14	14	12	11	9	7	6	8	6
10	13	17	17	16	12	12	9	8	13

Group the ages into five classes and construct a frequency table. What are the class boundaries? What are the class marks?

2.8. The number of children in the families of 37 students in an algebra class were as follows:

2	3	3	1	4	11	3	3	2	1
1	4	7	3	2	2	3	4	5	1
2	7	3	6	4	3	3	2	4	3
1	3	2	2	2	5	6			

Assuming there are 37 different families, construct a frequency distribution table of the number of children per family. (Do not group into classes.) Add a percentage column to the table and use it to write a short report on family size.

2.9. The body weights in grams of 62 male bobwhite quail trapped in Ohio in 1946–1948 are as follows:

210.4	203.8	189.4	196.4	179.6	195.1
217.9	198.3	217.3	215.6	220.2	181.4
198.5	189.7	185.9	198.4	174.7	177.0
170.7	189.8	181.6	187.9	194.6	184.3
216.2	212.6	230.3	173.6	180.7	183.3
192.5	201.3	176.6	174.3	163.4	199.7
191.8	166.5	194.8	202.5	191.5	184.3
174.8	190.4	191.4	194.9	222.2	195.8
177.2	184.3	194.9	202.0	205.2	194.9
187.8	235.8	202.0	212.6	198.5	212.8
233.9	187.4				

(a) Construct a frequency distribution table having the eight equal length classes 160.0–169.9, 170.0–179.9, ..., 230.0–239.9.
(b) Construct a "less than" cumulative relative frequency table.

2.10. The mileage death rates (i.e., number of deaths per 100,000,000 vehicle miles) for the 50 states in 1963 were as follows (sources were state traffic authorities):

6.8	5.6	4.4	5.2	3.5	6.0	6.4	6.0	6.2	5.4
5.6	3.1	6.0	7.0	5.3	4.6	4.5	6.0	6.0	4.8
6.7	3.8	5.1	6.7	5.2	7.4	6.4	4.3	5.2	6.4
6.5	6.0	5.8	4.5	7.2	5.4	6.7	2.8	5.7	5.4
5.2	6.6	5.3	4.5	5.5	3.5	4.9	8.0	6.4	7.4

(a) Construct a frequency distribution table with equal length classes where the first class has limits 2.5 and 3.2.
(b) Construct a "less than" cumulative relative frequency table.
(c) Construct a "greater than" cumulative relative frequency table.
(d) Use a cumulative distribution to find the proportion of states with mileage death rates less than 4.05; greater than 4.85.
(e) What is the approximate mileage death rate below which 75 percent of the values fall? Above which 50 percent of the values fall?

2.11. Construct a frequency table for the *types of vehicles* passing along a highway during the first hour after school.

2.12. Construct a frequency table for the color of hair of students in your classroom.

2.4. GRAPHIC METHODS OF PRESENTING DATA

We have already said that graphs may be used to display data in an organized way. A well-constructed graph is probably the quickest way to picture a set of data, but such a picture may also misinform if it is not prepared with care and with certain objectives in mind. Left to our own devices we tend to make diagrams that suit us. Thus, it is desirable that a few of the standard graphic forms be described in some detail.

For a quantitative variable, the most common graphic construction is the **histogram**. (It is probably the most useful as well.) We take the test scores of Table 2.3 and the grouping of Table 2.8 (which is easy to obtain from Table 2.4) to make the frequency histogram shown in Figure 2.1.

A **histogram** has values of the variable on the horizontal scale and frequencies on the vertical scale. For each class a rectangle is constructed with base equal to the class length and height determined from the class frequency. The *areas* of the rectangles must be proportional to the frequencies.

2.4. Graphic Methods of Presenting Data

Table 2.8. Frequency Distribution of Mathematics Test Scores

Class Limits	Class Boundaries	Frequency
40–49	39.5–49.5	4
50–59	49.5–59.5	15
60–69	59.5–69.5	44
70–79	69.5–79.5	46
80–89	79.5–89.5	14
90–99	89.5–99.5	2
	Total	125

If the bases are of equal width the heights of the rectangles are also proportional to the frequencies.

A histogram may be marked in several ways. Since the bases of rectangles extend from one boundary to the other, the boundary values are usually marked, but the class marks or even class limits may be used. The heights may be indicated by a vertical frequency axis (see Figure 2.1) or by frequencies placed on top of the rectangles (see Figures 2.1 and 2.2). The latter method is preferred for classes with unequal lengths or when a class overlaps zero. The maximum height of a histogram should not normally exceed its width.

To illustrate the construction of a frequency histogram with classes of unequal length we combine the fourth and fifth classes of Table 2.8 and leave the others as they appear. The new fourth class has boundaries 69.5 and 89.5 with

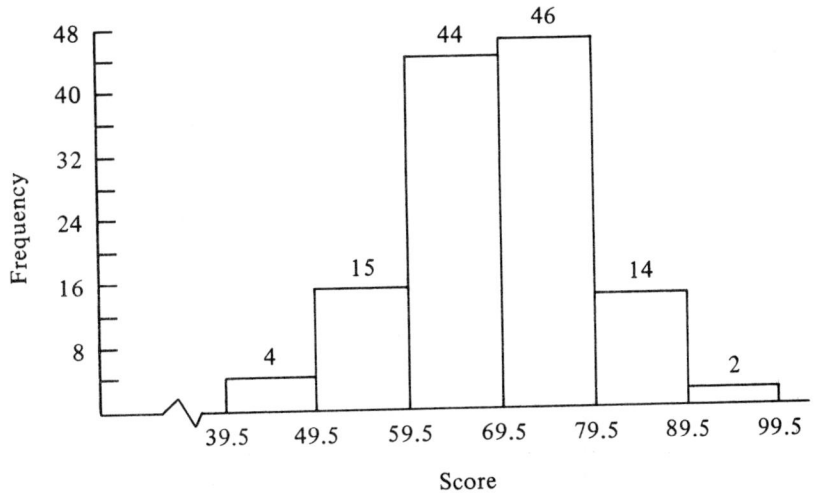

Figure 2.1. Frequency Histogram of Test Scores (Equal Intervals)

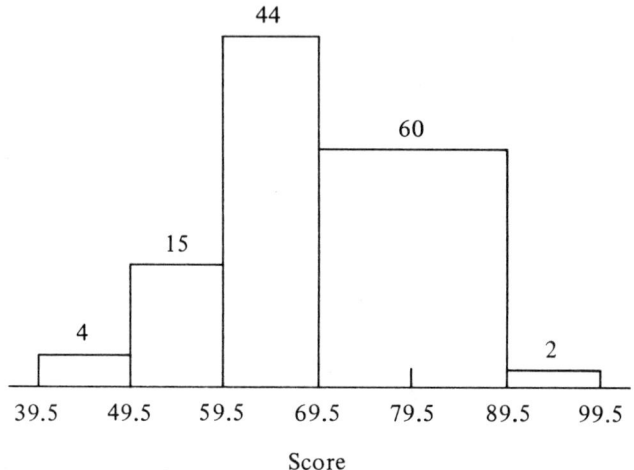

Figure 2.2. Frequency Histogram of Test Scores (Unequal Intervals)

frequency 60. The corresponding frequency histogram is shown in Figure 2.2. (The diagram in Figure 2.3 is not a histogram, but it illustrates a common error made in the construction of histograms with unequal length classes.) When the two classes of Table 2.8 were combined the length of the new class was doubled.

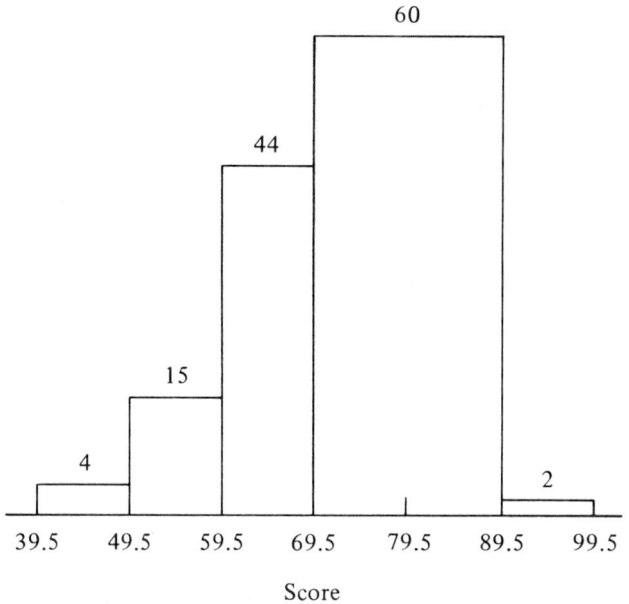

Figure 2.3. Wrong Construction for a Histogram

But the sum of the areas of the two rectangles over the interval from 69.5 to 89.5 must equal the area of the single rectangle over the same interval, since the total frequency is the same and, when viewing a histogram, the eye perceives *area*, not height alone. Thus, the height of the new rectangle must be one-half the combined heights of the first two rectangles. (In terms of numbers, the area of the two rectangles in Figure 2.1 is $10(46) + 10(14) = 10(60) = 600$. Thus, in order for the area of the corresponding new rectangle in Figure 2.2 to be 600 the height h must be such that $20h = 600$ or $h = 30$, which is one-half of $46 + 14$.)

Many times, relative frequencies (or proportions) are used instead of frequencies. For example, the comparison of two similar distributions may be made in terms of proportions. When relative frequencies replace frequencies in a histogram the area of a rectangle is generally made numerically equal to the relative frequency of its class. Thus, *the total area of a relative frequency histogram would be one.* (The reader should note that the total area of a frequency histogram is often numerically the same as the total frequency of the distribution.)

A histogram may be used to approximate the proportion of values that falls in a certain interval on the horizontal scale. (See Examples 3.10 and 3.11 for illustrations.) Also, a histogram may be used to approximate the area under a smooth curve for an interval on the horizontal scale, or a smooth curve may be used to approximate the area under a histogram (see Section 5.11).

Histograms may be constructed for variables whose values are integers (that is, for count data). (See Exercises 2.8, 2.13, and 2.14 for illustrations.) The base of each rectangle would generally be one unit in width. Hence, for each rectangle the height and area would be numerically the same, and we could use either.

If a class is open (as in Table 2.1) or if the variable is qualitative, a histogram cannot be constructed. However, a very similar diagram, called a **bar graph**, may be used. A typical bar graph is shown in Figure 2.4 (data from Table 2.6). The *heights* of the bars are proportional to the class (or category) frequencies and the widths of the bars are all the same. The unit of measure on the horizontal scale is not important (for a qualitative variable, numbers would not likely be used). The bars are equally spaced along the horizontal direction, and all bars have the same base line. Adjacent bars do not share a common boundary line as do rectangles of a histogram. *In a bar graph the heights of bars are to be used, but not the areas.*

A *second important graph* for a quantitative variable is known as **cumulative frequency polygon** or, for short, **frequency ogive**. A typical ogive graph is shown in Figure 2.5.

> The **class boundaries** of an **ogive** are located on the horizontal axis and the cumulative frequencies (or cumulative relative frequencies or cumulative percentages) on the vertical axis. Then a point is plotted above each class boundary a distance equal to the corresponding cumulative frequency, and all such adjacent pairs of points are connected by straight line segments.

26 Frequency Distributions

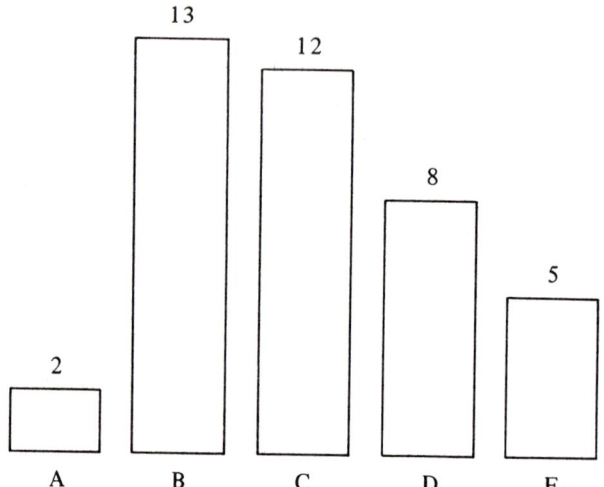

Figure 2.4. Bar Graph of Grades of Freshmen

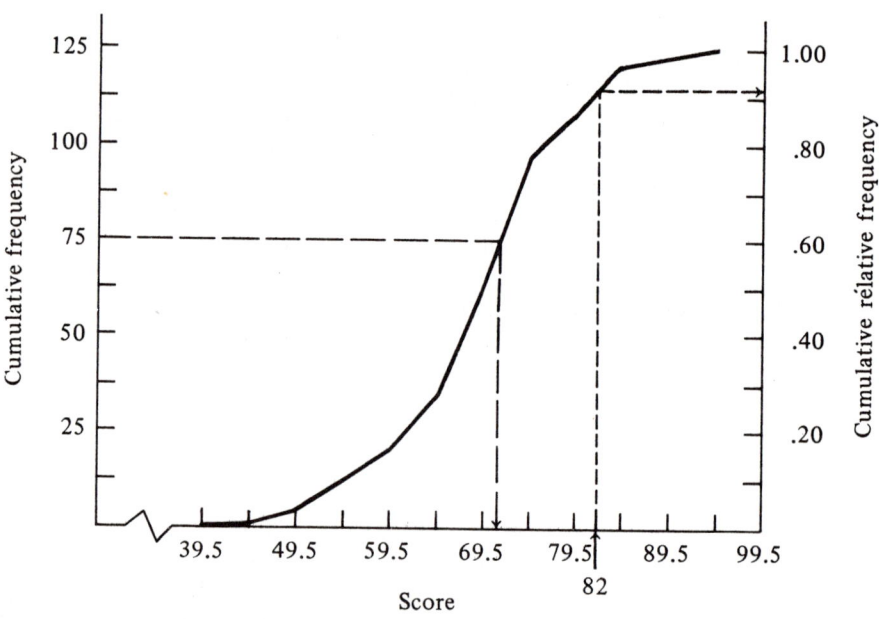

Figure 2.5. Ogive for Test Scores (data from Table 2.5)

Note that the ogive line graph starts on the horizontal axis at the lower boundary value of the first class, the cumulative frequency less than this boundary being zero. As in Figure 2.5, a second vertical scale is sometimes added on the right when both cumulative frequencies and cumulative relative frequencies are required.

If a "less than" relative frequency ogive is constructed one can find (1) the proportion of values below (less than) a fixed value of the variable or (2) the value below which a fixed proportion of the values of the variable fall. An illustration of (1) is on the right side of Figure 2.5. The proportion of scores less than 82 is approximately 92. (First the vertical line through 82 and the ogive is constructed. Then the horizontal line through the point of intersection on the ogive is extended to the relative frequency scale.) To illustrate (2), start at cumulative frequency 75 (or relative frequency 60). Draw a horizontal line to the ogive, and from the point of intersection on the ogive drop a perpendicular to the horizontal scale. The solution is approximately 71. Thus, we have found by construction that 71 is the value below which 75 (or 60 percent) of the scores fall. (See Exercise 2.21 for a problem of this type.)

A *third type of graph* for a quantitative variable, not as often used as the other two types, is the **frequency polygon** (see Figure 2.6).

For a **frequency polygon** locate class marks on the horizontal axis and frequencies (or relative frequencies) on the vertical axis. Then plot a point above each class mark a distance equal to the corresponding frequency, and connect by straight line segments all such adjacent pairs of points. One class with zero frequency is added on

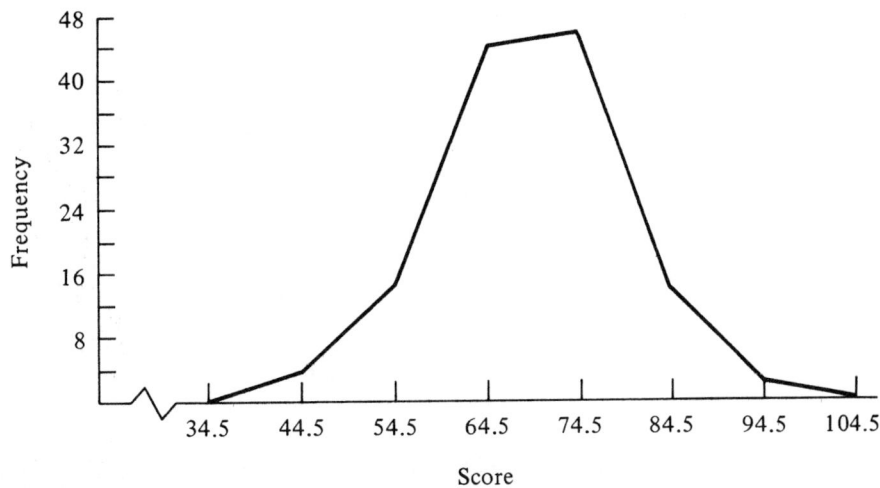

Figure 2.6. Frequency Polygon for Test Scores of Table 2.8

either end of the distribution to close the graph with the horizontal axis.

The primary purpose of the frequency (or relative frequency) polygon is to show the "shape" of the distribution. The relative frequency polygon is very useful when two or more distributions are to be compared. For example, several such graphs may be constructed on the same set of axes. In this way any differences in the "shapes" of the distributions immediately becomes clear. (One of the other two types of graphs should be used if details of the distributions are required.)

The diagrams discussed in this section have great visual appeal. There are other ways of presenting distributions that have greater "eye" appeal and that may be more effective for certain purposes. The "pie chart" and "picture-symbol graph" are two well-known methods which may be found in industrial reports, magazines, and other places. The variety of ways a distribution may be presented pictorially seems to be limited only by the artistic ability of the investigator.

2.5. POPULATION AND SAMPLE

In the first sections of this chapter we have called **values of a variable** such things as recordings of fact, data, categories, and observations. Also, we have presented some typical methods of describing the **distribution of a variable**. In following sections we normally use **observation** to mean recording of fact (or possible recording of fact)—such an observation may be real or conceptual. For example, each of the following is considered an observation: height of 71 inches, color of eyes is blue, smell of perfume, four children in a car, three errors on a printed page, first thought in the morning, or guess for a name. Variable and observation are basic terms in any study of statistics.

Two other basic words, already mentioned in Chapter 1 and which need clarification, are population and sample. Both words are used in a broader sense in scientific circles than outside. The nonscientist is likely to use the term population only when reference is to people, and has the tendency to equate "all people in given geographic boundaries" with "population." However, to a scientist

> A **population** is a collection of all elements (i.e., objects or observations) having some common attribute.

Hence, all horses in the state of Wyoming is a population of horses, all houses in Boston is a population of houses, the heights of all trees in a forest is a population of heights of trees, all possible outcomes of a toss of a coin is a population of outcomes, or all possible temperatures in an experiment is a population of temperatures. Yes, and all people in the state of Texas in 1970 is the population of Texas in 1970.

A **sample** is a subset of a population.

(In statistics, we usually think of a sample as having two or more values.) For example, if the population is the set of IQ scores of students at Thomas Jefferson High School in September 1970, then a sample may be the IQ scores of the girls in Algebra 2, Section 1. Similarly, if all fish in Lake Michigan on November 1, 1971, is a population, then all fish caught in Lake Michigan on that day is a sample.

The reader has probably already observed that all the fish in Lake Michigan might be a sample for a person who is interested in all fish in the five Great Lakes. Thus, it becomes clear that *a collection of observations may be considered either a population or a sample, depending on how we look at the collection.* If we are interested in making generalizations about the IQ's of all students in the state of Virginia, then all IQ scores at Thomas Jefferson High School would represent a sample. However, if we were not interested in making any generalization from the IQ scores at Thomas Jefferson High School, then the IQ scores could be looked at as a population.

The number of elements in a collection is called the **size of the collection**. Thus, the **size of a population** (or, briefly, **population**) may be either finite or infinite. For example, the size of the population of houses in Boston is finite, but the population of possible tosses of a coin is infinite. (Note that the size of a sample is likely to be finite.)

Clearly, a population (or a sample) has a distribution. We have already seen that the variable of a distribution may be numerical or nonnumerical. Not all numerical variables are quantitative. For example, a zip code or social security number or number on the jersey of a basketball player is like a name. However, most numerical variables are quantitative variables; nonnumerical variables along with the types of numerical variables mentioned in the last sentence are qualitative variables.

We may think of a numerical variable as being continuous or discrete. A **continuous variable** is one which can have any value between two distinct numbers; a **discrete variable** is one which can have only isolated values. The outcomes of a single toss of a die are values of a discrete variable since the only possible numbers of dots are 1, 2, 3, 4, 5, 6. The temperature of an oven is a continuous variable since its value might be any number between 200 and 500 degrees fahrenheit, say. (The reader should note that recordings of temperature are values of a discrete variable, but that actual temperatures are values of a continuous variable.)

Histograms and frequency polygons (as well as corresponding smooth curves) are very useful in showing the great variety of shapes of distributions of a quantitative variable. The smooth curve graphs of a continuous variable in Figure 2.7 show different shapes of distributions. Figures 2.7 (a) and (b) illustrate symmetric distributions, Figures 2.7(c) and (d) nonsymmetric distributions, and Figure 2.7(d) a distribution skewed to the right.

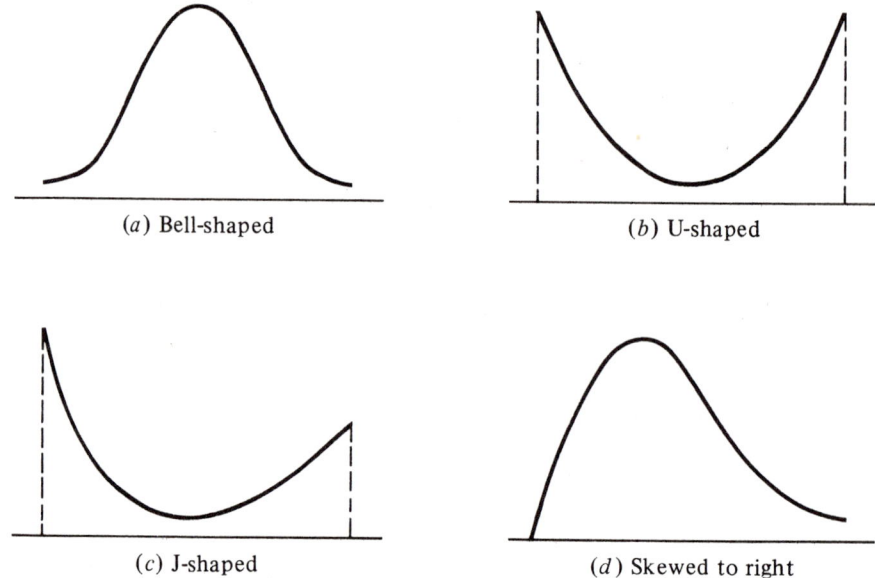

Figure 2.7. **Distribution Curves**

A distribution is **symmetric** if it can be folded along a vertical axis so that the two sides of the graph coincide; otherwise, the distribution is nonsymmetric.

Among the nonsymmetric distributions there are several types of interest. Two types are shown in Figure 2.7. A distribution with its one "tail" on the right is said to be **skewed** to the right; one with its one tail on the left is skewed to the left. (In Chapter 3 we introduce some ways to measure skewness.)

When we consider IQ scores of students, "student" is a variable and "score" is a variable which depends on the student variable. Similarly, in an experiment in which temperature of an oven is under consideration, we have an "oven" variable and a "temperature" variable which depends on the oven variable. In these examples, "student" and "oven" are called object variables, and "IQ" and "temperature" are called observation variables. Thus, in a statistical discussion, we might refer to a population of objects or to the corresponding resulting population of observations.

Statistics deals more with the population of observations than with the population of objects. However, it is very important that we carefully describe a population of objects, and, when sampling from such a population, give special attention to the method of sampling (see Chapter 6).

2.6. EXERCISES

2.13. Use the data of Exercise 2.8 to construct a
   (a) frequency polygon
   (b) "less than" cumulative frequency polygon

2.14. Use the data of Table 2.9 to construct a
   (a) bar chart for the total population
   (b) pie chart for the nonfarm population. *Hint:* A **pie chart** is a circle with pie-like sections which are proportional to the percentages of the categories.

Table 2.9. Families by Number of Own Children Under 18 Years Old, 1960*

Number of Children	Percentage of Population by		
	Farm	Nonfarm	Total
0	45.3	42.9	43.0
1	15.9	18.7	18.5
2	13.0	18.4	18.0
3	11.4	11.0	11.1
4 or more	14.4	9.0	9.4

*Department of Commerce, Bureau of the Census

2.15. Use the data of Exercise 2.10 to construct a
   (a) frequency histogram
   (b) "greater than" cumulative frequency polygon
   (c) Use part (b) to find the mileage death rate above which 10 percent of the mileage death rate values fall.

2.16. For the data of Exercise 2.7 draw a
   (a) relative frequency histogram
   (b) "less than" cumulative frequency polygon. *Hint:* Assume age is to be interpreted in the usual way. For example, a 6-year-old child is some age from 6 years and 0 days to 6 years and 365 days old (ignore leap year).
   (c) Use both part (a) and part (b) to find the value below which 50 percent of the ages fall.

2.17. Construct a "less than" cumulative relative frequency polygon for the weights of quail in Exercise 2.9. Use the ogive constructed to find the two values between which the middle 50 percent of the weights fall.

2.18. Classify each of the following variables in terms of quantitative or qualitative; object or observation; finite or infinite; discrete, continuous, or neither.
(a) degrees fahrenheit
(b) grams weight
(c) score on algebra test
(d) score in a basketball game
(e) color of eyes
(f) number of defective machine parts
(g) social security number
(h) suspension bridge
(i) rank in graduating class

2.19. Give three illustrations of a population of objects and a corresponding population of observations.

C2.20. The data of Table 2.10 were obtained from the Bureau of Census, Department of Commerce. A calculating machine should be used in parts (a), (b), and (c) since the percentage columns in the table are of no use in making the graphs.
(a) On the same graph, construct relative frequency polygons of school enrollment for 1920 and 1960 and write a short report in which the two distributions are compared.
(b) Construct a relative frequency histogram of school enrollment for the 1960 data.
(c) Construct a "less than" cumulative relative frequency polygon for the 1920 data. Use the graph to find the percentage of students under 15 years of age.

Table 2.10. School Enrollment by Age

	1920		1960	
Age	Number (in thousands)	Percentage of Population	Number (in thousands)	Percentage of Population
5	441	18.8	1764	44.8
6	1481	63.3	3175	83.2
7	1905	83.3	3636	97.0
8	2011	88.5	3524	97.8
9	1944	90.4	3388	98.0
10	2078	93.0	3395	97.9
11	1970	93.9	3388	97.8
12	2083	93.2	3478	97.5
13	1877	92.5	3391	96.9
14	1767	86.3	2605	95.3
15	1357	72.9	2589	92.9
16	1002	50.8	2437	86.3
17	642	34.6	2161	75.6
18	414	21.7	1258	50.6
19	253	13.8	742	32.8
20	148	8.3	512	23.5
Total	21373	64.3	41443	81.8

2.21. In 1964 the average hourly wage of production workers in the South was $1.90 and in the West it was $2.67. Table 2.11 shows the percent of workers falling in each class.

Table 2.11. Hourly Wage in March 1964 of Production Workers, Manufacturing Industries*

(Excludes Premium Pay for Overtime and for Work on Weekends, Holidays, and Late Shifts)

Hourly Wage	Percent of Workers	
	South	West
Under $1.25	1.8	0.1
1.25 to 1.29	17.0	1.7
1.30 to 1.39	11.3	1.9
1.40 to 1.49	8.4	1.7
1.50 to 1.59	7.8	2.4
1.60 to 1.69	5.6	1.9
1.70 to 1.79	5.0	2.7
1.80 to 1.89	4.2	2.9
1.90 to 1.99	3.3	2.8
2.00 to 2.19	7.1	7.7
2.20 to 2.39	5.7	9.4
2.40 to 2.59	5.2	11.4
2.60 to 2.79	4.4	11.4
2.80 to 2.99	3.7	10.2
3.00 and over	9.5	31.8
Total	100.0	100.0

*Department of Labor, Bureau of Labor Statistics

Before obtaining solutions, group these data so that classes have a common length of 0.20, except for the open classes at each end of each distribution.

(a) Use one set of axes to construct a percentage polygon of workers in the South and a percentage polygon for workers in the West. *Hint:* Let the lower limit of the first class be $1.20 and the upper limit of the last class be $3.19.

(b) Use one set of axes to construct a "less than" cumulative percentage polygon for workers in the South and a "less than" cumulative percentage polygon for workers in the West.

(c) Use these graphs to estimate the wage below which 25 percent of the workers fall in each section. If there are 2,913,000 workers in the South and 1,340,000 workers in the West, how many make wages which fall in the lowest quarter in each section of the country?

(d) What percentage of the workers in the South make wages below the 25 percent point of Western workers?

c2.22. The distribution of the population in the United States in 1960 is shown in Table 2.12. Let the upper limit of the last class be 89 years, for ease of computations. If you have a

Table 2.12. Population by Age and Sex in 1960 (in thousands)*

Age	Male	Female	Total
Under 5	10,352	10,013	20,364
5 to 9	9,572	9,254	18,825
10 to 14	8,595	8,314	16,910
15 to 19	6,814	6,651	13,465
20 to 24	5,558	5,554	11,112
25 to 34	11,323	11,588	22,909
35 to 44	11,873	12,350	24,223
45 to 54	10,143	10,438	20,581
55 to 64	7,561	8,066	15,628
65 and over	7,537	9,121	16,659
Total	89,328	91,349	180,676

*Department of Commerce, Bureau of the Census

calculating machine, construct cumulative relative frequency polygons and compare the distributions of male and female.

3

STATISTICAL MEASURES

3.1. INTRODUCTION

The tabular and graphic methods described in Chapter 2 give very little accurate quantitative information about sets of data. For example, we might observe that two relative frequency tables (or two histograms) differ, but find it impossible to say how much they differ. This illustrates one of many needs for some numerical descriptions of data.

Many numerical measures have been proposed. It is not always easy to decide which measure (or measures) should be used. The choice may depend on such things as the distribution of the data, the purpose of the investigation, or the amount of time required for computation. In this chapter, we are not concerned with such choices—we simply introduce a few of the measures which are most commonly used. Actually, we concentrate on two "measures of location" and two "measures of variation."

These four measures are defined in terms of a finite set of data denoted by

$$(3.1) \qquad y_1, y_2, \ldots, y_N$$

We do not distinguish between the terms "data from a population" and "data from a sample" at this time, since, for statistical description, it is not important. However, we denote measures of a set of data by lowercase Greek letters or by capital Latin letters, symbols used when data are from a population. That is, we use a finite population of size N and the notations associated with *any* population to discuss measures which characterize a set of data.

3.2. MEASURES OF LOCATION

The reader has probably noticed that observations have a tendency to cluster at some particular location on the scale of measurement. A measure of such a position is called a **measure of location**. Often the cluster point is near the "center." Thus, measures of location are also referred to as **measures of central position,** or **measures of central tendency,** or **measures of central value**.

The measure of location most commonly used in both applied and theoretical work is the **arithmetic mean**, or **mean**, as it is usually expressed. The mean is defined as the sum of all the values of a variable divided by the number of such values. If we denote N values by the symbols in Expression (3.1) and the mean by μ (Greek letter mu), then the mean is defined by

(3.2)
$$\mu = \frac{y_1 + y_2 + \cdots + y_N}{N}$$

(The mean of a **sample** of size n is denoted by $\bar{y}$ and defined by

$$\bar{y} = \frac{y_1 + y_2 + \cdots + y_n}{n}$$

It is to be understood that n is a positive integer and that $y_1, \ldots, y_n$ is some subset of values from a population. We will use the sample notation when it is necessary to distinguish between the population mean μ and a sample mean.)

Sometimes the layman uses the term "average" to refer to the mean. Unfortunately, "average" may also refer to other measures of location (for example, see Exercises 3.7 through 3.12), or, loosely, to such things as average beach, average basketball player, or average American. Thus, when reference is made to a specific measure of location, we use the appropriate technical name rather than the more general term "average."

EXAMPLE 3.1. The number of children in the families of 30 students in a statistics class were

2	3	3	1	4	11	3	3	2	1
1	4	7	3	2	2	3	4	5	1
2	7	3	6	4	3	3	2	4	3

Find the arithmetic mean.

Let $y_1 = 2, y_2 = 3, \ldots, y_{29} = 4, y_{30} = 3$. Then the mean number of children per family in the statistics class is

$$\mu = \frac{2 + 3 + \cdots + 4 + 3}{30} = \frac{102}{30} = 3.4$$

(If we considered the class to be a sample and used it to make an inference then the mean would be denoted by $\bar{y}$.)

No family has 3.4 children, but this is a number which, in a way, may typify the set of data. For example, in case 30 different families are involved, the mean times the number of families gives the total number of children in the 30 families. That is

$$30(3.4) = 102$$

or, in general

(3.3) $$N\mu = y_1 + y_2 + \cdots + y_N$$

Next, we observe that there is a large concentration of values near the mean. For example, 22 of the 30 observations have values 2, 3, 4, or 5. That is, $\frac{22}{30}$ or approximately 73 percent of the data values are within 1.6 units of the mean.

We notice that the mean is not in the center of the distribution. For there are 20 observations below and only 10 observations above the mean.

Thus, the mean may be a useful measure for some purposes and not very desirable for others. For example, the *median* is often a much better measure of the center (or middle) of a distribution.

The definition of the median, denoted by M, depends on whether N [of Expression (3.1)] is odd or even. First, arrange the data in increasing (or decreasing) order of magnitude. If N is odd, the **median** is the value of the $(N+1)/2$th ordered observation; if N is even, the **median** is the value of the mean of the $N/2$th and $[(N/2) + 1]$th ordered observations. (The reader should recall that we are now using the notation for a population median. A sample median is defined the same way, but is denoted by **m** whenever it is necessary to distinguish between the population median and a sample median.)

EXAMPLE 3.2. Find the median size of family for the data of Example 3.1. The ordered observations are

1	1	1	1	2	2	2	2	2	2
3	3	3	3	<u>3</u>	<u>3</u>	3	3	3	3
4	4	4	4	4	5	6	7	7	11

The two middle ordered observations, the 15th and 16th, are underlined and have values 3 and 3. Thus, the median is

$$M = \frac{3+3}{2} = 3$$

Clearly, when N is even, the median is a value such that no more than half the observations are less than M and no more than half are greater than M. In Example 3.2, 10 observations are below 3 and 10 are above 3, but 20 observations are equal to or less than 3 and 20 are equal to or greater than 3. Note that when the two middle ordered observations are different, the median is a value with one-half of the observations being greater and one-half being less. For example, the median of 4, 5, 8, 9, 11, 14, is 8.5 with three values greater and three values less.

When N is odd, the median always has the value of an observation, and less than half of the observations are smaller than M and less than half are greater than M. For example, the median of 7, 8, 10, 10, 22 is 10. Two values are less than the median and only one is larger.

The median is not affected as much as the mean by extreme values. This property is illustrated by Example 3.3 and is the primary reason why distributions of such things as "income," "debt," and "life of lightbulb" are usually measured by the median.

EXAMPLE 3.3. Compare the mean and median for the following sets of data of yearly income in thousands of dollars.

$$\text{Set A} = \{1, 5, 9, 9, 12\}$$
$$\text{Set B} = \{5, 9, 9, 12, 20\}$$
$$\text{Set C} = \{5, 9, 9, 12, 200\}$$

The reader may wish to check the following computations.

Set	Total	Mean	Median
A	36	7.2	9
B	55	11	9
C	235	47	9

For each set the median income of $9,000 is fairly typical—actually, four of five incomes are "near" the median. For Sets A and B the mean incomes of $7,200 and $11,000, respectively, give a fairly good indication of the typical income. But for Set C the mean of $47,000 is not representative of any income. Clearly, the value of the mean tends to the extreme value.

The mean of a distribution of a quantitative variable is unique, but the mean is not even defined for a qualitative variable. However, if the observations of a qualitative variable can be ordered, the median is defined. For example, the measurements of heights of 9 students may be unknown, but one can still determine which student has the median height. Also, it may be possible to order the "effectiveness" of several guest speakers and, thus, determine the speaker (or speakers) with median effectiveness.

In a large proportion of applied problems the mean is preferred to the median as a measure of location. This is due to the fact that in many situations the mean fluctuates less from sample to sample (see Exercise 3.6).

An investigator often needs to find a measure of location for combined distributions. For example, a school superintendent may wish to combine data for the six schools in the city, a labor supervisor may wish to combine information from several labor unions, or a census expert may wish to use information from the 50 states to make a report on the United States. We illustrate the principle with two very simple distributions.

EXAMPLE 3.4. Find the mean and median of the distribution obtained by combining the following two distributions.

Distribution 1:   9   11   11   14   15
Distribution 2:   2   4   5   7   9   10   12

The reader can show that the mean and median of Distribution 1 are $\mu_1 = \frac{60}{5} = 12$ and $M_1 = 11$, respectively, and for Distribution 2 they are $\mu_2 = \frac{49}{7} = 7$ and $M_2 = 7$.

The combined distribution is

2   4   5   7   9   9   10   11   11   12   14   15

The median is $M_c = (9 + 10)/2 = 9.5$ and the mean is $\mu_c = \frac{109}{12} \doteq 9.1$, where $\doteq$ denotes "approximately equal to."

If there are many numbers, an immense amount of work is required to obtain the appropriate measure of location from the combined distribution. Fortunately, the most useful measure of location, the combined mean μ_c, can be found by the formula

(3.4) $$\mu_c = \frac{N_1\mu_1 + N_2\mu_2}{N_1 + N_2}$$

where N_1 and N_2 denote the sizes of the respective distributions. [The extension of Formula (3.4) to three or more distributions is straightforward.] The use of Formula (3.4) represents a great saving in time, especially when μ_1 and μ_2 are already available. For Example 3.4 we use Formula (3.4) to verify the value already computed the longer way, that is

$$\mu_c = \frac{5(12) + 7(7)}{5 + 7} = \frac{109}{12} \doteq 9.1$$

No such formula exists for finding M_c. In order to find M_c one must always arrange the data of the combined distributions.

Many times two or more distributions are compared in terms of some measure of location. Such a measure should not be selected simply because it is easy to compute; instead it should be selected because it is considered the "best measure of location" for the particular investigation. The measure of location to be used for such comparisons should normally be selected before the data are collected. (See Exercise 3.12 for an illustration.)

3.3. EXERCISES

3.1. The numbers of students enrolled in the nine science classes at Anne Bates High School in 1970–1971 were 17, 43, 37, 40, 13, 35, 29, 38, 36.
(a) Find the mean and the median.
(b) Which measure of location do you prefer? Why?

3.2. In a United Fund drive the mean contribution of 132,760 families solicited was exactly $2.37.
  (a) What was the total amount contributed?
  (b) Suppose the mean contribution of $2.37 is accurate to one-half of one cent, what could be the total contribution?
  (c) Suppose the median contribution was $2.37 and no family gave $2.37, what can you say about the total contribution?

3.3. Use the 50 observations of Exercise 2.10 to find the mean and the median. (Note that the formula $(N + 1)/2$ can, if interpreted properly, be used to give the position of the median, for the $(50 + 1)/2 = 25.5$th largest value may be interpreted as "that value halfway between the $\frac{50}{2} = 25$th and the $\frac{50}{2} + 1 = 26$th values."

3.4. Find the mean and median of each of the following:
  (a) −7   −5   −3   −1   1   3   5   7
  (b) −8   −4   −2   −2   1   3   5   7
  (c) −13 −1   −1   −1   4   4   4   4
  (d) −700 −5   −3   −1   1   3   5   7
  (e) −7   −5   −3   −1   1   3   5   700
  (f) −700 −5   −3   −1   1   3   5   700
  (g) Decide whether each of the distributions in parts (a) through (f) is symmetrical or skewed, and if skewed how extreme and in which direction. Compare the values of the mean and median in terms of "shape" of each distribution.

3.5. In each of the following compute the mean and median whenever possible, and make whatever comparisons (or comments) seem pertinent:
  (a) The annual incomes (in 1966) of seven neighbors were $8,600; $9,200; $10,700; $9,500; $12,500; $11,950; and "over $20,000."
  (b) A certain type of manufactured item is packed in lots of 100. In 20 lots it was found that one lot had 10 defective items, each of two lots had one defective item, and each of 17 lots had no defective item.
  (c) Seven boys ran the 440-yard dash at a track meet and each finished. However, the timekeepers got so excited they forgot to start their watches.
  (d) In a cake-baking contest "symmetry and texture" were used as criterion to rank six cakes A, B, C, D, E, F. A panel of three judges gave the following order of preference, starting with the best:

  B   D   C   A   F   E

  (e) In a class of 22 students there were two A's, five B's, twelve C's, two D's, and one F.

E3.6. The measurements of a population of five objects are 1, 4, 7, 10, 13.
  (a) Draw all (10) samples of three measurements from this population and compute the mean ($\bar{y}$) and median (m) of each sample.
  (b) Make a frequency table of sample means ($\bar{y}$); of sample medians (m). Compare

the fluctuations (variations) of these two frequency tables. *Hint*: It may be desirable to draw frequency polygons.

(c) Replace the population measurement 13 by 43 and work parts (a) and (b).

E3.7. If each of N numbers has a positive value, the **geometric mean**, G, is defined as the positive Nth root of the product of the N numbers; that is

$$G = \sqrt[N]{y_1 y_2 \cdots y_N}$$

Find G for each of the following:

(a) 4   16
(b) 1 $\frac{1}{2}$ $\frac{1}{32}$
(c) 1.08   1.23   1.56   2.00

Hint: Note that the logarithm of the geometric mean is the arithmetic mean of the logarithms of the numbers.

The geometric mean is used primarily as an "average" of rates of change, ratios, or index numbers. For example, in part (c) the four numbers might indicate "cost of living indexes" in city A on January 1, 1950, 1955, 1960, and 1965, respectively. Then G would be a type of average cost-of-living index.

E3.8. If each of N numbers is different from zero, the **harmonic mean**, H, is defined as the reciprocal of the arithmetic mean of the reciprocals of the N numbers; that is

$$H = \frac{N}{\frac{1}{y_1} + \frac{1}{y_2} + \cdots \frac{1}{y_N}}$$

The harmonic mean is normally used in case the numbers are such things as time rates or cost rates.

Find the harmonic mean for (a) and (b).

(a) 4   16
(b) 1 $\frac{1}{2}$ $\frac{1}{32}$

(c) A motorist travels from Richmond to Washington at 40 miles per hour and returns at 60 miles per hour. What is the average speed for the round trip? Does the harmonic mean give the correct average speed? Show why. *Hint*: Use the formula which expresses distance as the product of rate by time.

(d) A homemaker spent $3.00 on one variety of apples costing 75 cents per dozen and another $3.00 on a second variety of apples costing 30 cents per dozen. Find the average price per dozen of all apples bought. Does the harmonic mean give the correct average price? Show why.

(e) A motorcycle travels up a hill at the rate of 30 miles per hour. At what speed must the motorcycle return in order for the average speed for the round trip to be 55 miles per hour?

E3.9. The **mode** of a set of data is that value or category (or values or categories) which occurs with maximum frequency. If the frequency of each value in a distribution is the same, the mode does not exist; otherwise there may be one mode or more than one mode. Once the data are arranged in a frequency table, the mode (or modes) may be determined by inspection. An advantage of the mode is that it is defined for both quantitative and qualitative variables.

Find the mode (or modes) for
(a) the data of Exercise 2.8
(b) the data of Exercise 2.9

E3.10. We define one more measure of location. The **mid-range** is the mean of the largest and smallest value of a set of data. Such a value may be computed rapidly and in some cases is a good estimate of the mean. Find the mid-range of the data of Exercise 2.10, and compare this value with the mean and median found in Exercise 3.3.

E3.11. Compute μ, G, and H in each of the following and compare their magnitudes by ranking (see Exercises 3.7 and 3.8 for definitions of G and H):
(a) 1   4
(b) $\frac{1}{2}$ $\frac{1}{4}$ $\frac{1}{8}$
*(c) a ay ay^2 (assume $a > 0$ and $y > 0$)

E3.12. Suppose a certain car distributor keeps hearing the complaint that his make of car, say Car A, uses a lot of gas. Suppose further that the manufacturer of Car A claims that his car uses less gas than either of the two closest competitive types of car, say Car B and Car C. The distributor decides to obtain facts to back up the manufacturer. Thus, he arranges to have one test driver take eighteen 1976 model cars, six of each type, over the same route and use the same kind of gas. The results in terms of consumption of gallons of gas were as follows:

Car A:	20.3	23.3	22.8	23.2	22.7	23.0
Car B:	24.2	20.8	22.7	21.6	21.3	22.4
Car C:	22.5	24.7	19.3	21.9	23.4	20.4

The reader can show that the means, to two decimal places, are

$$\mu_A = 22.55 \qquad \mu_B = 22.17 \qquad \mu_C = 22.03$$

Thus, using the mean as a basis of comparison, Car A is third best. Since the medians are

$$M_A = \frac{22.8 + 23.0}{2} = 22.90$$

$$M_B = \frac{21.6 + 22.4}{2} = 22.00$$

$$M_C = \frac{21.9 + 22.5}{2} = 22.20$$

it would appear that Car A does use more gas than the two closest competitors. But the distributor will not give up. After trying several other measures of location he discovers that the mid-range (see Exercise 3.10) values for Cars A, B, C are

$$\frac{20.3 + 23.3}{2} = 21.8$$

$$\frac{20.8 + 24.2}{2} = 22.5$$

$$\frac{19.3 + 24.7}{2} = 22.0$$

respectively. Thus, the distributor has "facts" to justify his claim that "on the average" Car A uses less gas than the two closest competitors. (Such misuse of statistics has led many to believe there are three kinds of lies: lies, damned lies, and statistics.)
(a) Verify the computations given above.
*(b) Compute and compare the three harmonic means.
(c) Which measure of location in this exercise do you think is best? Why?

The reader should recognize that the measures used for comparison should be selected before the data are obtained. Also, it is possible that any differences among these samples may be due to chance causes. Thus, it will be necessary for us to look carefully at variation among samples before making a definitive statement relating to such matters.

3.4. MEASURES OF LOCATION—GROUPING AND CODING

Section 3.2 concentrated on properties of the mean and median. Both of these measures of location are used so often that any abbreviation in amount of writing or in computation should be welcome. This section is for such purposes.
First, we write a compact formula for the mean as

(3.5) $$\mu = \frac{\Sigma y_i}{N}$$

where Σy_i denotes the sum $y_1 + y_2 + \cdots + y_N$. The Greek letter capital sigma Σ indicates that all y's in the context of the problem are to be added, and the subscript i indicates that y is to have the successive subscripts $1, 2, \cdots, N$. [Some writers omit i in Equation (3.5). We retain i to indicate "value of a variable" so that later we may easily distinguish a variable from a constant.]

In a set of 10 values we may express the sum of the first six values as $\sum_{i=1}^{6} y_i$. Similarly the expression $\sum_{i=7}^{10} y_i$ is an abbreviation for $y_7 + y_8 + y_9 + y_{10}$. Thus, in general

(3.6) $$\sum_{i=N_1}^{N_2} y_i = y_{N_1} + \cdots + y_{N_2}$$

where N_1 and N_2 denote integers with $N_1 < N_2$.

EXAMPLE 3.5. In Example 3.1 we found the mean number of children in 30 families by adding 30 numbers and dividing by 30. If the same data are arranged as in Table 3.1, the formula for the mean may be written as

$$\mu = \frac{y_1 f_1 + \cdots + y_8 f_8}{f_1 + \cdots + f_8} = \frac{\Sigma y_i f_i}{\Sigma f_i}$$

and the value of the mean is

$$\mu = \frac{1(4) + 2(6) + \cdots + 11(1)}{4 + 6 + \cdots + 1} = \frac{102}{30} = 3.4$$

In this case there are eight *different* values of y. Thus, the subscript takes integral values from 1 to 8 inclusive instead of values from 1 to 30 inclusive.

Table 3.1. Frequency Distribution of 30 Children in a Statistics Class

Number of Children per Family y_i	Frequency of Each Number f_i	$y_i f_i$
1	4	4
2	6	12
3	10	30
4	5	20
5	1	5
6	1	6
7	2	14
11	1	11
Totals	30	102

In general, let N values be placed in a frequency table. Suppose the resulting table has h distinct values, and suppose the frequency of the ith value is f_i. Then the mean is given by

(3.7) $$\mu = \frac{\Sigma y_i f_i}{N}$$

where $N = \Sigma_{i=1}^{h} f_i$ and $\Sigma y_i f_i = y_1 f_1 + y_2 f_2 + \cdots + y_h f_h$. Clearly, all values of the original data are retained. Thus

(3.8) $$\mu = \frac{\Sigma y_i}{N} = \frac{\Sigma y_j f_j}{N}$$

where different subscripts are used (in this equation) to indicate summation over

45 Measures of Location—Grouping and Coding

different numbers of terms. Note that Equation (3.8) could also be written as

$$\frac{\sum_{i=1}^{N} y_i}{N} = \frac{\sum_{i=1}^{h} y_i f_i}{N}$$

For a frequency table in which values have been *grouped into classes*, let y_i and f_i denote the **class mark** and **class frequency**, respectively, of the ith class. Then the mean for grouped data is given by Equation (3.7). Since all values in a class are assumed to have the same value, the class mark, an approximation is introduced. Hence, the mean for grouped data may not be the same as for the original data.

EXAMPLE 3.6. Compute the means for the data of Tables 2.3 and 2.4. For Table 2.3 we find that

$$\mu = \frac{68 + 84 + \cdots + 78}{125} = \frac{8{,}603}{125} = 68.824 \doteq 68.8$$

and for Table 2.4

$$\mu = \frac{42(1) + 47(3) + \cdots + 92(2)}{125} = \frac{8{,}580}{125} = 68.640 \doteq 68.6$$

The exact mean is $\frac{8{,}603}{125}$ and the approximate mean is $\frac{8{,}580}{125}$. Thus, the error introduced by grouping is $\frac{23}{125}$ in this case. The percentage error is $\frac{23}{8{,}603} 100\%$ $\doteq 0.27\%$. (Actually, the error is usually very small when care is exercised in selecting the classes.)

It should be observed that μ is not defined for grouped data when any class is open-ended, since the class marks cannot be found. For example, the mean cannot be found for the data of Table 2.1 since both the first and last classes are open.

EXAMPLE 3.7. The reader has probably already observed that the data in Table 2.4 could be coded (transformed) so as to find the mean with less effort. Thus, if 67 is subtracted from each class mark and the resulting difference divided by 5 the new u values are as indicated in Table 3.2. If we let μ_y and μ_u denote the means with respect to the y and u variables, respectively, it follows that

$$\mu_u = \frac{-5(1) + (-4)(3) + \cdots + 5(2)}{125} = \frac{41}{125}$$

and, by substituting in Equation (3.9), that

$$\mu_y = 67 + 5(\mu_u) = 67 + 5\left(\frac{41}{125}\right) = 68.64$$

Table 3.2. Frequency Distribution of Test Scores and Computation of the Mean

Class Mark y_i	Coded Data $u_i = (y_i - 67)/5$	Frequency f_i	$u_i f_i$
42	−5	1	−5
47	−4	3	−12
52	−3	7	−21
57	−2	8	−16
62	−1	16	−16
67	0	28	0
72	1	34	34
77	2	12	24
82	3	13	39
87	4	1	4
92	5	2	10
Totals		125	41

Theorem 3.1. If $y_1, y_2, \cdots, y_N$ are values of any set of data, and if these data are transformed to new data $u_1, u_2, \cdots, u_N$ by the formula

$$u_i = \frac{y_i - y'}{c} \quad (i = 1, \ldots, N)$$

where y' is any constant and c is any constant different from zero, then

(3.9) $$u_y = y' + c\mu_u$$

We prove Theorem 3.1 later. This theorem holds for both raw and grouped data. For grouped data with classes of equal length, the reader is advised to let y' be a class mark near the "center" of the distribution and to let c be the common class length. (The reader should work Example 3.7 with 67 replaced by 72, say.)

There are some important special cases of Theorem 3.1. When $c = 1$, $u_i = y_i - y'$ and $\mu_y = \mu_u + y'$. When $y' = 0$, $u_i = y_i/c$ and $\mu_y = c\mu_u$. Finally, when $c = 1/a$, $u_i = a(y_i - y')$ and $\mu_y = \mu_u/a + y'$. (In each case, observe that inverse operations were used to decode. In the third case, y_i was coded by first subtracting y' and then multiplying the result by a. Then μ_u was decoded by first dividing by a and then adding y'. Thus, *a general rule to use in finding μ_y is to apply inverse operations in reverse order to μ_u.*)

The frequency formula for the mean, Equation (3.7), is a special case of a more general formula. Sometimes we need to find a mean when the values of y are multiplied by numbers which are not frequencies. Let any real number w_i be the *weight* of y_i where $i = 1, \cdots, h$. Then the weighted mean μ_w is defined by

(3.10) $$\mu_w = \frac{\sum w_i y_i}{\sum w_i}$$

where $\sum w_i \neq 0$.

EXAMPLE 3.8. During one week a salesman bought 17.9 gallons of gasoline at 56.9 cents per gallon, 20.6 gallons at 50.9 cents per gallon, 13.4 gallons at 57.9 cents per gallon, and 18.6 gallons at 56.5 cents per gallon. Find the mean price paid per gallon.
By Equation (3.10) we find

$$\mu_w = \frac{(17.9)(56.9) + (20.6)(50.9) + (13.4)(57.9) + (18.6)(56.5)}{17.9 + 20.6 + 13.4 + 18.6}$$

$$= \frac{3893.81}{70.5} = 55.23 \text{ cents per gallon}$$

Also, Equation (3.10) may be considered a generalization of Equation (3.4), the formula for the mean of combined distributions. Suppose the N_i values of the ith distribution have mean μ_i ($i = 1, 2, \ldots, h$). If we let $w_i = N_i$ and $y_i = \mu_i$, the mean μ_c of the combined distribution is

(3.11) $$\mu_c = \frac{\sum N_i \mu_i}{\sum N_i}$$

EXAMPLE 3.9. Find the shooting percentage of a high school basketball team on which player A made 65 percent of 100 shots, player B 70 percent of 200 shots, player C 75 percent of 40 shots, player D 60 percent of 120 shots, and player E 40 percent of 30 shots.
Let $N_1 = 100$, $\mu_1 = 65$, $N_2 = 200$, $\mu_2 = 70$, $N_3 = 40$, $\mu_3 = 75$, $N_4 = 120$, $\mu_4 = 60$, $N_5 = 30$ and $\mu_5 = 40$. We find that

$$\mu_c = \frac{100(65) + 200(70) + 40(75) + 120(60) + 30(40)}{100 + 200 + 40 + 120 + 30}$$

$$= \frac{31900}{490} = 65.1 \text{ percent}$$

The median for grouped data (or for data for a continuous variable) may be found with the aid of a frequency distribution table or a frequency histogram. We describe the method for a histogram. (The reader should be able to give the procedure for a table.) The **median** is that value, sometimes denoted by $y_{.50}$, along the horizontal axis for which 0.50 of the area of the histogram is to the left. Before giving a formula we illustrate the procedure for finding the median of a set of data.

EXAMPLE 3.10. Find the median for the test score data of Figure 2.1.

Since the total frequency is 125, the median is a value along the horizontal axis of the histogram for which $\frac{125}{2} = 62.5$ units of area are to the left. The class which contains the median, called **median class**, has lower boundary $b = 59.5$ and upper boundary 69.5 since there are $F' = 4 + 15 = 19$ frequency units less than 59.5 but $F'' = 4 + 15 + 44 = 63$ frequency units less than 69.5. The median class has length $c = 10$ and frequency $f' = 44$. We require a value along the interval of the median class which is

$$\frac{62.5 - 19}{44} = \frac{43.5}{44}$$

of the distance from 59.5 to 69.5. That is, the median is

$$M = b + \left(\frac{0.50N - F'}{f'}\right)c$$

$$= 59.5 + \frac{43.5}{44}(10) \doteq 59.5 + 9.9 = 69.4$$

We have defined median in terms of "area to the left" and "lower boundary." Obviously, we could have defined median in terms of (1) "area to the left" and "upper boundary," (2) "area to the right" and "upper boundary," or (3) "area to the right" and "lower boundary." Also, note that a "less than" ogive frequency polygon may be used to find the median of a distribution.

The median is a particular measure from a class of measures known as **fractiles** or **quantiles**. Just as the median is that value below which $\frac{1}{2} = 0.50$ of the area of a histogram lies, we *define a fractile y_p to be that value of y below which a specific proportion p of the area of the histogram lies*. Most fractile values may be found by the method used in obtaining the median of Example 3.10. Thus, if b denotes the *lower boundary* of the class containing a fractile y_p, F' denotes the *total frequency* less than b, f' denotes the *frequency of the class containing y_p*, and c denotes the *length of the class containing y_p*, then

(3.12) $$y_p = b + \left(\frac{pN - F'}{f'}\right)c$$

Clearly, when $p = 0.50$, Equation (3.12) yields the formula used to find the median of Example 3.10. (When a fractile y_p falls on a boundary b, then

$$y_p = b$$

Another rare case is illustrated in Example 3.12.)

Several fractiles are of special interest. Some have particular names and are designated by special symbols. Those fractiles corresponding to .25, .50, .75 are called *first quartile, second quartile, third quartile* and are denoted by Q_1, Q_2, Q_3, respectively. Fractiles for .10, .20, $\cdots$, .90 are called *first decile, second* decile, $\cdots$, ninth decile and denoted by $D_1, D_2, \cdots, D_9$, respectively. Finally, fractiles for .01, .02, $\cdots$, .99 are called *first percentile, second percentile*, $\cdots$, *ninety-ninth percentile* and denoted by $P_1, P_2, \cdots, P_{99}$. Clearly, the median M, the second quartile Q_2, the fifth decile D_5, and the fiftieth percentile P_{50} represent the same value $y_{.50}$.

EXAMPLE 3.11. Find the first quartile for the test score data of Figure 2.1.

Since $p = .25$ and $N = 125$, we wish to find a value of y with $(.25)(125) = 31.25$ units of area to the left. Clearly, the first quartile Q_1 (also, denoted by P_{25} or $y_{.25}$) falls in the third interval from 59.5 to 69.5 since the cumulative frequency up to 59.5 is 19 and that up to 69.5 is 63. The interval containing Q_1 has length 10 and frequency 44. Thus, the first quartile is

$$Q_1 = 59.5 + \left(\frac{31.25 - 19}{44}\right) 10$$
$$\doteq 59.5 + 2.8 = 62.3$$

(It should be observed that 25 percent of the area is below and 75 percent of the area is above Q_1.)

Whenever a "gap" occurs in data we need to extend our rule for finding a fractile. We say there is a gap in the data if some class (between the first and last) has zero frequency or if some internal interval of the corresponding histogram has a rectangle with zero area. If a fractile y_p falls in such an interval, the rule given above would allow the fractile to be any value in the interval. In order to have a unique value, we define *the fractile*, in this case, to be that value which is p of the way across the interval from left to right.

EXAMPLE 3.12. Find the third quartile of the following minimum daily temperatures:

63 58 56 60 57 61 54 55 54 58 57 55

By arranging the temperatures we obtain the following frequency distribution:

Temperature	54	55	56	57	58	60	61	63
Frequency	2	2	1	2	2	1	1	1

Since a class is of unit length two gaps occur in the data. The third quartile Q_3 falls in the first gap from 58.5 to 59.5. The unique value of Q_3, determined from the definition of the last paragraph, is

$$Q = 58.5 + (0.75)(59.5 - 58.5)$$
$$= 58.5 + 0.75$$
$$= 59.25 \doteq 59.2 \text{ degrees}$$

3.5. EXERCISES

3.13. Check the computations in Example 3.6.

3.14. The heights in inches of 20 girls in an algebra class were

| 67 | 64 | 65 | 63 | 65 | 66 | 68 | 61 | 64 | 63 |
| 65 | 64 | 66 | 66 | 67 | 63 | 65 | 66 | 62 | 65 |

(a) Find the mean height by Equation (3.5) and by Equation (3.7).
(b) Find the median height from the histogram of arranged data.

3.15. (a) Make a frequency table for the age data of Exercise 2.7, but do not group. Find the mean by Equation (3.7).
(b) Find the mean for the data as they were grouped for Exercise 2.7.
(c) Find the percentage of error made by the grouping in (b).
(d) Use the table of (b) to find the median.

3.16. (a) In Example 3.7 let $u_i = (y_i - 72)/5$. First find μ_u, and then find μ_y by Theorem 3.1.
(b) In Example 3.7 let $u_i = y_i - 60$. Find μ_u and find μ_y by Theorem 3.1.

3.17. Table 3.3 shows the frequency distribution of grades of 100 students on a standardized test in mathematics.

Table 3.3. Grades of 100 Students

Class Limits	Frequency	Cumulative Frequency
45–49	2	2
50–54	3	5
55–59	0	5
60–64	7	12
65–69	19	31
70–74	20	51
75–79	14	65
80–84	15	80
85–89	12	92
90–94	3	95
95–99	5	100

(a) Find the mean grade by Equation (3.7).
(b) Find the mean grade by using the transformation $u_i = (y_i - 72)/5$.
(c) Find the median grade directly from the table.
(d) Find $y_{.05}, y_{.25}, y_{.65}$, and $y_{.90}$, the boundaries between consecutive grades F, D, C, B, A.

c3.18. (a) Find the median of the distribution obtained by combining the 225 test scores of the distributions of Table 2.4 and Exercise 3.17.
(b) Find the mean of the distribution of 225 test scores in part (a).

c3.19. (a) Use Table 2.11 to find the median hourly wage of workers in the South; of workers in the West.
(b) In Table 2.11, assume the first class mark to be $1.220 and the last class mark to be $3.095. Find the mean hourly wage of workers in the South; of workers in the West. Compare the means with the medians of part (a); the averages of Exercise 2.21. Explain the differences.

3.20. In a certain class a grade of B ranges from 84 to 91 inclusive. The teacher gives four tests and a final examination. The final examination "counts" twice as much as each of the four tests. Suppose student M makes 73, 90, 85, and 70 on the four tests.
(a) What is the lowest grade M can make on the examination and still get a B on the course?
(b) Is it possible for M to get an A on the course? Explain. Assume an A ranges from 92 to 100 inclusive.

*3.21. Suppose 100 students have a mean grade of 72.4, with the boys and girls in the group having mean grades of 70.6 and 78.7, respectively. How many students are boys?

3.22. The mean height of 32 girls is 63.4 inches. The mean heights of the boys and of the combined groups are 69.8 inches and 67.0 inches, respectively. How many boys are there?

*3.23. The 40 football players on team A have a mean weight which is 8 pounds more than that of the 50 players on team B. If the mean weight of the 90 players is 190.0 pounds, what is the mean weight of each team?

*3.24. Suppose the weight of a measurement is the square of its reciprocal. Find the weighted mean of 4, 5, and 6.

3.25. Use the data of Exercise 3.17 to compute and compare the following measures of location:

$$\mu \qquad \frac{Q_1 + Q_3}{2} \qquad \frac{D_2 + D_8}{2} \qquad \frac{P_{17} + P_{83}}{2}$$

3.26. In a symmetrical distribution with a single mode (see Exercise 3.9) the arithmetic mean, the median, and the mode are the same; in a skewed distribution they differ, the median falling between the other two. In fact, these three measures are approximately related by the equation

$$\mu - \text{mode} = 3(\mu - M)$$

Use the data of Exercise 3.17 to check this equation.

3.6. SUMMATION PROPERTIES

Sums of many types are found in statistics. The capital letter sigma, Σ, introduced in Section 3.4, is a useful abbreviation for a sum. A few important properties involving the **sigma notation**, sometimes called the **summation notation**, will be helpful in later work.

We have already learned that the sum of

$$y_1, y_2, \ldots, y_N$$

may be expressed as $\sum y_i$ or $\sum_{i=1}^{N} y_i$. The letter y is an arbitrary symbol for a variable, the number subscript on y designates the number of the observation, and the letter subscript i denotes an arbitrary positive integer from 1 to N inclusive. The letters u, v, w, y, z, and f are also used to denote variables, and j, k, l, and m are sometimes used to denote subscripts. Thus, the sum

$$y_1 + y_2 + \cdots + y_N$$

may also be expressed as

$$\sum_{j=1}^{N} y_j \quad \text{or} \quad \sum_{m=1}^{N} y_m \quad \text{or} \quad \sum_{i=0}^{N-1} y_{i+1} \cdots$$

The symbol following Σ may be any sort of mathematical expression. For example, we have already seen that when yf follows Σ in the expression $\sum_{i=1}^{h} y_i f_i$ the expanded form becomes $y_1 f_1 + y_2 f_2 + \cdots + y_h f_h$. Further, the sum $y_2 f_2 + y_3 f_3 + \cdots + y_7 f_7$ may be abbreviated as $\sum_{i=2}^{7} y_i f_i$.

Three useful properties of summation are

(3.13) $$\sum_{i=1}^{N} k = Nk$$

(3.14) $$\sum_{i=1}^{N} k u_i = k \sum_{i=1}^{N} u_i$$

(3.15) $$\sum_{i=1}^{N} (u_i \pm v_i) = \sum_{i=1}^{N} u_i \pm \sum_{i=1}^{N} v_i$$

It is to be understood that k is a constant and that both u_i and v_i may be functions of y_i and f_i. The proofs of these **summation identities** follow directly from the definition

(3.16) $$\sum_{i=1}^{N} y_i = y_1 + y_2 + \cdots + y_N$$

Let $y_i = k$ for each i. Substituting in Equation (3.16) gives

Summation Properties

$$\sum_{i=1}^{N} k = \overbrace{k + k + \cdots + k}^{N \text{ terms}} = Nk$$

Thus, Equation (3.13) holds. Further, if we substitute ku_i for y_i ($i = 1, 2, \ldots, N$) in Equation (3.16) we get

$$\sum_{i=1}^{N} ku_i = ku_1 + ku_2 + \cdots + ku_N$$
$$= k(u_1 + u_2 + \cdots + u_N)$$
$$= k \sum_{i=1}^{N} u_i$$

So Equation (3.14) holds. The proof of Equation (3.15) is left to the reader (see Exercise 3.32).

We now use these three summation identities to prove Equation (3.9) of Theorem 3.1. First we note that

$$u_i = \frac{y_i - y'}{c} = \frac{1}{c}(y_i - y') \quad i = 1, 2, \cdots, N$$

Then it follows that

$$\mu_u = \frac{\sum u_i}{N}, \text{ by definition of mean}$$

$$= \frac{\sum \frac{1}{c}(y_i - y')}{N}, \text{ by substitution}$$

$$= \frac{\frac{1}{c} \sum (y_i - y')}{N}, \text{ by Equation (3.14)}$$

$$= \frac{\frac{1}{c}[\sum y_i - \sum y']}{N}, \text{ by Equation (3.15)}$$

$$= \frac{1}{c}\left[\frac{\sum y_i}{N} - \frac{Ny'}{N}\right], \text{ by Equation (3.13) and algebra}$$

$$= \frac{1}{c}[\mu_y - y'], \text{ by definition of mean and algebra}$$

Hence

(3.17) $$c\mu_u = \mu_y - y'$$

Therefore

$$\mu_y = y' + c\mu_u$$

To further illustrate the three basic summation properties, we prove the following two very important identities:

$$(3.18) \qquad \sum_{i=1}^{h} (y_i - \mu) f_i = 0$$

$$(3.19) \qquad \sum_{i=1}^{h} (y_i - \mu)^2 f_i = \sum_{i=1}^{h} y_i^2 f_i - \frac{\left(\sum_{i=1}^{h} y_i f_i\right)^2}{N}$$

Proof of Equation (3.18):

$\sum (y_i - \mu) f_i = \sum (y_i f_i - \mu f_i)$, by algebra

$\qquad = \sum y_i f_i - \sum \mu f_i$, by Equation (3.15)

$\qquad = \sum y_i f_i - \mu \sum f_i$, by Equation (3.14)

$\qquad = \sum y_i f_i - \frac{\sum y_i f_i}{N} N$, by definition of mean and $\sum f_i = N$

$\qquad = 0$, by algebra

Thus Equation (3.18) holds.

Proof of Equation (3.19):

$\sum (y_i - \mu)^2 f_i = \sum (y_i^2 f_i - 2\mu y_i f_i + \mu^2 f_i)$, by algebra

$\qquad = \sum y_i^2 f_i - \sum 2\mu y_i f_i + \sum \mu^2 f_i$, by extension of Equation (3.15)

$\qquad = \sum y_i^2 f_i - 2\mu \sum y_i f_i + \mu^2 \sum f_i$, by Equation (3.14)

$\qquad = \sum y_i^2 f_i - 2\left(\frac{\sum y_i f_i}{N}\right) \sum y_i f_i + \left(\frac{\sum y_i f_i}{N}\right)^2 N$, by definition of mean and $\sum f_i = N$

$\qquad = \sum y_i^2 f_i - \frac{(\sum y_i f_i)^2}{N}$, by algebra.

Thus Equation (3.19) holds.

Occasionally we use a **double summation**. We illustrate properties with

$$(3.20) \qquad \sum_{i=1}^{3} \sum_{j=1}^{2} y_{ij}$$

We may expand Expression (3.20) by first writing either

$$(3.21) \qquad \sum_{i=1}^{3} \sum_{j=1}^{2} y_{ij} = \sum_{j=1}^{2} y_{1j} + \sum_{j=1}^{2} y_{2j} + \sum_{j=1}^{2} y_{3j}$$

or

$$(3.22) \qquad \sum_{i=1}^{3} \sum_{j=1}^{2} y_{ij} = \sum_{i=1}^{3} (y_{i1} + y_{i2})$$

Then the final result, obtained from either Equation (3.21) or (3.22), is

$$\sum_{i=1}^{3}\sum_{j=1}^{2} y_{ij} = y_{11} + y_{12} + y_{21} + y_{22} + y_{31} + y_{32}$$

Generally, the reader is advised to expand Expression (3.20) by way of Equation (3.22).

For a double summation such as Expression (3.20) we may interchange the order of summation. In general, if a and b are two positive integers equal to or greater than one we may write

(3.23) $$\sum_{i=1}^{a}\sum_{j=1}^{b} y_{ij} = \sum_{j=1}^{b}\sum_{i=1}^{a} y_{ij}$$

(That is, the order of summation is not important for a rectangular array of symbols.)

3.7. EXERCISES

3.27. Write each of the following without summation signs:

(a) $\sum_{i=1}^{3} y_i^2$ (b) $\sum_{j=2}^{4} (y_j - 1)$ (c) $\sum_{k=1}^{3} y_k^2$

(d) $\sum_{k=1}^{4} (y_k - \mu)^2$ (e) $\sum_{j=2}^{5} (y_j - 3)f_j$ (f) $\sum_{i=1}^{3} (ay_i + b)$

(g) $\left[\sum_{i=1}^{a} (y_i - \mu)\right]^2$ (h) $\sum_{i=1}^{3}\sum_{j=1}^{4} y_{ij} f_i$ (i) $\sum_{i=1}^{3}\sum_{j=1}^{n_i} y_{ij}^2$

3.28. Write each of the following with summation signs:
(a) $y_1^2 + y_2^2 + \cdots + y_9^2$
(b) $(y_1 - 4)f_1 + (y_2 - 4)f_2 + \cdots + (y_N - 4)f_N$
(c) $[(y_1 - 4) + (y_2 - 4) + (y_3 - 4)]^2$
(d) $[(y_1 - \mu)^2 f_1 + (y_2 - \mu)^2 f_2 + \cdots + (y_N - \mu)^2 f_N]/N$
(e) $y_{11} + y_{12} + y_{13} + y_{14} + y_{21} + y_{22} + y_{23} + y_{24}$
(f) $(y_{11} - \mu_1)^2 + (y_{12} - \mu_1)^2 + (y_{21} - \mu_2)^2 + (y_{22} - \mu_2)^2$
$+ (y_{31} - \mu_3)^2 + (y_{32} - \mu_3)^2$

3.29. If $y_1 = 3$, $y_2 = -2$, $y_3 = 0$, and $y_4 = 6$, find the value of each of the first four parts of Exercise 3.27.

3.30. If $y_1 = -3$, $y_2 = -1$, $y_3 = 2$, $y_4 = 3$, $y_5 = 4$, find the value of

(a) $\left[\sum_{i=1}^{5} (y_i + 1)\right]^2$ (b) $\sqrt{\sum_{j=2}^{4} y_j(y_j - 2)}$

(c) $\sum_{i=1}^{5} \frac{(y_i - \mu)^2}{5}$ (d) $\left[\sum_{i=3}^{5} \sqrt{y_i}\right]^2$

3.31. Use the following data to evaluate (a) through (g):

$y_{11} = 3 \quad y_{21} = 2 \quad y_{31} = -3 \quad y_{41} = 1$
$y_{12} = 1 \quad y_{22} = -1 \quad y_{32} = 3 \quad y_{42} = 2$
$y_{13} = 2 \quad y_{23} = -1 \quad y_{33} = 0 \quad y_{43} = -1$

(a) $\sum_{j=1}^{3} y_{ij}$ for each i ($i = 1, 2, 3, 4$)   (b) $\sum_{i=1}^{4} y_{ij}$ for each j ($j = 1, 2, 3$)

(c) $\sum_{i=1}^{4} \sum_{j=1}^{3} y_{ij}$ using (a)   (d) $\sum_{i=1}^{4} \sum_{j=1}^{3} y_{ij}^2$

(e) $\sum_{i=1}^{4} \left[\sum_{j=1}^{3} y_{ij} \right]^2$ (f) $\sum_{j=1}^{3} \left[\sum_{i=1}^{4} y_{ij} \right]^2$

(g) $\left[\sum_{i=1}^{4} \sum_{j=1}^{3} y_{ij} \right]^2$

3.32. Prove Equation (3.15). Note that Equation (3.15) may be extended to three or more terms. For example

$$\sum_i (u_i + v_i - w_i) = \sum_i u_i + \sum_i v_i - \sum_i w_i$$

3.33. Let $\mu_i = \dfrac{\sum_{j=1}^{N_i} y_{ij}}{N_i}$ for each i ($i = 1, \cdots, h$) and let

$$\mu_c = \dfrac{\sum_{i=1}^{h} \sum_{j=1}^{N_i} y_{ij}}{\sum_{i=1}^{h} N_i}$$

Prove Equation (3.11); i.e., prove $\mu_c = \dfrac{\sum_i N_i \mu_i}{\sum_i N_i}$

3.34. Prove: $\sum_{i=1}^{N} (y_i - \mu) = 0$

3.35. Prove: $\sum_{i=1}^{N} (y_i - \mu)^2 = \sum_{i=1}^{N} y_i^2 - \dfrac{\left(\sum_{i=1}^{N} y_i \right)^2}{N}$

Prove the identities in Exercises 3.36 through 3.40. Assume that there are N values of y and that the summation is over all values.

3.36. $\sum (y_i - \mu)^2 = \sum y_i^2 - N \mu^2$

E3.37. $\sum [(y_i - \mu)^2 + y_i(\mu - 1)] = \sum y_i^2 - N \mu$

E3.38. $\sum [y_j(y_j - \mu) + \mu^2] = \sum y_j^2$

E3.39. $\sum [y_i(y_i - \mu) + \mu - y_i] = \sum y_i^2 - \dfrac{(\sum y_i)^2}{N}$

E3.40. $\sum y_i(y_i - \mu)^2 - \mu \sum (\mu - 2y_i)y_i = \sum y_i^3$

*3.41. A measure of location, L, is sometimes defined to be a numerical description with the property that when the same number y' is subtracted from each observation then y' must also be subtracted from L. Note that we have already proved [see Equation (3.17) and let $c = 1$] that μ is a measure of location.

Use this definition to determine whether each of the following is a measure of location: median, harmonic mean, geometric mean, and $(Q_1 + Q_3)/2$.

3.8. MEASURES OF VARIATION

Two sets of data may have the same value for a measure of location, but still have quite different distributions. For example, the numbers 48, 51, 52, 50, 49 have a mean of 50 as do the numbers 35, 60, 60, 45, 50. But the numbers in the first case are much more closely clustered about the mean. We would like a measure of dispersion (or "variation" or "scatter") to reflect this property. That is, a measure of variation should be defined so that the greater the scatter in distributions the larger the number which measures variation.

The **range** of a distribution, defined as the positive difference in the two extreme values of a distribution, is sometimes a useful **measure of dispersion**. The two distributions of the last paragraph have ranges of $52 - 48 = 4$ and $60 - 35 = 25$, respectively. Clearly, the distribution with the larger dispersion has the larger range, but this is not always the case. For example, the distribution with values 48, 48, 52, 52 has the same range as the distribution with values 48, 50, 50, 52. Since these two distributions are not equally disperse it is clear that the range, as a measure of variation, does not distinguish between them. This is primarily because the values between the two extremes are not used in the computation of the range. Thus, the range is not used with most distributions. However, for a very small set of values, say from two to seven, the range is occasionally applied. The measure is attractive because it is easy to understand and easy to compute.

A natural way to measure scatter of a distribution (and use every observation) would be in terms of distance of observations from a measure of location. If the mean is the measure of location, then the distance of the ith observation y_i from the mean μ is $y_i - \mu$, and it is called the *deviation of y_i from the mean*. If the measure of location is the median M, then $y_i - M$ is the *deviation of y_i from the median*. We first discuss measures of variation (or dispersion) in terms of deviations from the mean.

For any set of data $y_1, y_2, \cdots, y_N$, the set of deviations from their mean

(3.24) $\qquad y_1 - \mu, y_2 - \mu, \cdots, y_N - \mu$

is a **distribution of deviations**. Thus, we might consider using the mean of these N

deviations as a measure of variation. But according to Exercise 3.34

$$\Sigma (y_i - \mu) = 0$$

and hence the mean of the deviations in Expression (3.24) is also zero. Since this property holds regardless of the amount of scatter in a distribution, the **mean of the deviations from the mean** is of no use in comparing the variation of two or more distributions.

Fortunately, we are actually interested in the **magnitudes** of the deviations; that is, in the absolute values

$$|y_1 - \mu|, |y_2 - \mu|, \cdots, |y_N - \mu|$$

So we consider the **mean of the absolute values of the deviations from the mean**, that is

(3.25) $$\frac{\Sigma |y_i - \mu|}{N}$$

Expression (3.25) is also called **mean deviation**. When data are used Expression (3.25) is relatively easy to compute, and its value increases as the scatter of a distribution gets larger. However, in theoretical work the mean deviation is difficult, if not impossible, to use due to the presence of absolute values. Thus, we shall generally ignore the mean deviation.

Fortunately, there is a way to avoid the use of absolute value signs and still retain the essential nature of Expression (3.25). To do this we find the mean of the squared deviations from the mean and then take the positive square root so that the measure of variation is expressed in terms of the unit of measure of the observations. Such a measure of variation is called the **standard deviation** and is denoted by σ. In symbols, the finite population $y_1, \cdots, y_N$ has standard deviation

(3.26) $$\sigma = \sqrt{\frac{\Sigma (y_i - \mu)^2}{N}}$$

(The symbol σ is the lowercase Greek letter sigma.) Even though the standard deviation is more difficult to compute than the mean deviation, it is theoretically quite satisfactory.

The square of the standard deviation, termed the **variance** and denoted by σ^2, is also used as a measure of variation. Thus the variance σ^2 is defined by

(3.27) $$\sigma^2 = \frac{\Sigma (y_i - \mu)^2}{N}$$

and is in terms of the square of the unit of measure of the variable. For example, if y is measured in terms of "feet," then σ^2 is in terms of "square feet" while σ is in terms of "feet."

EXAMPLE 3.13. Compare the standard deviations of the following distributions:

Distribution A: 1, 2, 3, 4, 5
Distribution B: 1, 1, 3, 5, 5

The mean of each distribution is 3. The standard deviation for Distribution A is

$$\sigma_A = \sqrt{\frac{(1-3)^2 + (2-3)^2 + (3-3)^2 + (4-3)^2 + (5-3)^2}{5}}$$
$$= \sqrt{2} \doteq 1.414$$

and the standard deviation of Distribution B is

$$\sigma_B = \sqrt{\frac{(1-3)^2 + (1-3)^2 + (3-3)^2 + (5-3)^2 + (5-3)^2}{5}}$$
$$= \sqrt{16/5} = \sqrt{3.2} \doteq 1.789$$

Clearly, the distribution with greater variation has the larger standard deviation. Since

$$\sigma_A^2 = 2 \quad \text{and} \quad \sigma_B^2 = 3.2$$

a similar statement can be made about the variances.

It may be of interest to compare the mean deviations of the two distributions of Example 3.13. The mean deviation for Distribution A is

$$\frac{|1-3| + |2-3| + |3-3| + |4-3| + |5-3|}{5} = \frac{6}{5} = 1.2$$

and the mean deviation for Distribution B is $8/5 = 1.6$. Again, we observe that the set with greater variation has larger mean deviation. The reader should note that when all data values are the same, there is no variation and both the standard deviation and the mean deviation are zero.

3.9. SOME APPLICATIONS OF THE STANDARD DEVIATION

We have already noted that the standard deviation may be used as a measure of dispersion of values about the mean. The smaller the value of σ, the closer the values of the distribution concentrate about the mean. Furthermore, if a large collection of observations has a distribution which is "bell-shaped" (like the one shown in Figure 2.7), it can be shown that about $\frac{2}{3}$ of the observations fall in the interval from $\mu - \sigma$ to $\mu + \sigma$; about $\frac{19}{20}$ fall in the interval from $\mu - 2\sigma$ to $\mu + 2\sigma$; and about $\frac{997}{1,000}$ fall in the interval from $\mu - 3\sigma$ to $\mu + 3\sigma$.

EXAMPLE 3.14. Find the proportion of test scores in Table 2.3 which fall in each of the intervals of the last paragraph.

From the frequency distribution of the grouped test scores (Table 2.4) we observe that the distribution is approximately bell-shaped. It can be shown that $\mu = 68.824$ and $\sigma = 9.297$ (see Exercise 3.61.). Thus, the interval from $\mu - \sigma$ to $\mu + \sigma$ becomes 59.5 — 78.1. By actual count we find that 88 observations fall in this interval. That is, $\frac{88}{125} = .704$ is the proportion of values which falls in an interval within one standard deviation of the mean. Also, 117 of 125 observations fall in the interval from $\mu - 2\sigma = 50.2$ to $\mu + 2\sigma = 87.4$, and 125 observations fall in the interval from $\mu - 3\sigma = 40.9$ to $\mu + 3\sigma = 96.7$. That is, the proportion of values in these intervals is 0.936 and 1.000, respectively. We see that these three proportions are fairly close to those (0.667, 0.950, 0.997) given above for the bell-shaped distribution. The proportion of values in each interval depends on a fairly large number of observations having a bell-shaped distribution.

There is a similar general property which does not depend on the "shape of the distribution" or "the number of observations." This property, known as **Chebyshev's theorem**, may be stated as "at least $[1 - (1/k^2)]$ is the proportion of observations which falls within k standard deviations of their mean" where k is some positive number. In particular, $1 - (1/2^2) = 0.75$ or more is the proportion of observations of *any distribution* which falls between $\mu - 2\sigma$ and $\mu + 2\sigma$. (Of course, this statement is much more conservative than the similar statement about the bell-shaped distribution because allowances must be made for all sorts of "wild" distributions.) The student might wish to experiment with certain specific distributions (see Exercise 3.47).

The mean and standard deviation are often used together to define another measure. If y is a measurement for a set of data, then

$$\frac{y - \mu}{\sigma}$$

is its corresponding **standard measurement**, and it measures how many standard deviations y deviates from the mean. The value of $(y - \mu)/\sigma$ is expressed in **standard units**.

A standard measurement is a so-called "pure number" or "number without dimension" or, mathematically, "abstract number." Thus, if y is expressed in inches, then μ and σ are expressed in inches. So $y - \mu$ is expressed in inches, but $(y - \mu)/\sigma$ is expressed in standard units. Such units of measurement are often useful in making comparisons.

EXAMPLE 3.15. On the first and second algebra tests Bob made grades of 90 and 88, respectively. In the absence of any further information about the grade distributions we might say that Bob's grades were essentially the same with the first test grade being only slightly better. With the additional information that the means on the first and second tests were 75 and 80, respectively, we find that Bob made 15 points more than the mean grade on the first test and only 8 points higher on the second test. Thus, it might appear that Bob did much better on the first test. But suppose we learn that the standard deviations on the first and second tests were 15 and 4, respectively. This places the grades 1 and 2 standard deviations above their respective means. Now, it appears that the second test grade was far superior. (Actually, in a class of 40, Bob made only the seventh highest grade on the first test and the highest grade on the second test.) Clearly, the dispersion of grades should be taken into account.

EXAMPLE 3.16. In a certain statistics course the final grade is determined from three tests and a final examination where the examination counts twice as much as each test. Use the data of Table 3.4 to determine whether Alice or Tom has the better final grade.

Table 3.4. Data for Example 3.16

	Mean Score	Standard Deviation	Alice's Score	Tom's Score
Test 1	70	10	80	90
Test 2	70	20	90	70
Test 3	70	5	75	90
Examination	75	15	90	75

The weighted mean scores of Alice and Tom are

$$\frac{80 + 90 + 75 + 2(90)}{1 + 1 + 1 + 2} = 85$$

and

$$\frac{90 + 70 + 90 + 2(75)}{1 + 1 + 1 + 2} = 80$$

respectively. Thus, ignoring information about the means and standard deviations, we would decide that Alice has the better grade by 5 points, and we might even record a grade of B for Alice and a grade of C for Tom.

In terms of standard scores, Alice has scores of $(80 - 70)/10 = 1$, 1, 1, and 1 with a weighted mean standard score of 1. Tom has standard scores of 2, 0, 4, and 0 with a weighted mean of 1.20. Thus, Tom's standing in class appears better than Alice's. (Actually, it can be shown that Alice's and Tom's weighted mean standard scores fall at the 84 and 88 percentiles, respectively.) Clearly, Alice has not earned a better letter grade than Tom.

3.10. EXERCISES

3.42. Compute the following measures of dispersion for the distribution with values

    13   16   12   18   13   17   16

(a) the mean deviation
(b) the range
(c) the standard deviation

3.43. Compute the following measures of dispersion for the distribution with values

    17   15   12   16   17   14   12   17

(a) the mean deviation
(b) the range
(c) the standard deviation

3.44. Decide which of the two distributions in Exercises 3.42 and 3.43 is more disperse (spread out). Then check each of the three measures of variation to determine if their values are consistent with your intuition. Write a summary statement.

3.45. Compute the variance for each of the following:

Distribution A:	6	14	4	6	16	2
Distribution B:	7	3	1	3	2	8
Distribution C:	6	9	21	24	9	3

Compare the dispersions of these distributions in terms of their variances. Write a summary statement, indicating whether you think the relative magnitudes of the variances should be taken seriously.

3.46. The arithmetic mean of the absolute value of the deviations of the observations from the median, denoted DM, is defined by

(3.28)
$$DM = \frac{\sum_{i=1}^{N} |y_i - M|}{N}$$

This measure of variation is called the **mean deviation about the median**.
(a) Use Equation (3.28) to find DM for the data of Exercise 3.42.
(b) It can be shown that the mean deviation about the median may also be computed by the formula

$$DM = \frac{\sum \text{larger half of values} - \sum \text{smaller half of values}}{N}$$

the median being excluded if N is an odd number. Use this formula to compute DM

for the data of Exercise 3.42. (The simplicity of this formula illustrates why the measure DM is sometimes preferred to the mean deviation about the mean.)
*(c) Prove the formula in part (b).

3.47. The four distributions in Table 3.5 have a common mean of 5.5. The standard deviations are 2.872, 2.062, 1.526, and 3.263, respectively.
(a) For each distribution, find the proportion of values between $\mu - 1.5\sigma$ and $\mu + 1.5\sigma$; between $\mu - 2\sigma$ and $\mu + 2\sigma$; and between $\mu - 2.5\sigma$ and $\mu + 2.5\sigma$.

Table 3.5. Distributions to be Used for Comparative Studies

I		II		III		IV	
y_i	f_i	y_i	f_i	y_i	f_i	y_i	f_i
1	5	1	1	1	1	1	10
2	5	2	3	2	1	2	5
3	5	3	5	3	1	3	5
4	5	4	7	4	7	4	0
5	5	5	9	5	15	5	0
6	5	6	9	6	15	6	5
7	5	7	7	7	7	7	5
8	5	8	5	8	1	8	10
9	5	9	3	9	1	9	5
10	5	10	1	10	1	10	5

(b) Compare the proportions found in part (a) with those given by Chebyshev's theorem. Which type of distribution seems to have proportions closest to those given by the theorem? Write a summary statement.

3.48. The final grade on a psychology course is determined from two tests prepared by the teacher, two standardized tests prepared by a testing service, one laboratory project, and a final examination. Table 3.6 shows the percent that each part counts toward the final grade, the mean and standard deviation of each part, and the numerical grades of three students. Use the information of Table 3.6 to determine the final grade (in standard units) of each student.

Table 3.6. Scores in a Psychology Course

	Percent of Final Grade	Mean Score	Standard Deviation	Score of Student		
				M	N	O
Test 1	15	75	10	95	85	100
Test 2	15	90	20	150	110	150
Test 3	15	500	100	600	660	500
Test 4	15	550	100	550	650	600
Project	15	35	4	51	31	47
Examination	25	150	40	210	190	170

3.49. In a summer school project the final mark of each student is based on grades received in courses in matrix algebra, programming, and statistics. Two tests and a final examination are given in both matrix algebra and statistics, but only a final examination is given in programming. All tests and examinations are based on a maximum of 100 points and an examination counts twice as much as a test. Use the data in Table 3.7 to determine which of the four top students should be the valedictorian.

Table 3.7. Summer Project Test and Examination Scores

	Mean Score	Standard Deviation	Scores of Student			
			W	X	Y	Z
Algebra Test 1	76	8	80	90	85	70
Algebra Test 2	74	10	80	90	85	70
Algebra Examination	75	10	80	90	85	70
Statistics Test 1	65	5	80	70	75	90
Statistics Test 2	70	4	80	70	75	90
Statistics Examination	60	5	80	70	75	80
Programming Examination	80	4	80	80	80	90

3.50. Standard measurements are small—generally ranging between -4 and $+4$. With scores (or grades) it is desirable that they be positive and larger. To adjust for this we introduce the **Z-score**, defined by

$$Z = 50 + 10u$$

where u is a standard score.
(a) For each student of Table 3.6 find all Z-scores and determine the weighted mean Z-score.
(b) Use Table 3.7 to work (a).

E3.51. Occasionally we need to compare the variation of two or more distributions when observations are expressed in different units of measure. One measure which is commonly used is the **coefficient of variation,** denoted by V, and defined by

$$V = \frac{\sigma}{\mu}$$

(Some writers call $100\sigma/\mu$ the coefficient of variation.) We observe that V is a measure of relative variation and is an abstract number. That is, V expresses the magnitude of the variation relative to the size of whatever is being measured in terms of a pure number.
(a) Compute the coefficient of variation for each of the distributions in Exercise 3.45, and make comparisons.
(b) Compare the coefficients of variation of the distributions in Exercise 3.47, and make comparisons.

E3.52. Compare the variation in weights for the following distributions:

Rats (in grams):   195 205 180 190 190 185 195 215 200 200
Cows (in pounds): 450 600 575 540 490 535 550 510

Hint: See Exercise 3.51.

3.11. MEASURES OF VARIATION—GROUPING AND CODING

If N values are arranged in a frequency table and there are h distinct values, then the variance is given by

(3.29) $$\sigma^2 = \frac{\sum_{i=1}^{h}(y_i - \mu)^2 f_i}{N}$$

where f_i denotes the frequency of y_i, the ith ordered value of y. Equation (3.29) gives exactly the same value of the variance as Equation (3.27), but requires less effort as a rule.

If the N values are grouped into h classes, and if y_i and f_i denote the class mark and class frequency, respectively, of the ith class, then Equation (3.29) gives the variance σ_g^2 for grouped data. It can be shown that σ_g^2 tends to be larger than σ^2 for any set of data and that the difference depends on the class length. (See Exercise 3.58 for an illustration.)

Since the standard deviation is the positive square root of the variance, it is not necessary that formulas for both be displayed. Thus, in the beginning, we restrict our attention to the variance.

In Example 3.13 the variances were easy to compute. This was mainly due to the presence of small integers and few values. Further, if μ had been rounded-off, troublesome approximations would have been introduced at an early stage in the computations. For these and other reasons the following computational form of the variance is normally used:

(3.30) $$\sigma^2 = \frac{N \sum y_i^2 f_i - (\sum y_i f_i)^2}{N^2}$$

Equation (3.30) is easy to derive from Equations (3.19) and (3.29) and will be left to the reader as an exercise (see Exercise 3.53).

EXAMPLE 3.17. Use Equations (3.29) and (3.30) to find the variance of the grouped test score data of Table 2.4. Compare the two methods of computing the variance.

In Example 3.6 we found the mean to be 68.64. In computing σ^2 by Equation (3.29) the mean is used in an intermediate step. Thus, we keep the two decimal places. (For if we rounded-off to 68.6 we would lose 0.04 for each of 125 values. The reader can show in Exercise 3.59 that this leads to a variance of 87.1120.) By the computations shown in Table 3.8 we find that the variance is

$$\sigma^2 \doteq \frac{10,888.8000}{125} \doteq 87.1104 \doteq 87.11$$

Table 3.8. Frequency Distribution of Grouped Test Scores and Computations for the Variance by Equation (3.29)

Class Mark y_i	Frequency f_i	Deviations $y_i - 68.64$	$(y_i - 68.64)^2$	$(y_i - 68.64)^2 f_i$
42	1	−26.64	709.6896	709.6896
47	3	−21.64	468.2896	1,404.8688
52	7	−16.64	276.8896	1,938.2272
57	8	−11.64	135.4896	1,083.9168
62	16	−6.64	44.0896	705.4336
67	28	−1.64	2.6896	75.3088
72	34	3.36	11.2896	383.8464
77	12	8.36	69.8896	838.6752
82	13	13.36	178.4896	2,320.3648
87	1	18.36	337.0896	337.0896
92	2	23.36	545.6896	1,091.3792
Totals	125			10,888.8000

Using Equation (3.30) and the computations shown in Table 3.9 we find the variance to be

$$\sigma^2 = \frac{125(599,820) - (8,580)^2}{(125)(125)}$$

$$= \frac{1,361,100}{15,625} = 87.1104 \doteq 87.11$$

In this case a decision about rounding-off is left to the very last step, and so any number of decimal places may be retained. Equation (3.30) is particularly useful with a desk calculator. In the absence of a desk calculator one of the following equivalent computational forms may be used to advantage:

(3.31) $$\sigma^2 = \frac{\sum y_i^2 f_i - \frac{(\sum y_i f_i)^2}{N}}{N}$$

or

(3.32) $$\sigma^2 = \frac{\sum y_i^2 f_i}{N} - \mu^2$$

Proofs of these equations will be left to the reader (see Exercise 3.53).

Table 3.9. Frequency Distribution of Grouped Test Scores and Computations for the Variance by Equation (3.30)

Class Mark y_i	Frequency f_i	$y_i f_i$	$y_i^2 f_i$
42	1	42	1764
47	3	141	6627
52	7	364	18928
57	8	456	25992
62	16	992	61504
67	28	1876	125692
72	34	2448	176256
77	12	924	71148
82	13	1066	87412
87	1	87	7569
92	2	184	16928
Totals	125	8580	599820

The above computations indicate why Equation (3.30) is better than Equation (3.29) for obtaining the variance of grouped data—there are fewer recordings, fewer calculations, and smaller rounding-off errors. But with each equation the computations are extensive. It is possible to shorten the work involved by coding (transforming) the data by the method of Theorem 3.2.

Theorem 3.2. If $y_1, y_2, \cdots, y_N$ are values of any set of data, and if these data are transformed to new data $u_1, u_2, \cdots, u_N$ by the formula

$$u_i = \frac{y_i - y'}{c} \quad (i = 1, \cdots, N)$$

where y' is any constant and c is any constant different from zero, then

(3.33) $$\sigma_y^2 = c^2 \sigma_u^2$$

and

(3.34) $$\sigma_y = c \sigma_u$$

It is to be understood that the subscripts u and y refer to the variables used in calculating the variance and the standard deviation.

The proof of Theorem 3.2 will be left to the reader (see Exercise 3.56). The theorem holds for both raw and grouped data. The constant added to (or subtracted from) each value of the variable does not affect the variance of the transformed data. That is, when $c = 1$, $u_i = y_i - y'$ and $\sigma_y = \sigma_u$. Sometimes it is desirable to state Equation (3.33) as

(3.35) $$\sigma_y^2 = c^2 \left(\frac{N \sum u_i^2 f_i - (\sum u_i f_i)^2}{N^2} \right)$$

in terms of the transformed values u_i.

EXAMPLE 3.18. Use Theorem 3.2 and Equation (3.30) to find the variance of the grouped test score data of Table 2.4.

If we let $y' = 67$ and $c = 5$, we obtain the values shown in Table 3.10. Substituting the totals of Table 3.10 in Equation (3.35) we obtain

$$\sigma_y^2 = 5^2 \left(\frac{125(449) - (41)^2}{(125)^2} \right) = 87.1104 \doteq 87.11$$

which is the value obtained earlier from Table 3.8. Of course we could use Equation (3.30) to find

$$\sigma_u^2 = \frac{125(440) - (41)^2}{125^2} \doteq 3.4844$$

and then substitute this value in Equation (3.33) to obtain

$$\sigma_y^2 \doteq 5^2(3.484) = 87.11$$

The standard deviation is then found to be $\sqrt{87.11} = 9.33$.

Table 3.10. Frequency Distribution of Grouped Test Scores and Computations for the Variance by Coding

Class Mark y_i	Frequency f_i	$u_i = \frac{y_i - 67}{5}$	$u_i f_i$	$u_i^2 f_i$
42	1	−5	−5	25
47	3	−4	−12	48
52	7	−3	−21	63
57	8	−2	−16	32
62	16	−1	−16	16
67	28	0	0	0
72	34	1	34	34
77	12	2	24	48
82	13	3	39	117
87	1	4	4	16
92	2	5	10	50
Totals	125		41	449

3.12. APPROXIMATIONS IN CALCULATIONS

In most of the calculations of this chapter we have found it necessary to "round-off" numbers. However, as we shall shortly observe, the usual rules for rounding-off numbers are not always adequate. When only one or two mathematical operations are required, the "rounded-off" answer is usually satisfactory; but when a long series of operations are performed, there is obviously more opportunity to make sizeable errors. Thus, we should take a close look at the "rounding-off" methods.

First, we review rules for rounding-off numbers in case measurements are made by an instrument which has equally spaced markings. Examples of such instruments are meter sticks, thermometers, scales, and electric meters. *If the object being measured falls closest to one marking we use that marking as our measurement; if it falls halfway between two markings we use the one (of two) which is designated by an even number.* Thus, if the mercury *falls halfway between* 72 and 73 degrees fahrenheit, we record the temperature as 72 degrees or, if the width of a table top is halfway between 1.23 and 1.24 meters, we record the width as 1.24 meters. But when the weight of a rat is *closer* to 195.3 grams than 195.4 grams we record the weight as 195.3 grams. In each case, the **true measurement** is the measurement recorded plus or minus one-half the smallest unit of measurement. Thus, a recording of 72 degrees means that the true temperature is between $72 - 0.5$ and $72 + 0.5$ a recording of 1.24 meters means that the true table width is between $1.24 - 0.005$ and $1.24 + 0.005$, and so forth. In these illustrations we say the measurement of temperature is accurate to two significant digits, of table width to three significant digits, and of weight of a rat to four significant digits.

There are well-known rules for rounding-off the product or sum of two or more numbers. We first consider the product of two numbers. The rule may be stated as follows: *If each of two numbers has* m *significant digits, then the product must be rounded-off (by the above rule) so that it has* m *digits,* sometimes erroneously referred to as *m* **significant digits**. For example, according to this rule the product of 1.3 and 1.7 is 2.2 since $(1.3)(1.7) = 2.21$ is closer to 2.2 than 2.3, but the true product of the measurements represented by 1.3 and 1.7 (as we shall soon see) might be quite different from 2.2. Furthermore, we often act as though the product 2.2 is accurate to two significant digits. That is, we make the false claim that the true product is between 2.15 and 2.25.

Let us examine this property more closely. Since both 1.3 and 1.7 are assumed to have two significant digits, then 1.3 represents a number between 1.25 and 1.35 and 1.7 represents a number between 1.65 and 1.75. Thus, the true product should be a number between

$$(1.25)(1.65) = \underline{2.0625} \text{ and } (1.35)(1.75) = \underline{2.3625}$$

Clearly, *the true product cannot be expressed in terms of two significant digits*—for the two significant digit number 2.2 indicates that the true product is between 2.15 and 2.25. Further, since the one significant digit number 2 indicates that the true product is between 1.5 and 2.5, *the true product cannot be expressed in terms of one significant digit*. In the first case the set of values from 2.15 to 2.25 does not include all possible true products; in the second case the set of values from 1.5 to 2.5 includes too many. This illustrates why a single approximate number, expressed in terms of significant digits, cannot be used to represent accurately the true product of two numbers.

Next, we briefly discuss an example of the usual rounding-off rule for the addition of two approximate numbers. The rule would state, if both 1.3 and 1.7 have two significant digits, then the sum $1.3 + 1.7 = 3.0$ has two significant digits. That is, the rule indicates that the true sum is between 2.95 and 3.05. However, the true sum is actually between

$$1.25 + 1.65 = 2.90 \text{ and } 1.35 + 1.75 = 3.10$$

Thus, we observe that the sum cannot accurately be expressed in terms of a single approximate number.

Similar statements could be made about the difference or quotient of two numbers. Actually, any careful treatment of the sum, product, difference, or quotient of two numbers would require that two extreme values be determined. Extensions to more than two numbers or more than one operation would require a similar treatment. Some of the essential features of such a treatment are made clearer in the following paragraphs.

Suppose an *approximate value of a number*, or, more briefly, **approximate number**, is a number which differs from the *true value* by some restricted small amount. Clearly, an approximate number may be expressed in several ways. Regardless of the way we write an approximate number, it is important that we understand that such a number represents an interval (or range) within which the true value of the number is located. If we let y_l denote the largest possible value and y_s the smallest possible value in the interval, then one useful way to express an approximate number y is the form

$$\begin{bmatrix} y_l \\ y_s \end{bmatrix}$$

An approximate number y expressed in this form is termed a **range number**, and the two extreme values y_l and y_s are the **components of the range number**.

Let

$$\begin{bmatrix} a_l \\ a_s \end{bmatrix} \text{ and } \begin{bmatrix} b_l \\ b_s \end{bmatrix}$$

denote range numbers of the two approximate numbers a and b, respectively. Then basic operations on two range numbers are defined by the following four equations:

(3.36) $$\begin{bmatrix} a_l \\ a_s \end{bmatrix} + \begin{bmatrix} b_l \\ b_s \end{bmatrix} = \begin{bmatrix} a_l + b_l \\ a_s + b_s \end{bmatrix}$$

(3.37) $$\begin{bmatrix} a_l \\ a_s \end{bmatrix} - \begin{bmatrix} b_l \\ b_s \end{bmatrix} = \begin{bmatrix} a_l - b_s \\ a_s - b_l \end{bmatrix}$$

$$
(3.38) \qquad \begin{bmatrix} a_I \\ a_s \end{bmatrix} \times \begin{bmatrix} b_I \\ b_s \end{bmatrix} = \begin{bmatrix} a_I \times b_I \\ a_s \times b_s \end{bmatrix}
$$

$$
(3.39) \qquad \begin{bmatrix} a_I \\ a_s \end{bmatrix} \div \begin{bmatrix} b_I \\ b_s \end{bmatrix} = \begin{bmatrix} a_I \div b_s \\ a_s \div b_I \end{bmatrix}
$$

The reader should note that Equation (3.36) gives the sum of two approximate numbers even though they are not given in terms of significant digits. Similar statements can be made about Equations (3.37), (3.38), and (3.39).

EXAMPLE 3.19. Suppose 2.1, 1.4, 1.9, and 2.3 are approximate numbers, each with two significant digits. Determine the range number for each of the following:

(a) $(2.1)(2.3) - (1.4)(1.9)$ (b) $2.1 \div 1.4$
(c) $(2.1 + 1.4 + 1.9 + 2.3) \div 4$ (d) $(2.1)^6$

From the definition given in Equation (3.38) it follows that

$$
(2.1)(2.3) = \begin{bmatrix} 2.15 \\ 2.05 \end{bmatrix} \begin{bmatrix} 2.35 \\ 2.25 \end{bmatrix} = \begin{bmatrix} 5.0525 \\ 4.6125 \end{bmatrix}
$$

$$
(1.4)(1.9) = \begin{bmatrix} 1.45 \\ 1.35 \end{bmatrix} \begin{bmatrix} 1.95 \\ 1.85 \end{bmatrix} = \begin{bmatrix} 2.8275 \\ 2.4975 \end{bmatrix}
$$

Thus, by Equation (3.37), the answer for part (a) is

$$
(2.1)(2.3) - (1.4)(1.9) = \begin{bmatrix} 5.0525 - 2.4975 \\ 4.6125 - 2.8275 \end{bmatrix} = \begin{bmatrix} 2.5550 \\ 1.7850 \end{bmatrix}
$$

By Equation (3.39), the solution for part (b) is

$$
2.1 \div 1.4 = \begin{bmatrix} 2.15 \\ 2.05 \end{bmatrix} \div \begin{bmatrix} 1.45 \\ 1.35 \end{bmatrix} = \begin{bmatrix} 2.15/1.35 \\ 2.05/1.45 \end{bmatrix}
$$

$$
= \begin{bmatrix} 1.592592\cdots \\ 1.413793\cdots \end{bmatrix} \doteq \begin{bmatrix} 1.593 \\ 1.413 \end{bmatrix}
$$

where the components of the range are rounded-off so that they define a range of values including the exact components which appear in the next-to-the-last step of the calculations. By repeated application of Equation (3.36) we find that

$$
2.1 + 1.4 + 1.9 + 2.3 = \begin{bmatrix} 2.15 \\ 2.05 \end{bmatrix} + \begin{bmatrix} 1.45 \\ 1.35 \end{bmatrix} + \begin{bmatrix} 1.95 \\ 1.85 \end{bmatrix} + \begin{bmatrix} 2.35 \\ 2.25 \end{bmatrix} = \begin{bmatrix} 7.90 \\ 7.50 \end{bmatrix}
$$

In part (c), assume the number 4 is exact. Thus, by Equation (3.39) it follows that

$$
(2.1 + 1.4 + 1.9 + 2.3) \div 4 = \begin{bmatrix} 7.90 \\ 7.50 \end{bmatrix} \div \begin{bmatrix} 4 \\ 4 \end{bmatrix} = \begin{bmatrix} 7.90/4 \\ 7.50/4 \end{bmatrix} = \begin{bmatrix} 1.975 \\ 1.875 \end{bmatrix}
$$

Finally, for part (d) we use Equation (3.38) repeatedly to find

$$(2.1)^6 = \begin{bmatrix} 2.15 \\ 2.05 \end{bmatrix}^6 = \begin{bmatrix} (2.15)^6 \\ (2.05)^6 \end{bmatrix} = \begin{bmatrix} 98.771297640625 \\ 74.220378765625 \end{bmatrix} \doteq \begin{bmatrix} 98.78 \\ 74.22 \end{bmatrix}$$

where the components of the range are rounded-off as for part (b).

EXAMPLE 3.20. Compute the values in Example 3.19 by the usual rounding-off methods, and compare the results with those found for Example 3.19.

Applying the rounding-off rules, we have for parts (a), (b), (c), (d), respectively

$$(2.1)(2.3) - (1.4)(1.9) = 4.83 - 2.66 = 2.17 \doteq 2.2$$
$$2.1 \div 1.4 = 1.5$$
$$(2.1 + 1.4 + 1.9 + 2.3) \div 4 = 1.925 \doteq 1.9$$
$$(2.1)^6 = 85.766121 \doteq 86$$

In terms of range numbers these four approximate numbers are, for parts (a), (b), (c), (d), respectively

$$\begin{bmatrix} 2.25 \\ 2.15 \end{bmatrix} \quad \begin{bmatrix} 1.55 \\ 1.45 \end{bmatrix} \quad \begin{bmatrix} 1.95 \\ 1.85 \end{bmatrix} \quad \begin{bmatrix} 86.5 \\ 85.5 \end{bmatrix}$$

and the corresponding true range numbers are

$$\begin{bmatrix} 2.555 \\ 1.785 \end{bmatrix} \quad \begin{bmatrix} 1.593 \\ 1.413 \end{bmatrix} \quad \begin{bmatrix} 1.975 \\ 1.875 \end{bmatrix} \quad \begin{bmatrix} 98.78 \\ 74.22 \end{bmatrix}$$

In each part, a two digit number does not adequately represent the true number because many possible values of the true number fall outside the indicated range of values of the corresponding approximate number. The reader should take special notice of the answers for part (d). See Dwyer [8] for further details.

The discussion on range numbers was given to emphasize some reasons why the rounding-off rules should be applied with caution. Two reasons to favor the rounding-off rules are (1) that an answer can be simply expressed by a single number, and (2) in a series of calculations numbers which overestimate and numbers which underestimate a set of true values behave in such a way that the errors tend to cancel (balance out) each other.

In certain calculations it seems desirable to carry one more digit in the solution than is found in the original data. For example, suppose ten measurements on the same object are

8.3 8.3 8.3 8.4 8.3 8.3 8.4 8.4 8.3 8.3

Clearly, the true measurement of the object is neither 8.3 nor 8.4, but some value in between. The mean of the measurements is

$$\frac{83.3}{10} = 8.33$$

and would appear to be better (more accurate) than either 8.3 or 8.4. This is usually the case when the "average" of several values is obtained.

Thus, *when computing characteristics like the mean, median, standard deviation, and mean deviation we follow the rule of rounding-off the solution to one more digit than is found in the numerically smallest number of the original data.* In most other calculations involving a few numbers we apply the usual rules of rounding-off. When a single number is raised to a large power, range numbers should be obtained or the number of "significant digits" in the solution should be decreased [see part (d) of Example 3.19].

In computing measurements which characterize a set of data, any intermediate computations may be carried out as far as one wishes. As a rule, from two to four more digits should be used. For example, when the standard deviation is computed by the defining formula [see Equation (3.26)], the mean μ should be recorded to two more digits than the original data.

3.13. EXERCISES

3.53. Prove Equations (3.30), (3.31), and (3.32).

3.54. Use Equation (3.30) with $f_i = 1$ to find each σ^2 in Exercise 3.45.

3.55. The weights of 40 girls are given in Table 3.11. Compute the variance to one decimal place by
(a) Equation (3.29)
(b) Equation (3.31)
(c) Theorem 3.2

Table 3.11. Weights of 40 Girls

Class Limits (in Pounds)	Frequency
90–99	3
100–109	6
110–119	11
120–129	14
130–139	4
140–149	2

3.56. Prove Theorem 3.2.

3.57. Use the data of Exercise 2.8 to compute
(a) the variance from a frequency table, and
(b) the variance from a grouped frequency table.

E3.58. When measurements are computed from a grouped frequency distribution, errors are introduced as a result of assuming the values of a class are concentrated at the class mark. Some writers have suggested corrections for certain measures of characteristics of specific types of distributions. For example, Sheppard has suggested that $c^2/12$ should be subtracted from the variance computed from grouped data with equal length classes, provided the distribution is bell-shaped and the common class lengths c are sufficiently small. For some grouped frequency distributions the corrected variance

$$\sigma_c^2 = \sigma^2 - \frac{c^2}{12}$$

is not necessarily an improvement over σ^2 and should be used with discretion.

Use Sheppard's correction to compute the variance for the data of Exercise 3.57 and compare with the correct variance computed in part (a) of Exercise 3.57. For further discussion see Mills [16].

C3.59. Use Equation (3.29) and the mean value 68.6 to find the variance of the grouped test score data of Table 2.4.

3.60. (a) By coding, find the variance of the data of the grouped frequency table constructed for Exercise 2.10(a).
(b) Find the proportion of weights within one standard deviation of the mean; within two standard deviations of the mean; and within three standard deviations of the mean.

C3.61. Compute the standard deviations for the distributions of Table 3.5.

3.62. Suppose 1.8, 2.1, 3.3, 2.7, and 1.5 are approximate numbers, each with two significant digits. Determine the components of the range number of each of the following:
(a) $1.8 + 2.1 + 3.3$
(b) $1.8 + 2.1 - 3.3$
(c) $(1.8)(2.1) + 3.3$
(d) $2.7 \div 1.5$
(e) $(1.5)^6$
(f) $(2.1 \div 1.8) \div 1.5$

3.63. Compare each range number obtained in Exercise 3.62 with the corresponding approximate number found by the usual rules of rounding-off.

E3.64. If the true value of y is in the interval from $y' - e$ to $y' + e$, we say y' represents y. Sometimes the set of values in the interval from $y' - e$ to $y' + e$ is denoted by $y'(e)$ where it is understood that e denotes the maximum error when y' represents y. We call $y'(e)$ the **approximate-error form** of a number y. For example, the approximate-error

75 Exercises

number 13(0.3) means that the true value is between $13 - 0.3 = 12.7$ and $13 + 0.3 = 13.3$. Evaluate each of the following by first changing to range numbers, and then express your solution in approximate-error form:

(a) $18(0.5) + 23(0.5) - 11(0.5)$
(b) $18(0.3) + 23(0.8) - 11(0.4)$
(c) $1.6(0.15) \div 0.9(0.32)$
(d) $6(0) \div 3.0(0.7)$
(e) $\sqrt{72.9(1.4)}$

4

DISCRETE PROBABILITY

4.1. INTRODUCTION

The word "probability" is used by people in almost every segment of our society; it is applied in different contexts and with different meanings. In certain loose contexts the word is understood by most; in any specific context it is apparently understood by relatively few.

Over the years probability has been the subject of considerable controversy. Today, even though it is possible to give a sound axiomatic definition of probability, there still exist disputes on its meaning and use. There are those who insist on a numerical measure for probability. Others claim that it is foolish to insist that probability necessarily be expressed numerically. We will not get involved in the controversy.

We consider only numerical probabilities. The *basic concepts are developed only for the discrete case*. However, most of the basic properties (definitions and rules) introduced for the discrete case apply to the continuous case as well.

4.2. SAMPLE SPACE

We think of probability in terms of real or conceptual "experiments," where "experiments" are discussed in terms of their "outcomes" or in terms of "events" which are collections of "outcomes." We accept "experiment" and "outcome" as undefined words which are commonly understood.

EXAMPLE 4.1. The instruction "think of a positive integer between 0 and 11.2" may be considered a conceptual experiment with outcomes

$$1, 2, 3, 4, 5, 6, 7, 8, 9, 10, 11.$$

In the general discrete case, we let each outcome in an experiment correspond to exactly one of the elements of the set

$$S = \{o_1, o_2, \cdots, o_n, \cdots\}$$

Sample Space

The set S is called a **sample space** associated with an experiment provided (1) each element of S denotes an outcome of the experiment, and (2) any outcome of the experiment corresponds to exactly one element of S.

EXAMPLE 4.2. The finite sample space for the experiment $\mathcal{E}_1$ of Example 4.1 is

$$S_1 = \{1, 2, 3, 4, 5, 6, 7, 8, 9, 10, 11\}$$

But

$$S' = \{1, 2, 3, 4, 5, 6, 7, 8, 9, 10, 11, 12\}$$

is not a sample space for the experiment since the element 12 in S' is not an outcome of the experiment, and

$$S'' = \{1, 2, 3, 4, 6, 7, 8, 9, 10, 11\}$$

is not a sample space for the experiment since the experimental outcome 5 is not an element of S''.

An experiment may involve more than one "trial." For example, in the experiment $\mathcal{E}_2$ with instruction "toss a coin twice" there are two trials. Suppose the only outcomes on each trial of the experiment are "head" and "tail" which we denote by H and T, respectively. Then the outcomes on each trial are elements of the set

$$R = \{H, T\}$$

and the outcomes of $\mathcal{E}_2$ are HH, HT, TH, and TT, where the first symbol of each pair denotes outcome on trial one and the second symbol denotes outcome on trial two. Thus, the sample space for $\mathcal{E}_2$ is

$$S_2 = \{HH, HT, TH, TT\}$$

Since HH is an abbreviated notation for the ordered pair (H,H), HT an abbreviated notation for (H,T), etc., we may also express S_2 as

$$S_2 = \{(H,H), (H,T), (T,H), (T,T)\}$$

or

$$S_2 = R \times R$$

where $R \times R$ denotes the **Cartesian product** of R and R.

The experiment $\mathcal{E}_3$ with instruction "toss two coins, a nickel and a dime" has the same sample space as $\mathcal{E}_2$, "toss a coin twice." In this case, we may let the outcomes of either coin in $\mathcal{E}_3$ correspond to the outcomes of trial one in $\mathcal{E}_2$ so that the outcomes of the remaining coin in $\mathcal{E}_3$ corresponds to the outcomes of trial two in $\mathcal{E}_2$.

There are many experiments whose sample spaces may be expressed in terms of Cartesian product sets. The outcomes of trials may be different or the number of trials may exceed two.

EXAMPLE 4.3. Specify the sample space for the experiment $\mathcal{E}_4$ in which "a coin is tossed with a die."
Letting

$$R_1 = \{H, T\} \quad \text{and} \quad R_2 = \{1, 2, 3, 4, 5, 6\}$$

we may write the Cartesian product set as

$$S_4 = \{(H, 1), (H, 2), \cdots, (H, 6), (T, 1), (T, 2), \cdots, (T, 6)\}$$

or as

$$S_4 = \{(x, y) | x \in R_1 \text{ and } y \in R_2\}$$

where the symbol $\in$ denotes "belongs to" and the bar | is read as "where." Note that the sample space could also be written as

$$S_4 = \{(1, H), (1, T), (2, H), (2, T), \cdots, (6, H), (6, T)\}$$

There are experiments whose sample spaces are not easily expressible in terms of Cartesian product sets. Three fairly typical examples follow.

EXAMPLE 4.4. Four similar disks are in a box. The symbol 1 is written on one, 2 on a second, 3 on a third, and 4 on a fourth. Let $\mathcal{E}_5$ be the experiment in which two disks are drawn one after the other and without the first being replaced in the box. The sample space may be expressed as

$$S_5 = \{(1,2), (1,3), (1,4), (2,1), (2,3), (2,4), (3,1), (3,2), (3,4), (4,1), (4,2), (4,3)\}$$

EXAMPLE 4.5. Consider the experiment $\mathcal{E}_6$ in which "a coin is tossed and then a die is tossed only in case a head shows on the coin." The possible outcomes are

$$T, H1, H2, \cdots, H6$$

Thus, the sample space may be expressed as

$$S_6 = \{T, (H,1), (H,2), (H,3), (H,4), (H,5), (H,6)\}$$

EXAMPLE 4.6. Let $\mathcal{E}_7$ be the experiment in which a coin is tossed until a head appears. The sample space is

$$S_7 = \{H, TH, TTH, TTTH, \cdots\}$$

If the outcomes are numbers (or are expressed in terms of numbers), the sample space may be conveniently represented by a diagram. Thus, for example, the collection of twelve dots of Figure 4.1 represents the sample space S_5. The diagram illustrates why the elements (or outcomes) of a sample space are also termed **sample points.**

In experiment $\mathcal{E}_2$ the sample space may be expressed as

$$S_2 = \{(1,1), (1,0), (0,1), (0,0)\}$$

where each number denotes the "number of heads" on a single trial. The collection of sample points in Figure 4.2 shows another representation of the sample space S_2. Sometimes Figure 4.3 is erroneously used to represent the sample space of experiment $\mathcal{E}_2$. In many situations Figure 4.3 is permissible, but there are some in which information is lost because the single circled point in Figure 4.3 corresponds to the two circled points in Figure 4.2. Thus, unless

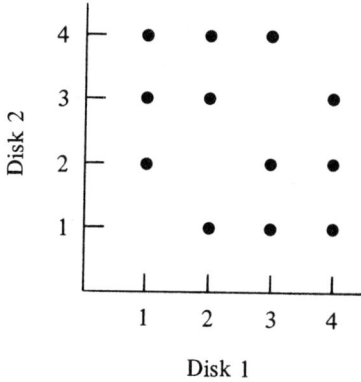

Figure 4.1.   Sample Space of Example 4.4

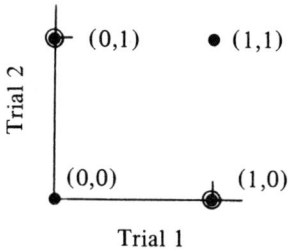

Figure 4.2.   Sample Space for Experiment $\mathcal{E}_2$

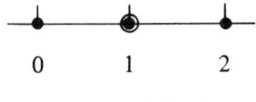

Number of Heads

Figure 4.3. Sample Space for Experiment $\mathcal{E}_2$

otherwise stated, in describing the sample space of an experiment we shall use the more detailed representation illustrated in Figure 4.2.

A second type of figure, known as a **tree diagram**, may be used to illustrate a sample space. In Figure 4.4 the seven terminal points to the right represent the sample space of Example 4.5.

Since Figures 4.1 and 4.2 are two-dimensional constructions, the corresponding sample spaces are often said to be two-dimensional. Clearly, we could construct three-dimensional sample spaces by the methods illustrated in Figures 4.1, 4.2, and 4.3. But it would be impossible to make similar constructions of sample spaces of higher dimensions. However, the tree diagram may be used to make constructions of sample spaces of any dimension.

We have seen that a discrete sample space may be either finite or infinite, and that it may be described in words, by set notation, by points in a dot diagram, or by a tree diagram. There are other ways to describe a discrete sample

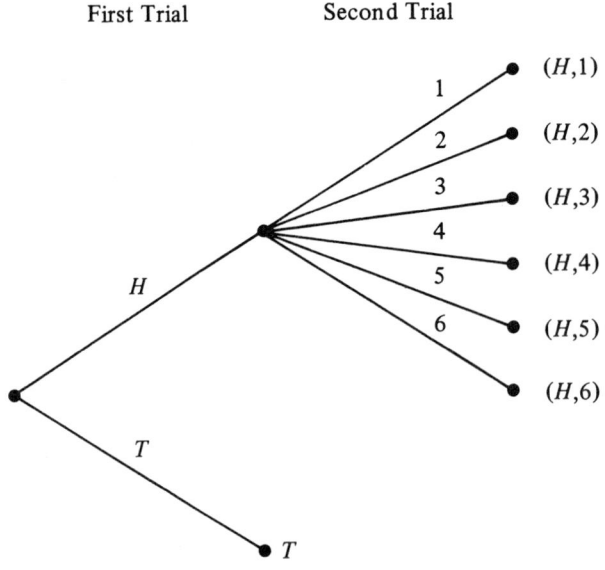

Figure 4.4. Tree Diagram of Example 4.5

space or, for that matter, a continuous sample space. All such methods may be summarized in the following formal definition:

> The **sample space** S for an experiment is the set of points representing the possible outcomes of the experiment.

4.3. EXERCISES

4.1. In each case specify the sample space for the indicated experiment.
   (a) A coin is tossed three times.
   (b) Two persons answer a question. Assume the only responses are "yes," "no," and "don't know."
   (c) Three objects are simultaneously drawn from a box of four distinguishable but similar objects.
   (d) A coin is tossed at most four times, but not after the second head occurs.
   (e) A study of families with three children is made and the sexes of the children are recorded.
   (f) Two letters are simultaneously selected from the word "level."
   (g) Two cards are dealt from an ordinary deck of cards.
   (h) A letter is chosen four times from the word "boy."
   (i) Three different people are selected from a city of 100,000 inhabitants.

4.2. Construct a diagram for parts (a), (b), (c), (d), (e), and (f) of Exercise 4.1.

4.3. Construct a tree diagram for the sample space of Example 4.6.

4.4. A regularly scheduled seminar in statistics has seven members. If fewer than four members are present the seminar is automatically cancelled. What is the sample space for the automatic cancellation of the seminar, assuming order of members to be ignored.

4.5. A committee of two is selected from five students, a, b, c, d, and e.
   (a) Specify the sample space S for the experiment and construct a diagram.
   (b) Specify the subset of S in which b is selected.
   (c) Specify the subset of S in which both b and c are not selected.

4.6. Two boys and two girls are in room A and one boy and three girls are in room B. Specify the sample space for the experiment in which a room is selected and then a person. Construct a diagram of the sample space.

4.4. EVENTS

We have seen that a *sample space* is used as the mathematical counterpart of an *experiment* and its *elements* correspond to *outcomes* in the experiment. Before we discuss probability the term "event" should be introduced.

82 Discrete Probability

An **event** is a subset of a sample space.

From this definition we observe that an event E, in mathematical terms, is a subset of elements of a set S or, in experimental terms, is a subset of outcomes of a sample space. We say an "event E occurs" provided the specific result of a single performance of an experiment is an element of E. Briefly, event E occurs provided a specific experimental outcome belongs to E.

EXAMPLE 4.7. The sample space for a single toss of a die is

$$S = \{1, 2, 3, 4, 5, 6\}$$

The subset

$$E = \{1, 2, 3\}$$

may be described as the event that the die shows fewer than four dots. Thus, if as a result of a single performance of the experiment two dots appear on the top side of a die, the event E occurs since 2 belongs to E. (Also, the event E would have occurred had the die shown either 1 dot or 3 dots. E would not have occurred had the die shown either 4 or 5 or 6 dots.)

We use capital letters to denote events. (When the number of events is large, subscripts are added.) It is possible for an experiment to have a large number of events. As a matter of fact it can be shown that when a set contains n elements, then 2^n events are possible.

EXAMPLE 4.8. Write all events which can be defined for the experiment $\mathcal{E}_3$ with instruction "toss two coins, a nickel and a dime."
We write the sample space as

$$S_3 = \{HH, HT, TH, TT\}$$

where the first symbol of each pair indicates the outcome for the nickel and the second the outcome for the dime. The events may be written as

$E_1 = \{HH\}$ $E_2 = \{HT\}$
$E_3 = \{TH\}$ $E_4 = \{TT\}$
$E_5 = \{HH, HT\}$ $E_6 = \{HH, TH\}$
$E_7 = \{HH, TT\}$ $E_8 = \{HT, TH\}$
$E_9 = \{HT, TT\}$ $E_{10} = \{TH, TT\}$
$E_{11} = \{HH, HT, TH\}$ $E_{12} = \{HH, HT, TT\}$
$E_{13} = \{HH, TH, TT\}$ $E_{14} = \{HT, TH, TT\}$
$E_{15} =$ null event $E_{16} = S$

where the null event contains no elements of S. Thus, when $n = 4$ there are $2^4 = 16$ possible events, including the null event and the sample space itself.

Events which contain a single element of S are called **simple events** or **elementary events**. Any event which does not contain any element of S is called a **null event** or **empty event** and is denoted by ∅. Any event which is not empty or simple is a **compound event**. The collection of all events defined on S is called an **event space**. Thus, in Example 4.8, the events E_1, E_2, E_3, and E_4 are simple, event E_{15} is null, and all other events are compound. The collection of all 16 events is an event space.

We have defined and discussed null event in terms of the elements of a specific sample space. However, it is not always done this way. For example, the set of all blue-eyed people is null relative to the sample space of Example 4.8. In such a context all that is required is that no element of S belong to ∅. Similar statements may be made about each event E.

It is possible to express any compound event in terms of simple events. Also, many practical problems in probability require that we be able to combine two or more compound events (see Section 4.5). Before discussing such topics, we consider Example 4.9.

EXAMPLE 4.9. Identify and combine some of the events of Example 4.4 in which the experiment was to "draw two disks, one after the other."

First, since the sample space (see Figure 4.1) contains 12 outcomes, it is possible to define $2^{12} = 4{,}096$ events. Suppose we let A denote the event that the number on the first disk is larger than the one on the second disk, B denote the event that the number of the second disk is greater than 2, C denote the event that the sum of the numbers on the two disks is 6 or 7, and D denote the event that the number on the second disk is twice that on the first disk. These four events are easy to visualize with the aid of Figure 4.5. We observe that

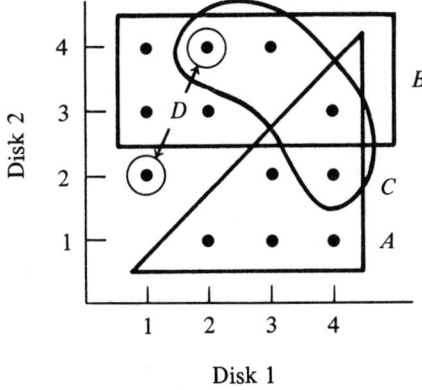

Figure 4.5. Events for Example 4.4

event A includes the six points inside the triangular boundary, event B the six points inside the rectangular boundary, event C the four points inside the oval boundary, and event D the two points inside the two circular boundaries. Also, the four events may be written as

$$A = \{(2,1), (3,1), (3,2), (4,1), (4,2), (4,3)\}$$
$$B = \{(1,3), (2,3), (4,3), (1,4), (2,4), (3,4)\}$$
$$C = \{(2,4), (3,4), (4,2), (4,3)\}$$
$$D = \{(1,2), (2,4)\}$$

Other events are expressible in terms of the four events A, B, C, and D. We note four such examples. First, the event F that the number on the first disk is smaller than the number on the second disk is the event "not A." Second, the event G that the number on the first disk is larger than the one on the second disk *and* the number on the second disk is greater than 2 is that part of S which is common to events A and B. Third, the event H that the number on the first disk is larger than the one on the second disk *or* the number on the second disk is greater than 2 *or* the number on the second disk is twice that on the first disk is the same as the sample space. Fourth, the event J that the number on the first disk is larger than the number on the second disk *and* the number on the second disk is twice that on the first disk is the same as the null event. That is, briefly

$$F = \text{not } A$$
$$G = \text{set of elements common to } A \text{ and } B = \{(4,3)\}$$
$$H = \text{set of elements in } A \text{ or } B \text{ or } D = S$$
$$J = \text{set of elements common to } A \text{ and } D = \emptyset$$

4.5. COMBINATIONS OF EVENTS

For the student with some knowledge of sets this section might not be necessary, but it probably will be useful as a review. For others the material should be straightforward enough. If more details or more illustrative methods are required, please check any standard reference on sets or see your teacher.

Complement. By definition, the *complement of any event* A *with respect to a sample space* S is that subset $\bar{A}$ of all outcomes of S which are not contained in A. $\bar{A}$ is an event of S and is typically read as "the complement of A" or as "A bar" or as "not A." In Example 4.9, $\bar{A} = F$ and

$$\bar{B} = \{(1,2), (2,1), (3,1), (3,2), (4,1), (4,2)\}$$

Intersection. Let A and B be any two events in a sample space S. By definition, the *intersection of* A *and* B, denoted by $A \cap B$ (and read also as

"A cap B" or "A and B"), is that event of S each of whose elements belong to both A and B. In Example 4.9,

$$A \cap B = G = \{(4,3)\} \quad \text{and} \quad A \cap D = J = \emptyset$$

Mutually Exclusive. When any two events A and B of S do not have an element in common, that is, when their intersection is the null event, we say A *and* B *are mutually exclusive events.* Thus, A and D in Example 4.9 are mutually exclusive since $A \cap D = \emptyset$. In any sample space, an event A and its complement $\bar{A}$ are mutually exclusive since $A \cap \bar{A} = \emptyset$.

Union. Let A and B be any two events in a sample space S. By definition the *union of* A *and* B, denoted by $A \cup B$ (and read also as "A cup B" or as "A or B"), is that event of S each of whose elements belong to A, to B, or to both. In Example 4.9,

$$A \cup B \cup D = H = S$$

and

$$C \cup D = \{(1,2), (2,4), (3,4), (4,2), (4,3)\}$$

In everyday usage a statement of the form "A or B" may apply the word "or" in two different ways, in either the *exclusive sense* or the *inclusive sense.* The statement "Joe expects to win or lose the tennis match" is an example of the exclusive sense, for either Joe "wins" or "loses" but not both. The statement "Joe expects to visit his father or mother" is an example of the inclusive sense, for Joe may visit his father, his mother, or both his parents. In our use of the statement "A or B" we always mean "or" in the inclusive sense. That is "$A \cup B$" always means "an element belongs to A, to B, or to both." If we ever wish to use "A or B" in the exclusive sense we shall use the statement "either A or B."

Intersection is a binary operation. That is, the operation intersection combines any two events A and B in S to form another event $A \cap B$ in S. The definition may be extended in two ways to three events A, B, and C in that order. Since $A \cap B$ and $B \cap C$ are events in S, $(A \cap B) \cap C$ and $A \cap (B \cap C)$ are also events in S. As an illustration consider B, C, and D of Example 4.9. First, we find that

$$B \cap C = \{(2,4), (3,4), (4,3)\} \quad \text{and} \quad C \cap D = \{(2,4)\}$$

So

$$(B \cap C) \cap D = \{(2,4), (3,4), (4,3)\} \cap \{(1,2), (2,4)\} = \{(2,4)\}$$

and

$$B \cap (C \cap D) = \{(1,3), (2,3), (4,3), (1,4), (2,4), (3,4)\} \cap \{(2,4)\} = \{(2,4)\}$$

Thus, $(B \cap C) \cap D$ and $B \cap (C \cap D)$ are two representations of the same event $\{(2,4)\}$ in S. That is, in this case

$$(B \cap C) \cap D = B \cap (C \cap D)$$

In fact, for any three events A, B, and C in an arbitrary sample space S it can be demonstrated that the associative property for intersection of sets holds. That is

(4.1) $$A \cap (B \cap C) = (A \cap B) \cap C$$

Since Equation (4.1) holds, we evaluate an expression like $X \cap Y \cap Z$ by first grouping adjacent pairs of sets. That is, $X \cap Y \cap Z$ is equivalent to $X \cap (Y \cap Z)$ or to $(X \cap Y) \cap Z$. More generally, we evaluate an expression like $E_1 \cap E_2 \cap E_3 \cap E_4 \cap E_5$ by grouping adjacent pairs of sets. For example, we might evaluate by first writing

$$E_1 \cap E_2 \cap E_3 \cap E_4 \cap E_5 = (E_1 \cap E_2) \cap [(E_3 \cap E_4) \cap E_5]$$

Of course, any other pair-wise grouping of adjacent sets would give the same event.

Union is a binary operation. Thus, by an argument similar to the above, we can demonstrate that the associative property for union of sets holds. That is,

(4.2) $$A \cup (B \cup C) = (A \cup B) \cup C$$

Generalizations of Equation (4.2) allow us to evaluate expressions such as $A_1 \cup A_2 \cup A_3 \cup A_4 \cup A_5 \cup A_6$ by grouping adjacent pairs of sets.

EXAMPLE 4.10. We have already noted that 4,096 events can be defined for the sample space of Example 4.9, and, in Section 4.4, we introduced the eight events, A, B, C, D, $\bar{A}$, G, S, and $\emptyset$. There are 4,088 more events. Fortunately, it is not necessary to list all 4,096 events or to give them names. It is possible to express all 4,096 events in terms of the 12 simple events

$$E_1 = \{(1,2)\} \quad E_2 = \{(1,3)\} \quad E_3 = \{(1,4)\}$$
$$E_4 = \{(2,1)\} \quad E_5 = \{(2,3)\} \quad E_6 = \{(2,4)\}$$
$$E_7 = \{(3,1)\} \quad E_8 = \{(3,2)\} \quad E_9 = \{(3,4)\}$$
$$E_{10} = \{(4,1)\} \quad E_{11} = \{(4,2)\} \quad E_{12} = \{(4,3)\}$$

For example

$$A = E_4 \cup E_7 \cup E_8 \cup E_{10} \cup E_{11} \cup E_{12}$$
$$C = E_6 \cup E_9 \cup E_{11} \cup E_{12}$$
$$D = E_1 \cup E_6$$
$$\emptyset = E_1 \cap E_2$$

When a sample space is finite, any compound event can be expressed as the union of two or more simple events and the empty event can be expressed as the intersection of any two simple events. Note that simple events are always mutually exclusive. (Of course, there are other ways to express all events in a sample space in terms of a small number of events, but the above method is straightforward and always works the same way.)

4.6. VENN DIAGRAMS

Expressions of events and relationships among events may often be easily displayed by special types of drawings called **Venn diagrams.** In such diagrams we let the sample space S be represented by a rectangle, while events are represented by closed regions within the rectangle, usually circles or parts of circles. Two or more events are shown as intersecting regions unless it is known that their intersection is empty.

EXAMPLE 4.11. In Figure 4.6 the circle represents event A, and the shaded region the complement of A; i.e., $\bar{A}$. Clearly, the complement of the shaded region is A. That is,

(4.3) $$\bar{\bar{A}} = A$$

Note: Venn diagrams may be used for both discrete and continuous cases. For discrete sample spaces outcomes may be represented as points and the constructions are made so that no outcome is on a boundary.

EXAMPLE 4.12. In Figure 4.7 event A is represented by the circle to the left and event B by the circle to the right. Thus, $A \cap B$ is represented by that part which is shaded by horizontal lines. The complement of B is that part of the rectangle which is outside the circle for B. Thus, $A \cap \bar{B}$ is represented by that part which is shaded by vertical lines. From the diagram it is clear that

$$(A \cap \bar{B}) \cup (A \cap B) = A \quad \text{and} \quad (A \cap \bar{B}) \cap (A \cap B) = \emptyset$$

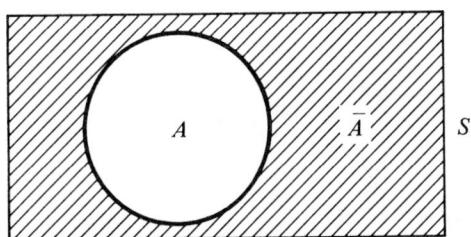

Figure 4.6. Venn Diagram

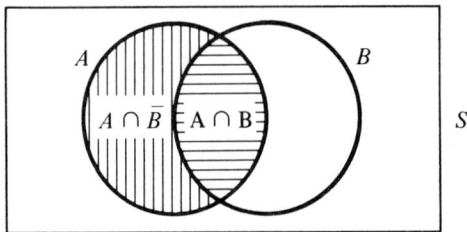

Figure 4.7. Venn Diagram

EXAMPLE 4.13. In Figure 4.8 the union of A and B, $A \cup B$, is represented by the shaded region. It should be clear that $A \cup B$ can be expressed as the union of three mutually exclusive events (i.e., disjoint regions) $A \cap \bar{B}$, $A \cap B$, $\bar{A} \cap B$. That is

$$A \cup B = (A \cap \bar{B}) \cup (A \cap B) \cup (\bar{A} \cap B)$$

Thus, we see that with the aid of a Venn diagram, two events which are not mutually exclusive can be expressed as the union of three mutually exclusive events.

EXAMPLE 4.14. In Figure 4.9(a), the shaded region represents the event $\overline{A \cap B}$. In Figure 4.9(b), the region shaded by horizontal lines represents event $\bar{A}$ and that region shaded by vertical lines represents event $\bar{B}$. Thus, $\bar{A} \cup \bar{B}$ is the event represented by single shading or double shading. By comparing Figures 4.9(a) and 4.9(b) we see that

(4.4) $$\overline{A \cap B} = \bar{A} \cup \bar{B}$$

since the parts shaded are identical.

EXAMPLE 4.15. We next wish to verify that

(4.5) $$A \cup (B \cap C) = (A \cup B) \cap (A \cup C)$$

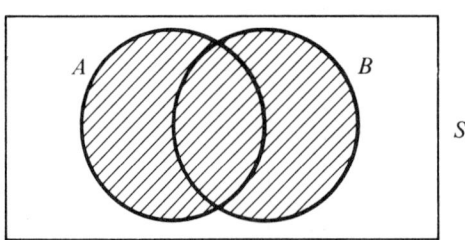

Figure 4.8. Venn Diagram

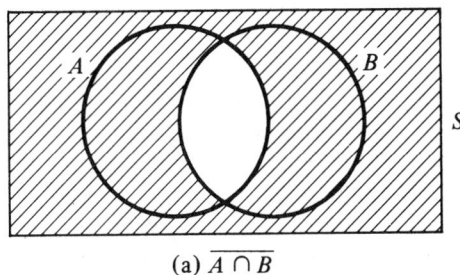

(a) $\overline{A \cap B}$

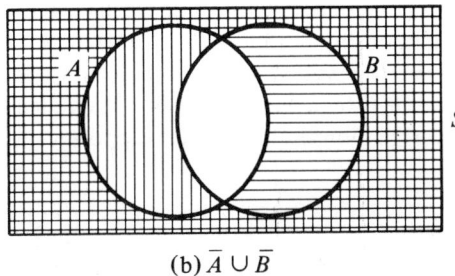

(b) $\overline{A} \cup \overline{B}$

Figure 4.9. Venn Diagrams to Verify Equation (4.4)

Equation (4.5) is one of two distributive laws (the other one is stated in Exercise 4.16). In Figure 4.10(a), event A is represented by the region shaded by horizontal lines, and $B \cap C$ is represented by the region shaded by vertical lines. Thus, the event $A \cup (B \cap C)$ is represented by the region which has single or double shading, the boundary being marked by a heavy line. In Figure 4.10(b), $A \cup B$ is represented by horizontal lines and $A \cup C$ by vertical lines. Thus, the event $(A \cup B) \cap (A \cup C)$ is represented by that region where there is

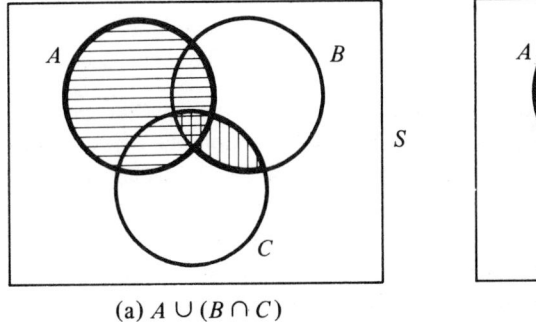

(a) $A \cup (B \cap C)$

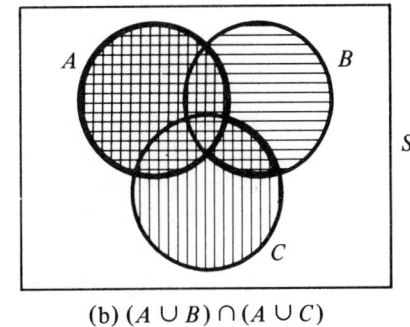

(b) $(A \cup B) \cap (A \cup C)$

Figure 4.10. Venn Diagram to Verify Equation (4.5)

only double shading, the boundary being marked by a heavy line. By comparing Figures 4.10(a) and 4.10(b) we see that Equation (4.5) holds, since the parts bounded by heavy lines are identical.

We have required that Venn diagrams be drawn so that regions intersect unless the events they represent are known to be mutually exclusive. Thus, in the general case, two regions are drawn so that the sample space is partitioned (divided) into four disjoint (mutually exclusive) regions; three regions are drawn so that the sample space is partitioned into eight disjoint regions. Readers should satisfy themselves that no diagram can be drawn with four intersecting regions which partitions the sample space into sixteen regions—thus, we have reason for limiting Venn diagrams to three circles.

Any relationship involving one, two, or three events can always be verified by a Venn diagram. If a relationship has more than three events a Venn diagram cannot be applied directly. However, a similar relationship with three events can often be found and verified by a Venn diagram. Then algebra of sets may be used to complete the verification. For example, suppose we wanted to prove that

(4.6) $$A \cup (B \cap C \cap D) = (A \cup B) \cap (A \cup C) \cap (A \cup D)$$

Since Equation (4.6) is a fairly obvious extension of Equation (4.5), we would first verify Equation (4.5) by a Venn diagram and then use $B \cap C \cap D = B \cap (C \cap D)$ in an algebraic proof. Actually, once the student has mastered the algebra of relationships involving one, two, or three events any extension to a finite number of events is not usually difficult.

4.7. EXERCISES

4.7. A die is tossed one time. Let A denote the event "die shows an even number," B denote the event "die shows an odd number," C denote the event "die shows a number less than three," and D denote the event "die shows a six." List any of these events which are
(a) mutually exclusive
(b) simple
(c) complementary
(d) compound
(e) If a one appears on the toss, which, if any, of the events occur?

4.8. Three coins are tossed one time. Let A denote the event "three heads show," B denote the event "two heads and one tail show," C denote the event "three tails show," and D denote the event "a head shows on the first coin." List any of these events which are
(a) mutually exclusive
(b) simple
(c) complementary
(d) compound
(e) If one head appears, which, if any, of the events occur?
(f) List and name all simple events for this experiment.
(g) Use the simple events named in part (f) to list all compound events containing exactly two elements.

4.9. In a sample poll, adults of voting age were asked "If the election were today would you vote for candidate X?" Assume the only responses to be "yes," "no," and "don't know." Let A denote the event "male responds with yes or no," B denote the event "the response is yes," C denote the event "female responds," D denote the event "female responds no *or* male responds don't know," and E denote the event "male responds *and* response is no." List any of these events which are

(a) mutually exclusive (b) simple
(c) complementary (d) compound
(e) Suppose no female in the poll responds with "don't know," what is the resulting sample space S' of possible outcomes?
(f) Let events A, B, C, D, and E be defined as before. In S' give a concise verbal description of each of the following: $B \cup D$, $B \cup \bar{C}$, $\overline{B \cup C}$, $\bar{B} \cap C$.

4.10. A red die is made so that opposite faces have the same number of dots, and three adjacent faces show 1, 2, and 3 dots, respectively. A white die is made like the red die. These two dice are tossed together. An outcome is designated as (x,y) where x and y denote number of dots on white die and red die, respectively. Let S denote the sample space of the experiment, A the event that "the sum of the dots on the two faces is odd," B the event that "the number of dots on the white die is odd," and C the event that "the number of dots on the red die is even."

(a) List the elements of the events S, A, B, and C.
(b) Find $A \cap B$, $A \cap \bar{B}$, and $\overline{A \cup B}$. Give a concise verbal description of each of these events.
(c) Let $D = \{(2,1), (2,2), (2,3)\}$ $E = \{(1,2), (2,1)\}$
$F = \{(1,1), (1,3), (3,1), (3,3)\}$ $G = \{(2,3)\}$
Give a concise verbal description of each event.
(d) If possible, express each of the events in (c) in terms of some or all of the symbols A, B, C, $\bar{\ }$.

4.11. Suppose A, B, and C are any events of a sample space S, and suppose the experiment for S is performed one time. Write an expression (use only the symbols A, B, C, $\cup$, $\cap$, $\bar{\ }$, and grouping symbols) for the event that of A, B, and C

(a) at least one occurs (b) only B occurs
(c) B and C occur but not A (d) all three occur
(e) none occurs (f) exactly two occur
(g) at most two occur

4.12. A box contains five similar balls. In a game the first player may draw 1, 2, 3, or 4 balls and the second player may draw 1, 2, or 3 balls from those remaining after the first player draws. Let S denote the sample space, A denote the event "the total number of balls drawn is 4," B denote the event "player two draws more balls than player one," C denote the event "player one draws at least 3 balls," and D denote the event "the players draw the same number of balls." List the sample points which belong to

(a) each of the five events defined above
(b) the complement of each event in (a)

(c) $\bar{A} \cap B$ and $A \cap \bar{B}$
(d) $(\bar{A} \cap B) \cup (A \cap \bar{B})$
(e) $\overline{A \cup B}$ and $\bar{A} \cap \bar{B}$
(f) $A \cap (B \cup C)$ and $(A \cap B) \cup (A \cap C)$
(g) $\overline{B \cup C \cup D}$

4.13. Use two Venn diagrams to verify that $\overline{A \cup B} = \bar{A} \cap \bar{B}$.

4.14. Use two Venn diagrams to verify that $A \cup (A \cap B) = A$.

4.15. Use Venn diagrams to verify: $A \cap (B \cup C) = (A \cap B) \cup (A \cap C)$.

4.16. Use Venn diagrams to verify: $\bar{A} \cap (B \cup C) = (\bar{A} \cap B) \cup (\bar{A} \cap C)$.

4.17. In a small seminar at a state university, Adam is an undergraduate physics major, Bob a graduate nonphysics major, Carol a graduate physics major, Dot a graduate nonphysics major, Earl an undergraduate nonphysics major, Fred a graduate physics major, and Gay an undergraduate physics major. Specify the event that the student is
(a) an undergraduate
(b) a male graduate
(c) an undergraduate physics major
(d) a female or a physics major
(e) either an undergraduate or a physics major

4.18. In an experiment an outcome is a positive integer from 1 to 9 inclusive. Three events of interest are $A = \{1, 3, 5, 7, 9\}$, $B = \{1, 4, 7\}$, and $C = \{1, 5, 9\}$.
(a) Use a Venn diagram to indicate the sample points in the eight mutually exclusive (disjoint) regions of the form $A \cap B \cap C$, $A \cap B \cap \bar{C}$, $\cdots$, $A \cap \bar{B} \cap \bar{C}$, $\bar{A} \cap \bar{B} \cap \bar{C}$.
(b) Indicate the sample points in each of the following regions:

$A \cap B$ $\quad$ $A \cap C$ $\quad$ $\overline{A \cup B}$ $\quad$ $\overline{A \cup B \cup C}$ $\quad$ $\bar{A} \cup B$ $\quad$ $\bar{A} \cap B$

(c) Express each of the following sets in terms of one or more of the events, A, B, C:

$\{1,3\}$ $\quad$ $\{3,4\}$ $\quad$ $\{5,7,9\}$ $\quad$ $\{1,2,6,8\}$

4.19. Let A, B, and C denote "French," "German," and "Russian," respectively. In a fraternity of 32 boys two are enrolled in A, B, and C, 4 in A and B, 6 in A and C, 7 in B and C, 11 in A, 14 in B, and 12 in C.
(a) How many boys are not enrolled in any of these languages?
(b) How many boys are enrolled in only one of these languages?

*4.20. In a group of 500 students, 135 are women, 345 are undergraduates, and 400 are from out of state. What is the smallest number of undergraduate men from out of state? Explain.

4.8. PROBABILITY: FINITE NUMBER OF EQUALLY LIKELY OUTCOMES

The word "probability" and its synonyms are widely used today. Some common expressions are "there is a 60 percent *chance* of rain today," "the machine makes *about* 3 defective *out of* every 100 parts," "*almost everybody* seemed to enjoy the show," and "in a toss of two dice the *probability* is $\frac{1}{6}$ that a seven will show." One question that immediately comes to mind is, "what, if anything, do all these statements have in common?" Clearly, they all express a degree of certainty (or uncertainty), and the assumption is that they are based on factual (or experimental) information. Some of the statements are expressed in numerical terms while others are not; some seem to be more plausible than others.

We attempt to come to grips with the notions expressed in the above and numerous other similar statements of uncertainty *by describing numerical probability for a finite sample space*. Furthermore, we take the simplest possible route to introduce the basic ideas of probability. Later we make the concept more inclusive.

Our first definition is of historical interest and is very useful in experiments such as games of chance. As a matter of fact, the definition was first used on games with dice, roulette, and cards. The *classical definition of probability* is as follows:

> If an experiment can result in N different (mutually exclusive) and equally likely outcomes, n of which correspond to the occurrence of an event E, then the **probability of the event E** is $\frac{n}{N}$ or briefly

(4.7) $$P(E) = \frac{n}{N}$$

Probability is an illustration of a set function. A *set function* is a rule which assigns one number to each event (set) of an event space. Another number assigned to each event could simply be the number n of outcomes in the event. This *number function* is also an illustration of a set function.

EXAMPLE 4.16. Use the classical definition to find the probability of obtaining (a) less than four on a single toss of a die, (b) a spade in a single draw from a deck of playing cards, and (c) a total of six on one toss of a pair of dice.

For (a), if the die is symmetrical and tossed vigorously it is reasonable to assume each of the six outcomes is equally likely. The event E can occur with the appearance of any one of the three outcomes 1, 2, 3. Thus, $N = 6$, $n = 3$ and the required probability is

$$P(E) = \frac{3}{6} = \frac{1}{2}$$

In part (b), if the deck of cards is well-shuffled, then on the draw it is reasonable to assume that each of the 52 outcomes is equally likely. Let A denote the event "a spade is drawn." There are 13 spades. Thus,

$$P(A) = \frac{13}{52} = \frac{1}{4}$$

In part (c), there are 36 outcomes in the sample space and five of them, (1, 5), (2, 4), (3, 3), (4, 2), (5, 1), lead to the occurrence of the event. On a vigorous toss of two well-balanced dice, we assume the 36 outcomes to be equally likely. Thus, the required probability is $\frac{5}{36}$.

Since the complement of E contains all the outcomes of S which are not in E, it follows from the definition of probability that

$$P(\bar{E}) = \frac{N-n}{N} = 1 - \frac{n}{N} = 1 - P(E)$$

That is, the *probability that* E *will not occur* is given by

(4.8) $$P(\bar{E}) = 1 - P(E)$$

If the probability of drawing a spade from a deck of playing cards is $\frac{1}{4}$, then by Equation (4.8) the probability of drawing a non-spade is $1 - \frac{1}{4} = \frac{3}{4}$. Since the probability of drawing a non-spade is 3 times greater than the probability of drawing a spade we say the *odds in favor* of a non-spade are 3 to 1, or, what is the same thing, the *odds against* spades are 3 to 1. Similarly, we could say the *odds in favor* of spades are 1 to 3 or the *odds against* non-spades are 1 to 3. Of the four statements on odds the third one is typically preferred since it expresses odds in favor of the original event. Generally, the *odds in favor of an event* E *may be defined as the ratio of the probability that the event will occur to the probability that the event will not occur* where the ratio is reduced to simple numbers (like $\frac{1}{4}$ to $\frac{3}{4}$ is reduced to 1 to 3).

If the odds in favor of an event E are a to b, then

(4.9) $$P(E) = \frac{a}{a+b} \quad \text{and} \quad P(\bar{E}) = \frac{b}{a+b}$$

For example, suppose the odds in favor of "a total of six on one toss of a pair of dice" are 5 to 31, then

$$P(\text{a total of six on one toss of a pair of dice}) = \frac{5}{5+31} = \frac{5}{36}$$

which agrees with the solution to Example 4.16(c).

The sample space S is an event. Thus, by definition

(4.10) $$P(S) = \frac{N}{N} = 1$$

Further, since the null event ∅ is the complement of S, by Equation (4.8) it follows that $P(\varnothing) = 1 - P(S) = 1 - 1 = 0$. That is,

(4.11) $$P(\varnothing) = 0$$

In each case, the number of outcomes n of an event E is between 0 and N inclusive. Hence $\frac{n}{N}$ is between 0 and 1 inclusive. So for any event E in a sample space S,

(4.12) $$0 \leq P(E) \leq 1$$

If the probability of an event E is 1, we say E is *certain to occur*; if the probability of E is 0, we say not E is certain to occur. For all other events, the magnitude of the probability of an event E measures the degree of certainty (confidence) we have in the occurrence of E. For example, if $P(E) = 0.99$, we have a very high degree of certainty E will occur when the experiment is performed; if $P(E) = 0.05$, we have moderately low confidence E will occur.

When outcomes of an experiment are equally likely, as they are assumed to be in this section, (1) probability and relative frequency of an event are the same number, and (2) the probability of every simple event is $1/N$. In later sections we see that these two properties do not necessarily hold for other cases.

4.9. PROBABILITY OF TWO OR MORE EVENTS: EQUALLY LIKELY OUTCOMES

Many times it is desirable that we be able to find the probability of a combination of two or more events in terms of probabilities of the single events. (In earlier sections, we have seen some of the types of expressions involving several events.) For example, we might wish to find "probability of an ace or a king" in terms of "probability of an ace" and "probability of a king."

Suppose A and B are two mutually exclusive events of a sample space S of an experiment which has N different and equally likely outcomes. Further, suppose the events A and B can occur in a and b ways, respectively. Then the event $A \cup B$ can occur in $a + b$ ways since A and B are disjoint events (i.e., $A \cap B = \varnothing$). By the definition of probability

$$P(A) = \frac{a}{N}, \ P(B) = \frac{b}{N}, \text{ and } P(A \cup B) = \frac{a+b}{N}$$

Thus

$$P(A) + P(B) = \frac{a}{N} + \frac{b}{N} = \frac{a+b}{N}$$

and, by substitution

(4.13) $$P(A \cup B) = P(A) + P(B)$$

whenever A and B are mutually exclusive events.

Equation (4.13) can easily be extended to any number of events. Suppose $A_1, A_2, \cdots, A_k$ are k mutually exclusive events of an experiment. Then

(4.14) $$P(A_1 \cup A_2 \cup \cdots \cup A_k) = P(A_1) + P(A_2) + \cdots + P(A_k)$$

where $k = 2, 3, \cdots, N$. The proof follows from Equation (4.13) by repeated use of the associative property of union of events (see page 86).

When $A_1, A_2, \cdots, A_k$ are mutually exclusive events of an experiment and

$$A_1 \cup A_2 \cup \cdots \cup A_k = S$$

we say the sample space has been partitioned into k events (subsets). *If events $A_1, A_2, \cdots, A_k$ form a partition of S*, then from Equations (4.10) and (4.14) it follows that

(4.15) $$P(A_1) + P(A_2) + \cdots + P(A_k) = 1$$

Earlier we learned that when a sample space is finite each compound event can be expressed as the union of simple events. Now we learn how the probability of a compound event can be obtained from the probabilities of simple events. Suppose

(4.16) $$B_1 = \{o_1\}, B_2 = \{o_2\}, \cdots, B_n = \{o_n\}$$

denote the simple events of a finite sample space

$$S = \{o_1, o_2, \cdots, o_n\}$$

Let E be any compound event. Then E is the union of two or more of the simple events in Expression (4.16). Since the simple events are mutually exclusive it follows from Equation (4.14) that

> *The probability of any event* E *equals the sum of the probabilities of the simple events of* E.

EXAMPLE 4.17. Use the sample space S shown in Figure 4.11 to illustrate the properties of this section.

Let $A_1 = \{(1, 1)\}$, $A_2 = \{(1, 1), (2, 2)\}$, $A_3 = \{(1, 2), (2, 1)\}$,
$A_4 = \{(1, 1), (1, 2), (3, 1)\}$, and $A_5 = \{(2, 2), (3, 1), (3, 2)\}$

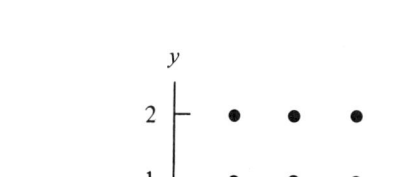

Figure 4.11. Finite Sample Space

Assume each of the six different outcomes of S to be equally likely. By the classical definition of probability

$$P(A_1) = \frac{1}{6} \qquad P(A_2) = P(A_3) = \frac{2}{6} \qquad P(A_4) = P(A_5) = \frac{3}{6}$$

The events A_1 and A_3 are mutually exclusive and

$$A_1 \cup A_3 = \{(1, 1), (1, 2), (2, 1)\}$$

Thus, by the classical definition of probability

$$P(A_1 \cup A_3) = \frac{3}{6} = \frac{1}{2}$$

But

$$P(A_1) + P(A_3) = \frac{1}{6} + \frac{2}{6} = \frac{3}{6} = \frac{1}{2}$$

Since these last two expressions have the same value, Equation (4.13) is verified. On the other hand A_1 and A_2 are *not mutually exclusive* and

$$A_1 \cup A_2 = \{(1, 1), (2, 2)\}$$

Thus

$$P(A_1 \cup A_2) = \frac{2}{6} = \frac{1}{3} \quad \text{and} \quad P(A_1) + P(A_2) = \frac{1}{6} + \frac{2}{6} = \frac{3}{6} = \frac{1}{2}$$

So, in this example,

(4.17) $$P(A_1 \cup A_2) < P(A_1) + P(A_2)$$

Thus, Equation (4.13) does not hold for events which are *not mutually exclusive*.
 The events A_1, A_3, and A_5 form a partition of S since they are mutually exclusive (i.e., no pair of events has an outcome in common) *and* $A_1 \cup A_3 \cup A_5$

$= S$. Using the definition of classical probability

$$P(A_1) + P(A_3) + P(A_5) = \frac{1}{6} + \frac{2}{6} + \frac{3}{6} = 1$$

and we have verified Equation (4.15).

Let $B_1 = A_1$, $B_2 = \{(1, 2)\}$, $B_3 = \{(2, 1)\}$, $B_4 = \{(2, 2)\}$,
$B_5 = \{(3, 1)\}$, $B_6 = \{(3, 2)\}$, and $E = \{(1, 1), (2, 2), (3, 2)\}$

By the classical definition of probability, each of these simple events has probability $\frac{1}{6}$ and $P(E) = \frac{3}{6} = \frac{1}{2}$. Since

$$P(B_1) + P(B_4) + P(B_6) = \frac{1}{6} + \frac{1}{6} + \frac{1}{6} = \frac{3}{6} = \frac{1}{2} \quad \text{and} \quad E = B_1 \cup B_4 \cup B_6$$

we have demonstrated that

$$P(E) = P(B_1) + P(B_4) + P(B_6)$$

which is a special case of Equation (4.14).

EXAMPLE 4.18. A card is drawn from a well-shuffled deck of 52 playing cards. What is the probability of drawing a black face card (jack, queen, king) or a diamond?

Let A denote the event "black face card" and B denote the event "a diamond." We want to find $P(A \cup B)$. There are 6 black face cards and 13 diamonds, and we assume that each of 52 cards is equally likely. Since A and B are mutually exclusive events,

$$P(A \cup B) = P(A) + P(B) = \frac{6}{52} + \frac{13}{52} = \frac{19}{52}$$

As a check we compute $P(A \cup B)$ directly. The event $A \cup B$ contains $6 + 13 = 19$ different cards. Thus

$$P(A \cup B) = \frac{19}{52}$$

and Equation (4.13) is again verified.

The reader might not yet see any advantage in using Equations (4.13), (4.14), and (4.15)—at least, there appears to be none in the illustrations. However, two things should be kept in mind. The illustrations were relatively simple, and we have introduced only the simplest concepts in preparation for others to come. Soon it should be obvious that these equations as well as their extensions are very useful.

Under the restriction that events A and B be mutually exclusive we have seen that $P(A \cup B)$ is equal to the *sum* of the probabilities, $P(A)$ and $P(B)$. Thus, it is natural to ask if there are conditions under which $P(A \cap B)$ is equal to the *product* of the probabilities, $P(A)$ and $P(B)$. Example 4.19 is given to suggest what the answer might be.

EXAMPLE 4.19. Refer to the four similar disks of Example 4.4. Again we draw two disks, but this time the first disk is returned and the four disks thoroughly mixed before the second disk is drawn. That is, the two disks are *drawn with replacement*. The resulting sample space is shown in Figure 4.12. Let A denote the event "disk 1 shows a number greater than 2," and B denote the event "disk 2 shows a number less than 2." Thus, $A \cap B$ denotes the event "disk 1 shows a number greater than 2 *and* disk 2 shows a number less than 2." Briefly, the events are

$$A = \{(3, 1), (3, 2), (3, 3), (3, 4), (4, 1), (4, 2), (4, 3), (4, 4)\}$$
$$B = \{(1, 1), (2, 1), (3, 1), (4, 1)\}$$
$$A \cap B = \{(3, 1), (4, 1)\}$$

Assuming the 16 outcomes are equally likely we find that

$$P(A) = \frac{8}{16} = \frac{1}{2} \qquad P(B) = \frac{4}{16} = \frac{1}{4} \qquad P(A \cap B) = \frac{2}{16} = \frac{1}{8}$$

Observe that

$$P(A)P(B) = \frac{1}{2} \times \frac{1}{4} = \frac{1}{8}$$

Thus, in this illustration

(4.18) $$P(A \cap B) = P(A)P(B)$$

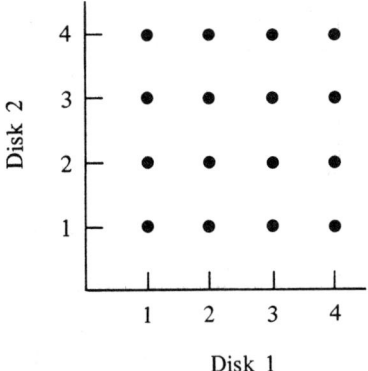

Figure 4.12. Sample Space for Example 4.19

What are the conditions under which Equation (4.18) holds? Or does the property always hold? In answer to the second question we return to Example 4.4 where the sample is *drawn without replacement* and the reduced sample space is the same as in Figure 4.12, except that the diagonal points, (1, 1), (2, 2), (3, 3), (4, 4), are missing. Defining events A and B as in Example 4.19, we have

$$A = \{(3, 1), (3, 2), (3, 4), (4, 1), (4, 2), (4, 3)\}$$
$$B = \{(2, 1), (3, 1), (4, 1)\}$$
$$A \cap B = \{(3, 1), (4, 1)\}$$

Now, assuming the 12 outcomes are equally likely we find that

$$P(A) = \frac{6}{12} = \frac{1}{2} \qquad P(B) = \frac{3}{12} = \frac{1}{4} \qquad P(A \cap B) = \frac{2}{12} = \frac{1}{6}$$

Since

$$P(A)P(B) = \frac{1}{2} \times \frac{1}{4} = \frac{1}{8} \neq \frac{1}{6} = P(A \cap B)$$

it is clear that Equation (4.18) does not always hold.

In Example 4.19 the conditions for the second draw were identical to those for the first draw. That is, the result of the second draw was not influenced by what occurred on the first draw. One could say that the second draw was "independent" of the first draw. On the other hand, for the sample space of Example 4.4, the result of the second draw depended on what occurred on the first draw. That is, the second draw was "dependent" on the first draw. This leads us to the following formal definition:

Events A and B are **independent** if and only if

(4.19) $$P(A \cap B) = P(A)P(B)$$

Statement (4.19) means (1) if A and B are independent, then $P(A \cap B) = P(A)P(B)$, and (2) if $P(A \cap B) = P(A)P(B)$, then A and B are independent. When $P(A \cap B) \neq P(A)P(B)$ we say **events A and B are dependent.**

The reader should not confuse the term "independent" with that of "mutually exclusive." Independent is defined in terms of "probability of events" whereas mutually exclusive is defined in terms of "events." Perhaps the distinction can be made clear by the following:

Theorem 4.1. If $P(A) \neq 0$, $P(B) \neq 0$, and events A and B are independent, then A and B have at least one outcome in common.

PROOF OF THEOREM 4.1. Suppose $A \cap B = \emptyset$. Then $P(A \cap B) = 0$. Since A and B are independent, $P(A)P(B) = P(A \cap B)$. Thus $P(A)P(B) = 0$, and it follows that $P(A) = 0$ or $P(B) = 0$. But this contradicts the condition that $P(A) \neq 0$ and $P(B) \neq 0$. Thus, our assumption $A \cap B = \emptyset$ is in error.

Since either $A \cap B = \emptyset$ or $\mathbf{A} \cap B \neq \emptyset$, we must conclude that $A \cap B \neq \emptyset$. That is, the intersection is not empty. So A and B have at least one outcome in common.

Mutually exclusive events A and B *never have* an outcome in common, but independent events *do have* an outcome in common, provided each event contains an outcome. Clearly, "independent" and "mutually exclusive" do not have the same meaning.

Extending the concept of independence to three or more events is not trivial. For it is possible to have A and B independent, A and C independent, and B and C independent and still to have $A \cap B$ and C, say, not independent or to have A, B, and C not independent. That is, it is possible for three events A, B, and C to be "pair-wise independent" and not be "mutually independent." Further, it is possible to have three-way independence and not have pair-wise independence. Thus, it is not enough to say A, B, and C are mutually independent whenever

$$P(A \cap B \cap C) = P(A)P(B)P(C)$$

(Illustrations of these two cases are given in Example 4.20.) This leads to the following formal definition:

Events A, B, and C are **mutually independent** if and only if

(4.20)
$$P(A \cap B) = P(A)P(B), P(A \cap C) = P(A)P(C),$$
$$P(B \cap C) = P(B)P(C), \text{ and } P(A \cap B \cap C) = P(A)P(B)P(C)$$

That is, A, B, and C are mutually independent (or, simply, independent) if and only if the multiplication rule holds for all combinations of two or three events. Thus, in the general case, we say $E_1, E_2, \cdots, E_n$ are mutually independent if and only if the multiplication rule holds for all combinations of two or more events. When events fail to be mutually independent they are said to be *dependent*. It will be left as an exercise for the reader to show that there are $2^n - n - 1$ equations in the definition of mutual independence of n events.

EXAMPLE 4.20. Use the Venn diagrams of Figure 4.13 to illustrate cases where three events are not mutually independent, but are either pair-wise independent or three-way independent. (The digits in the diagrams indicate the number of equally likely outcomes in each set.)

For Figure 4.13(a) we find that

$$P(A) = P(B) = P(C) = \frac{8}{16} = \frac{1}{2}$$

$$P(A \cap B) = P(A \cap C) = P(B \cap C) = \frac{4}{16} = \frac{1}{4}$$

$$P(A \cap B \cap C) = \frac{1}{16}$$

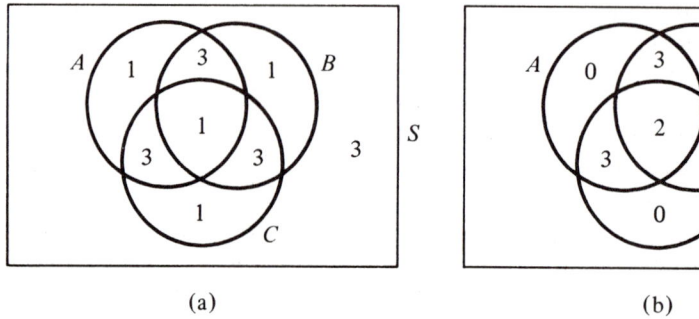

Figure 4.13. Venn Diagrams for Example 4.20

and therefore

$$P(A)P(B) = \frac{1}{2} \times \frac{1}{2} = \frac{1}{4} = P(A \cap B)$$

$$P(A)P(C) = \frac{1}{2} \times \frac{1}{2} = \frac{1}{4} = P(A \cap C)$$

$$P(B)P(C) = \frac{1}{2} \times \frac{1}{2} = \frac{1}{4} = P(B \cap C)$$

$$P(A)P(B)P(C) = \frac{1}{2} \times \frac{1}{2} \times \frac{1}{2} = \frac{1}{8} \neq P(A \cap B \cap C)$$

Thus, there is pair-wise independence, but three-way dependence. So the events A, B, and C are dependent.

For Figure 4.13(b) we find that

$$P(A) = P(B) = P(C) = \frac{8}{16} = \frac{1}{2}$$

$$P(A \cap B) = P(A \cap C) = P(B \cap C) = \frac{5}{16}$$

$$P(A \cap B \cap C) = \frac{2}{16} = \frac{1}{8}$$

$$P(A)P(B) = \frac{1}{2} \times \frac{1}{2} = \frac{1}{4} \neq P(A \cap B)$$

and

$$P(A)P(B)P(C) = \left(\frac{1}{2}\right)^3 = \frac{1}{8} = P(A \cap B \cap C)$$

Thus, there is pair-wise dependence, but there is three-way independence. So the events are dependent.

EXAMPLE 4.21. Construct a Venn diagram with three independent events. Let the events be A, B, and C. Further, let

$$P(A) = \frac{1}{2} = \frac{12}{24} \quad P(B) = \frac{1}{3} = \frac{8}{24} \quad P(C) = \frac{1}{4} = \frac{6}{24}$$

$$P(A \cap B) = P(A)P(B) = \frac{1}{6} = \frac{4}{24}$$

$$P(A \cap C) = P(A)P(C) = \frac{1}{8} = \frac{3}{24}$$

$$P(B \cap C) = P(B)P(C) = \frac{1}{12} = \frac{2}{24}$$

$$P(A \cap B \cap C) = P(A)P(B)P(C) = \frac{1}{24}$$

Since the least common denominator is 24, we let the sample space have 24 different and equally likely outcomes. Then the assignments are made as follows: Place 1 outcome in $A \cap B \cap C$; $4 - 1 = 3$ outcomes in $A \cap B \cap \bar{C}$; $3 - 1 = 2$ outcomes in $A \cap \bar{B} \cap C$; $2 - 1 = 1$ outcomes in $\bar{A} \cap B \cap C$; $12 - 3 - 2 - 1 = 6$ outcomes in $A \cap \bar{B} \cap \bar{C}$; etc. All the entries are shown in the Venn diagram of Figure 4.14 and should be checked.

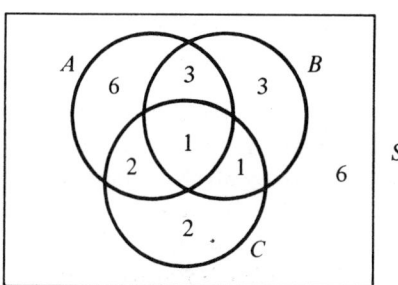

Figure 4.14. Venn Diagram for Example 4.21

4.10. EXERCISES

4.21. If a die is tossed, what is the probability that an even number shows?

4.22. One student is chosen from a class of 25 students with 8 boys. What is the probability that a girl is chosen?

4.23. If a pair of dice is tossed, what is the probability that the total on the two faces is a 6? a 10? an even number?

4.24. (a) If $P(A) = \frac{5}{9}$, what are the odds in favor of A?
(b) If $P(B) = \frac{13}{17}$, what are the odds in favor of $\bar{B}$?

4.25. If the odds in favor of event A are 7 to 11, find $P(\bar{A})$.

4.26. Let I be the event that a car traveled on an interstate highway, and X be the event that a car traveled over 60-miles per hour. State in words each of the following:
(a) $P(X)$ (b) $P(\bar{I})$ (c) $P(I \cup X)$
(d) $P(I \cap X)$ (e) $P(I \cap \bar{X})$ (f) $P(\bar{X} \cup \bar{I})$

4.27. In a family of four children, what is the probability that three or four are girls? Assume the probability of a boy being born is 0.50.

4.28. One card is drawn from a well-shuffled deck of 52 playing cards. What is the probability of getting a card with an odd number?

4.29. A student is selected at random from the students of a four-year college. What would be wrong, if anything, in saying the probability of obtaining a senior is $\frac{1}{4}$?

4.30. Four similar disks, numbered 1, 2, 3, 4 are placed in a box and thoroughly mixed. If the disks are selected one after the other until all are drawn, what is the probability that the numbers, in order, are 4, 3, 2, 1? What is the probability of a specific different order?

4.31. A person with a blindfold on places four equal sized volumes of a set on a shelf. What is the probability the four books will be out of order?

E4.32. In gambling, when Mr. X gives Mr. Y 3 to 2 odds it means that if event A is successful then Mr. X pays $3 for every $2 bet by Mr. Y. Of course, if event A is not successful then Mr. Y loses $2 for every $3 put down by Mr. X. In the case just cited one could argue that Mr. Y believes that $P(A) = 2/(3 + 2) = 0.40$.
    In a football game your friend offers to give 5 to 2 odds that your team will win, but you refuse. When your friend offers 10 to 3 odds you accept. Based on the argument of the first paragraph of this exercise, between what two values would you estimate "the probability of success of your team" to be?

4.33. Suppose A and B are mutually exclusive events with $P(A) = 0.6$ and $P(B) = 0.3$. Find
(a) $P(\bar{A})$ and $P(\bar{B})$ (b) $P(A \cup B)$
(c) $P(\bar{A} \cap \bar{B})$ (d) $P(A \cap \bar{B})$

4.34. Two dice are tossed one time. Find the probability that
(a) The total on the two faces is not 3
(b) Each die shows only a 3 or a 4
(c) Each die shows 5 or more
(d) At least one die shows fewer than 3
(e) Both dice show fewer than 3
(f) Only one die shows fewer than 3

4.35. A coin is tossed twice. Let E_1, E_2, E_3 denote events "the two tosses are different," "a head occurs on the second toss," "a tail occurs on the first toss," respectively. Which of the following groups of events are independent?
(a) E_1 and E_2
(b) E_1 and E_3
(c) E_2 and E_3
(d) $E_1, E_2,$ and E_3

4.36. A coin is tossed three times. Let A_1 and A_2 denote events "tails on two coins" and "tails on three coins," respectively.
(a) Are $\bar{A}_1$ and $\bar{A}_2$ independent events?
(b) Find an event B such that A_1 and B are independent.

4.37. A die is tossed three times. Assume the outcomes of the individual tosses are independent.
(a) What is the probability that an even number appears on the first toss, a number greater than 3 on the second toss, and a 2 on the third toss?
(b) What is the probability that the sum of the first two tosses shows a seven *and* the third toss shows a three?

4.38. A pair of dice, one red and one white, is tossed. Let events A, B, C, D denote "a number greater than 4 appears on the white die," "a number less than 4 appears on the red die," "a total of 7 appears on two dice," "the number is less than 3 on the white die *and* the number is less than 4 on the red die," respectively.
(a) Which pair (pairs) of events is mutually exclusive?
(b) Which pair (pairs) of events is independent?
(c) Find the probability of the union of each pair of mutually exclusive events in part (a).
(d) Find the probability of either A or B.

4.39. What are the probabilities of the events in Exercise 4.18(b)?

4.40. Suppose a sample space is
$$S = \{1, 2, 3, 4, 5, 6, 7, 8, 9\}$$
Let
$$A = \{1, 2, 3\} \qquad B = \{4, 5, 6\} \qquad C = \{6, 7, 8\}$$
(a) Events A and B are mutually exclusive. Which, if any, of the following pairs are mutually exclusive?
A and $\bar{B}$ $\qquad$ $\bar{A}$ and B $\qquad$ $\bar{A}$ and $\bar{B}$

(b) Events B and C are independent. Which, if any, of the following pairs are independent?
B and $\bar{C}$ $\qquad$ $\bar{B}$ and C $\qquad$ $\bar{B}$ and $\bar{C}$

4.41. In a sample space with 4 equally likely outcomes 1, 2, 3, 4 let $A = \{1, 2\}, B = \{2, 3\}$ and $C = \{1, 3\}$. Are events A, B, C independent?

c4.42. In a sample space with 30 equally likely outcomes 1, 2, $\cdots$, 30 let

$$A = \{1, 3, 5, \cdots, 29\}$$
$$B = \{5, 7, 10, 14, 15, 20, 21, 25, 28, 30\}$$
$$C = \{3, 6, 7, 9, 12, 13, 14, 18, 21, 24, 27, 28\}$$

(a) Are the three events A, B, C independent?
(b) Are the three events A, B, $\bar{C}$ independent?
(c) Are the three events A, $\bar{B}$, $\bar{C}$ independent?

4.43. Construct a Venn diagram in which the only reason events A, B, and C fail to be independent is that A and C are dependent.

4.44. Construct a Venn diagram in which the only reason events A, B, and C fail to be independent is that A, B, and C are dependent.

4.11. CONDITIONAL PROBABILITY: EQUALLY LIKELY OUTCOMES

The probability of an event depends on the sample space. In earlier sections of this chapter the sample space was always specified or clearly understood; in future work this might not be the case. Thus, when reference is made to the probability of an event A in sample space S, we may use the notation $P(A|S)$, read "probability of event A given sample space S." The vertical bar is read "given." [When there is no confusion, we continue to use $P(A)$ in place of $P(A|S)$.]

A few illustrations should make clear the importance of specifying the sample space. The probability of a three on a single toss of a die is $\frac{1}{6}$, but if we have the added information that an odd number shows on the face of the die, then the probability is $\frac{1}{3}$. On a single draw from a well-shuffled deck of cards the probability of a spade is $\frac{13}{52} = \frac{1}{4}$, but if we know that the card drawn is black, the probability is $\frac{13}{26} = \frac{1}{2}$.

From the above illustrations, we conclude that the probability of an event in the original sample space may differ from that in a subset of the sample space, called a new sample space. A new sample space is also termed a **reduced sample space**. Since the reduced sample space is defined by an added condition beyond that required to define the sample space, probabilities of events in the reduced sample space are called **conditional probabilities**. Example 4.22 should clarify the ideas and symbols of conditional probability.

EXAMPLE 4.22. In an experiment with instruction "toss a well-balanced die" the sample space is

$$S = \{1, 2, 3, 4, 5, 6\}$$

Three events are $A = \{1, 3, 5\}$, $B = \{1, 2, 3\}$, and $C = \{1, 6\}$. Assume each outcome in S is equally likely.

In sample space S

$$P(A|S) = P(A) = \frac{3}{6} = \frac{1}{2}$$

Suppose B is the reduced sample space. Since the outcomes in S were equally likely, they are equally likely in B. Thus, the conditional probability of event A relative to event B is

$$P(A|B) = \frac{2}{3}$$

since only two of the outcomes in A (namely, 1 and 3) are also in the reduced sample space B, i.e., $A \cap B = \{1, 3\}$. Suppose C is the reduced sample space. The outcomes in C are equally likely and the conditional probability of A relative to C is

$$P(A|C) = \frac{1}{2}$$

since event A contains only one of the outcomes in the reduced sample space $\{1, 6\}$, that is, $A \cap C = \{1\}$.

Note that $P(A) = P(A|C)$, but $P(A) \neq P(A|B)$. In words, the proportion of outcomes in A relative to S is *the same* as the proportion of outcomes in A relative to C, but the proportion of outcomes in A relative to S is *not the same* as the proportion of outcomes in A relative to B. With the classical definition of probability the phrase "proportion of outcomes in A relative to S" is interchangeable with the phrase "probability of A relative to S," etc.

The conditional probability of an event is still probability of that event. Thus, all the properties of probability which we have studied still hold for conditional probabilities, provided the reduced sample space is used.

There are times when it is desirable that we be able to relate probabilities relative to a reduced sample space with those relative to the original sample space. Example 4.23 introduces such relationships.

EXAMPLE 4.23. Use the sample space and events of Example 4.22 to find conditional probabilities in terms of probabilities relative to S.

First, by the classical definition of probability

$$P(A) = \frac{3}{6} \qquad P(B) = \frac{3}{6} \qquad P(C) = \frac{2}{6}$$

$$P(A \cap B) = P(\{1, 3\}) = \frac{2}{6}$$

$$P(A \cap C) = P(\{1\}) = \frac{1}{6}$$

From Example 4.22 and the above probabilities we have

$$P(A|B) = \frac{2}{3} = \frac{\frac{2}{6}}{\frac{3}{6}} = \frac{P(A \cap B)}{P(B)}$$

$$P(A|C) = \frac{1}{2} = \frac{\frac{1}{6}}{\frac{2}{6}} = \frac{P(A \cap C)}{P(C)}$$

Further, from the definition of conditional probability and the above probabilities relative to S we find

$$P(B|A) = \frac{2}{3} = \frac{\frac{2}{6}}{\frac{3}{6}} = \frac{P(A \cap B)}{P(A)}$$

$$P(C|A) = \frac{1}{3} = \frac{\frac{1}{6}}{\frac{3}{6}} = \frac{P(A \cap C)}{P(A)}$$

Note that $P(C) = P(C|A)$, but $P(B) \neq P(B|A)$.

The property verified four times in Example 4.23, often stated as the *definition* of conditional probability, is desirable for more general kinds of sample spaces. However, for our purposes the property is

Theorem 4.2. Let A and B be any two events in a finite sample space S. Then

(4.21) $$P(A|B) = \frac{P(A \cap B)}{P(B)} \text{ provided } P(B) \neq 0$$

PROOF OF THEOREM 4.2. Since the outcomes in S are assumed to be equally likely, those in B are also equally likely. Let m, n, and N denote the number of different outcomes in A, B, and S, respectively, and $n \neq 0$. Further, let p denote the number of outcomes in A which are also in B. Then

$$P(A|B) = \frac{p}{n} = \frac{p/N}{n/N} = \frac{P(A \cap B)}{P(B)}$$

and Theorem 4.2 holds.

Equation (4.21) is seldom used to compute conditional probabilities in finite sample spaces. Its usefulness comes (1) in relating independent events to conditional probability, and (2) in computing $P(A \cap B)$. In this connection it is not difficult to show (see Exercise 4.58) the following four useful properties:

Property 1. If $P(A|B) = P(A)$, $P(A) \neq 0$, and $P(B) \neq 0$, then $P(B|A) = P(B)$

Property 2. If A and B are independent events with $P(A) \neq 0$ and $P(B) \neq 0$, then $P(A|B) = P(A)$ and $P(B|A) = P(B)$

Property 3. If $P(A) \neq 0$, $P(B) \neq 0$, and $P(A|B) = P(A)$ [or $P(B|A) = P(B)$], then A and B are independent events

Property 4. For any two events A and B $P(A \cap B) = P(A)P(B|A) = P(B)P(A|B)$

For the equally likely case, the restrictions $P(A) \neq 0$ and $P(B) \neq 0$ are of no practical consequences. For example, $P(B)$ must be different from zero to talk sensibly about the probability of A given B.

For Example 4.22 we know that $P(A) = P(A|C)$. Thus, by Property 3, A and C are independent events. Since $P(A) \neq P(A|B)$ it follows that A and B are dependent events.

It is informative to note in Example 4.23 that $P(A \cap C) = P(A)P(C)$ and $P(A \cap B) \neq P(A)P(B)$. However, $P(B)P(A|B) = \frac{3}{6} \times \frac{2}{3} = \frac{2}{6} = P(A \cap B)$, and Property 4 is verified.

EXAMPLE 4.24. Classroom A contains 7 girls and 23 boys, and classroom B, next door, contains 2 girls and 38 boys. If a classroom is selected at random (each room has the same chance of being selected) and then a student is selected at random, what is the probability that the student is a girl?

The following steps should be straightforward:

$$P(\text{girl}) = P(\text{classroom A and girl or classroom B and girl})$$
$$= P(\text{classroom A and girl}) + P(\text{classroom B and girl})$$
$$= P(\text{classroom A}) \times P(\text{girl}|\text{classroom A})$$
$$+ P(\text{classroom B}) \times P(\text{girl}|\text{classroom B})$$
$$= \frac{1}{2} \times \frac{7}{30} + \frac{1}{2} \times \frac{2}{40} = \frac{17}{120}$$

(The use of tree diagrams in problems of this type may simplify the explanation of conditional probabilities.)

Property 4 can be extended to the intersection of any number of events. For example, if A, B, and C are any events in sample space S, then one of the six forms is

(4.22) $$P(A \cap B \cap C) = P(A)P(B|A)P(C|A \cap B)$$

since

$$P(A \cap B \cap C) = P[(A \cap B) \cap C] = P(A \cap B)P(C|A \cap B)$$
$$= P(A)P(B|A)P(C|A \cap B)$$

Other forms of Equation (4.22) will be left as an exercise (see Exercise 4.56) for the student.

EXAMPLE 4.25. In three draws from a well-shuffled deck of playing cards, find (a) the probability of drawing three queens, and (b) the probability of drawing an ace and two kings.

In (a), let Q denote event "queen is drawn" and let subscript denote number of the draw. Then the required probability is

$$P(Q_1 \cap Q_2 \cap Q_3) = P(Q_1) \times P(Q_2|Q_1) \times P(Q_3|Q_1 \cap Q_2)$$
$$= \frac{4}{52} \times \frac{3}{51} \times \frac{2}{50} = \frac{1}{5{,}525}$$

In (b), let A denote event "ace is drawn," K denote event "king is drawn," and subscript denote number of the draw. The required event occurs when one of the three mutually exclusive events

$$A_1 \cap K_2 \cap K_3,\ K_1 \cap A_2 \cap K_3,\ \text{and}\ K_1 \cap K_2 \cap A_3$$

occurs. Thus, the required probability is P(ace and two kings)

$$= P[(A_1 \cap K_2 \cap K_3) \cup (K_1 \cap A_2 \cap K_3) \cup (K_1 \cap K_2 \cap A_3)]$$
$$= P(A_1 \cap K_2 \cap K_3) + P(K_1 \cap A_2 \cap K_3) + P(K_1 \cap K_2 \cap A_3)$$
$$= P(A_1)P(K_2|A_1)P(K_3|A_1 \cap K_2) + P(K_1)P(A_2|K_1)P(K_3|K_1 \cap A_2)$$
$$+ P(K_1)P(K_2|K_1)P(A_3|K_1 \cap K_2)$$
$$= \frac{4}{52} \times \frac{4}{51} \times \frac{3}{50} + \frac{4}{52} \times \frac{4}{51} \times \frac{3}{50} + \frac{4}{52} \times \frac{3}{51} \times \frac{4}{50}$$
$$= 3\,\frac{4 \times 4 \times 3}{52 \times 51 \times 50} = \frac{6}{5{,}525}$$

We have a general method (where independence need not be assumed) for computing the probability of the intersection of any number of events in terms of products of probabilities of individual events. To make our set of basic formulas complete we need a general method (where events need not be mutually exclusive) for computing the probability of the union of any number of events in terms of sums of probabilities of events. Such an equation for two events, is stated in

Theorem 4.3. For any two events A and B in sample space S

(4.23) $$P(A \cup B) = P(A) + P(B) - P(A \cap B)$$

PROOF OF THEOREM 4.3. Think of the probability of $A \cup B$ as the sum of the probabilities of the simple events (outcomes) in $A \cup B$. Then $P(A) + P(B)$ is the sum of the probabilities of outcomes in A plus the sum of the probabilities of outcomes in B. Since A and B are not mutually exclusive, $P(A) + P(B)$ includes the probabilities of outcomes in the intersection $A \cap B$ twice. If $P(A \cap B)$ is subtracted once we have

$$P(A \cup B) = P(A) + P(B) - P(A \cap B)$$

The formula for the probability of the union of two events is typically expressed as in Equation (4.23). Often the following form is useful in computations:

(4.24) $$P(A \cup B) = P(A) + P(B) - P(A)P(B|A)$$

Statements for the probability of the union of more than two events may be found in Exercise 4.66.

EXAMPLE 4.26. A well-balanced die is tossed twice. Let A denote the event "number less than 3 on first toss," and B denote the event "number greater than 3 on second toss." Find $P(A \cup B)$.

From the classical definition of probability

$$P(A) = \frac{12}{36} = \frac{1}{3} \qquad P(B) = \frac{18}{36} = \frac{1}{2} \qquad P(B|A) = \frac{6}{12} = \frac{1}{2}$$

Thus, by Equation (4.24)

$$P(A \cup B) = \frac{1}{3} + \frac{1}{2} - \frac{1}{3} \times \frac{1}{2} = \frac{2+3-1}{6} = \frac{2}{3}$$

The required probability may also be found as follows:

$$P(A \cup B) = 1 - P(\overline{A \cup B}) = 1 - P(\bar{A} \cap \bar{B})$$
$$= 1 - P(\bar{A})\,P(\bar{B}|\bar{A}) = 1 - \frac{24}{36} \times \frac{12}{24} = \frac{2}{3}$$

4.12. THREE DEFINITIONS OF PROBABILITY

The basic properties of probability have already been presented. The classical definition was used to develop these properties, but once we have them we find they hold for the broader definitions of probability presented in this section. There are times when each of three definitions of probability is useful. The primary object in introducing them is to bring out the strong and weak points of each definition—no one definition is entirely adequate.

The primary advantage of the classical definition is its simplicity. Whenever it is applicable, probability of an event is found by enumeration. However, there are some obvious disadvantages. In many experiments there are more than a finite number of outcomes. So any extension of this definition becomes difficult or impossible because of division by N, the number of outcomes. Further, the expression "equally likely" is troublesome from at least two points of view. First, it is not a reasonable assumption in many experiments, and second, probability is *defined* in terms of "equally likely." Thus, the definition may be considered circular. Finally, such a definition requires that all possible outcomes be determined in advance. For example, the classical concept is inadequate to determine the probability of death on the Pennsylvania Turnpike or of an unborn baby's being female or of a person going blind.

Scientists find a second definition most useful for it gives estimates of probabilities of events in any situation where repeated observations can be made. The *relative frequency definition* of probability is as follows:

(4.25)   Suppose an experiment is performed T times, and the event E occurs t times. Let the relative frequency $\frac{t}{T}$ of the event E in T performances of the experiment by denoted by $R(E)$. Then the probability $P(E)$ of the event E is the limit approached by $R(E)$ as T increases indefinitely, it being assumed a limit value exists.

This is an *operational definition*, a type often used in research involving observations, and it is useful because it enables the research worker to find satisfactory estimates of probability. (The "limit" of this definition is not a mathematical limit.)

The relative frequency concept of probability is based on observational evidence, and the classical concept is not. For example, in the classical concept we say the probability of a one occurring on a single toss of a die is $\frac{1}{6}$ without even tossing the die, but according to the relative frequency concept we would need to toss the die many times before arriving at a satisfactory estimate of probability. Several difficulties are immediately apparent. How large is "many times"? When is an estimate "satisfactory"? Why an "estimate" of probability? Why not find "probability"? Most of these questions just do not seem very important when we (1) consider that an estimate can be made almost as good as we choose by letting the number of tosses increase, and (2) remember that people have worked a long time with approximations to (or estimates of) numbers such as π, e, and $\sqrt{2}$. Of course, the real advantage is that *estimates of probability can be found* in a wide variety of problems; if repeated observations can be made, then an estimate of probability can be obtained. For example, if we found that 2,542 peas out of 10,000 of a certain strain were pale green or yellow, we would estimate the probability as $0.2542 \doteq 0.25$ that peas of this strain are pale green or yellow. Or, if we found in a certain town Q that of the 73,461 people who lived to be 40-years of age only 12,967 were alive on their 70th birthday, we would say that the probability of a 40-year-old person of town Q living to be 70 is approximately $\frac{12,967}{73,461} \doteq 0.177$. Example 4.27 should shed further light on the relative frequency approach to probability.

EXAMPLE 4.27. Toss a lopsided quarter and record the proportion of heads (a) every 10 tosses for 20 times, (b) every 100 tosses for 20 times, and (c) every 1,000 tosses for 20 times. Use the information to discuss the relative frequency definition of probability.

The results of the experiments are shown in Table 4.1. In the groups with 10 tosses the relative frequency of heads varies between 0.30 and 0.80; in the groups with 100 tosses it varies between 0.51 and 0.68; and in the groups with 1,000 tosses it varies between 0.590 and 0.616. Clearly, these results suggest that (1) the relative frequencies cluster about a fixed value of approximately 0.60, and (2) the variation about the central value gets smaller as the groups get larger. Thus, it would appear that the probability of heads for the particular quarter of the experiment may be taken as 0.60.

Table 4.1. Relative Frequency of Heads per Group

Group	Number of Tosses		
	10	100	1000
1	.60	.57	.603
2	.50	.55	.591
3	.60	.61	.593
4	.30	.59	.593
5	.60	.51	.613
6	.50	.61	.599
7	.70	.56	.608
8	.60	.63	.596
9	.80	.60	.601
10	.60	.66	.616
11	.60	.68	.592
12	.70	.63	.601
13	.80	.62	.590
14	.50	.62	.607
15	.70	.65	.604
16	.40	.66	.595
17	.80	.55	.604
18	.50	.62	.605
19	.50	.55	.594
20	.60	.54	.603

EXAMPLE 4.28. Use the numbers of heads obtained in part (a) of Example 4.27 to illustrate the limit property of the relative frequency definition of probability.

The computations are illustrated in Table 4.2, and the decreasing variation about some probability is shown in Figure 4.15. As the total number T of tosses gets larger, the relative frequencies get closer and closer to some value—the value could be 0.60. At least it would appear that 0.60 is a satisfactory estimate of the probability. The reader can use the larger number of tosses in parts (b) and (c) of Example 4.27 to obtain (see Exercise 4.69) estimates of the probability of heads for a particular coin.

114 Discrete Probability

Table 4.2. Relative Frequency as Number of Tosses Accumulates

Number of Tosses, T	10	20	30	40	...	190	200
Number of Heads, t	6	11	17	20	...	113	119
Relative Frequency of Heads	.600	.550	.567	.500	...	.595	.595

If an experiment can be repeated a large number of times, according to the relative frequency definition, the probability of an event is the *proportion of times the event will occur in the long run*. But suppose an experiment cannot be repeated? How then does one estimate the probability of an event?

EXAMPLE 4.29. How does one estimate the probability that Molly Brown will make an A in first semester calculus at a large university?

Since the event can happen only once, the relative frequency concept of probability does not seem to apply. Strictly speaking, this is the case. However, there is a sense in which the concept does apply. Of the hundreds of students who have taken first semester calculus with Molly's teacher during the last few years, suppose 6 percent made an A. Then we may take 0.06 as an estimate of the probability that Molly will make an A.

In general, *if a specific experiment is not repeatable, we take the proportion of successes in similar experiments as the estimate of probability of success in the specific experiment*. The problem is to determine which experiments to allow as similar experiments. In Example 4.29 it might have been more meaningful to take into account such things as sex of student, year in the university, and aptitude for mathematics.

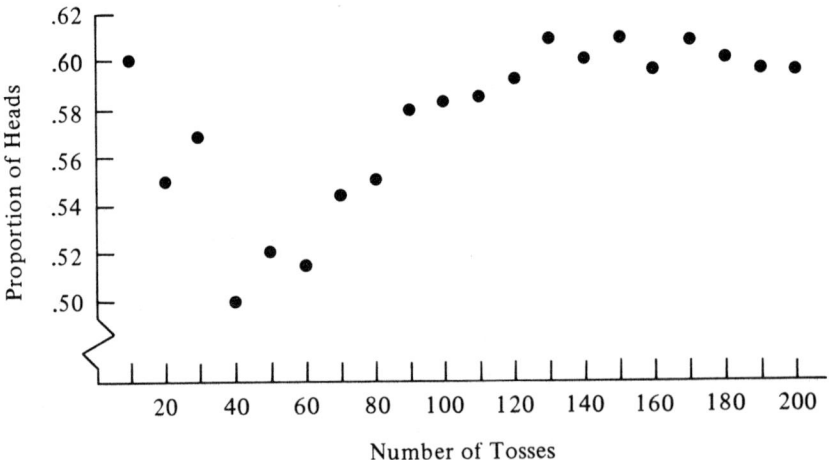

Figure 4.15. Relative Frequency of Heads as Number of Tosses Increases

In case similar experiments do not exist, then an educated guess should be made of the probability of an event. This is what is usually done when a new venture or adventure is undertaken. In such situations, another ingredient, based on chance, adds complications to the problem of obtaining reliable estimates of probability.

All properties of probability derived from the classical definition of probability are consistent with the following rules for assigning the probabilities to an event:

Rule I. Construct a finite sample space S of all possible outcomes where $S = \{o_1, o_2, \ldots, o_N\}$

Rule II. Assign equal probabilities to the simple events. That is, let $P(\{o_i\}) = 1/N$ where $i = 1, 2, \cdots, N$

Rule III. Let the probability of any event E in S be the sum of the probabilities of the simple events contained in E

Common-sense knowledge that all simple events of a sample space cannot always have the same probabilities plus the relative frequency definition of probability lead us to restate Rule II as

Rule II'. Assign probability p_i to simple event $\{o_i\}$ subject to the restrictions

(a) $p_i \geq 0$ for $i = 1, 2, \cdots, N$

(b) $\sum_{i=1}^{N} p_i = 1$

Clearly, Rule II satisfies the restrictions of Rule II' and is a special case of Rule II'. That is, *Rule II' is a generalization of Rule II*. Hence, an acceptable *mathematical definition of finite discrete probability* is:

(4.26)   Use Rules I, II', and III to assign the probability of any event E. This is a generalization of the classical definition of probability.

It can be shown that all the properties of probability already stated (and summarized in Section 4.13) can be derived from Rules I, II', and III. (For a mathematical definition of probability stated in terms of events rather than simple events see Exercise 4.81.) Thus, once it is clear how probabilities of separate events can be found from simple events, we should automatically know how to find probabilities of combinations of two or more events.

EXAMPLE 4.30. Possible outcomes to a certain question are "yes," "no," and "don't know," and the corresponding simple events are denoted by Y, N, and D, respectively. If it is known that $P(Y) = 0.40$ and $P(N) = 0.45$, what are the probabilities of the other six events?

By Rule II'(b), $P(Y) + P(N) + P(D) = 1$. Thus,
$$P(D) = 1 - P(Y) - P(N) = 1 - 0.40 - 0.45 = 0.15$$
By Rule III, $P(Y \cup N) = P(Y) + P(N) = 0.40 + 0.45 = 0.85$,
$$P(Y \cup D) = P(Y) + P(D) = 0.40 + 0.15 = 0.55, \text{ and}$$
$$P(N \cup D) = P(N) + P(D) = 0.45 + 0.15 = 0.60$$

Finally, $P(S) = 1$ and $P(\emptyset) = 0$ for every sample space.

EXAMPLE 4.31. If $S = \{a, b, c\}$, $P(\{a, b\}) = 0.5$, and $P(\{b, c\}) = 0.7$, find the probability of each simple event.

By Equation (4.23) we know that

$$P(\{a, b\} \cup \{b, c\}) = P(\{a, b\}) + P(\{b, c\}) - P(\{a, b\} \cap \{b, c\})$$

Thus, by substitution

$$P(S) = 0.5 + 0.7 - P(\{b\}) \text{ or } 1 = 1.2 - P(\{b\}) \text{ or } P(\{b\}) = 0.2$$

Since

$$P(\{a, b\}) = P(\{a\}) + P(\{b\}) \text{ and } P(\{b, c\}) = P(\{b\}) + P(\{c\})$$

we have by substitution

$$0.5 = P(\{a\}) + 0.2 \text{ and } 0.7 = 0.2 + P(\{c\})$$

Thus

$$P(\{a\}) = 0.3 \text{ and } P(\{c\}) = 0.5$$

An experiment with a discrete sample space can have either a finite or a countably infinite number of outcomes. (When there are *countably infinite* outcomes in the sample space there are as many outcomes as there are positive integers.) We have already considered probability of events for finite sample spaces. For theoretical purposes, we sometimes need to find probability of events in a sample space with countably infinite outcomes. Whenever this is the case we need to extend Rule II'(b) so that the restriction is

$$\text{Rule II'(b')} \quad \sum_{i=1}^{\infty} p_i = 1$$

The other rules remain the same. (See Section 4.13 for summary.) An example should show that restriction II'(b') does not cause any difficulty.

EXAMPLE 4.32. In Example 4.6 the sample space is

$$S_7 = \{H, TH, TTH, TTTH, \cdots\}$$

Show that Rule II'(b') is satisfied.

Since an outcome on any toss of the coin is independent of the outcome on any other toss, and since we may use the classical definition on each outcome, we have, assuming the coin is fair,

$$p_1 = P(\{H\}) = \frac{1}{2} \qquad p_2 = P(\{TH\}) = P(\{T\} \cap \{H\}) = \frac{1}{2} \times \frac{1}{2} = \left(\frac{1}{2}\right)^2,$$

$$p_3 = P(\{TTH\}) = \frac{1}{2} \times \frac{1}{2} \times \frac{1}{2} = \left(\frac{1}{2}\right)^3, \ldots, p_n = \left(\frac{1}{2}\right)^n, \ldots$$

Thus

$$\sum_{i=1}^{\infty} p_i = p_1 + p_2 + p_3 + \cdots$$

$$= \frac{1}{2} + \left(\frac{1}{2}\right)^2 + \left(\frac{1}{2}\right)^3 + \cdots$$

$$= \frac{\frac{1}{2}}{1 - \frac{1}{2}}, \text{ by the formula for the sum of the terms in an infinite geometric progression}$$

$$= 1$$

Thus, restriction II'(b') does hold, along with Rules I, II'(a), and III.

For the student who does not know about the geometric progression let

$$s_1 = p_1, \ s_2 = p_1 + p_2, \ s_3 = p_1 + p_2 + p_3, \ \ldots$$

For Example 4.32

$$s_1 = \frac{1}{2}, \ s_2 = \frac{1}{2} + \frac{1}{4} = \frac{3}{4}, \ s_3 = \frac{1}{2} + \frac{1}{4} + \frac{1}{8} = \frac{7}{8} = \frac{2^3 - 1}{2^3},$$

$$\ldots, \ s_n = \frac{1}{2} + \frac{1}{4} + \frac{1}{8} + \cdots + \left(\frac{1}{2}\right)^n = \frac{2^n - 1}{2^n} = 1 - \frac{1}{2^n}, \ \ldots$$

Thus, as n gets larger, $1/2^n$ gets closer and closer to zero, and $1 - (1/2^n)$ gets closer and closer to 1. This means that the probability of the set of all outcomes after some sufficiently large n is nearly zero.

The properties resulting from the mathematical definition of probability [Statement (4.26)] do not require that the sample space be discrete. *The sample space could be continuous* (or part discrete and part continuous) *and the properties of probability would still hold.* However, the number of sample points and the number of events would no longer be countable. Hence, we would expect some extra difficulties in assigning probabilities and in the calculus of probability. For example, it would not be fruitful to work with simple events for they would always have probability zero since a ratio like $1/N$ approaches zero as N approaches infinity. Instead, we would work from the axiomatic definition of Exercise 4.81.

4.13. SUMMARY OF PROBABILITY

For any discrete sample space, finite or countably infinite, the following three rules may be used to assign probabilities to events:

Rule I. Construct a sample space S of all possible outcomes.

Rule II'. Assign probability p_i to simple event $\{o_i\}$ subject to the two restrictions
(a) $p_i \geq 0$ for each i
(b) for finite sample space, $\sum_{i=1}^{N} p_i = 1$
or
(b') for countably infinite sample space, $\sum_{i=1}^{\infty} p_i = 1$

Rule III. Let the probability of any event E in S be the sum of the probabilities of the events contained in E

The following properties of probability can be derived from Rules I, II', and III in ways similar to those already presented for the equally likely case (we assume event notation is understood):

(4.8) $\qquad P(\bar{E}) = 1 - P(E)$

(4.10) $\qquad P(S) = 1$

(4.11) $\qquad P(\emptyset) = 0$

(4.12) $\qquad 0 \leq P(E) \leq 1$

When events are mutually exclusive

(4.13) $\qquad P(A \cup B) = P(A) + P(B)$

(4.14) $\qquad P(A_1 \cup A_2 \cup \cdots \cup A_k) = P(A_1) + P(A_2) + \cdots + P(A_k)$

When events $A_1, A_2, \cdots, A_k$ form a **partition** of S

(4.15) $\qquad P(A_1) + P(A_2) + \cdots + P(A_k) = 1$

(4.19) $\begin{cases} \text{Two events } A \text{ and } B \text{ are independent if and only if} \\ P(A \cap B) = P(A)P(B) \end{cases}$

(4.20) $\begin{cases} \text{Three events } A, B, \text{ and } C \text{ are mutually independent (or} \\ \text{independent) if and only if} \\ P(A \cap B) = P(A)P(B), P(A \cap C) = P(A)P(C), \\ P(B \cap C) = P(B)P(C), \text{ and } P(A \cap B \cap C) = P(A)P(B)P(C) \end{cases}$

For any events A and B,

(4.21) $\qquad P(A \mid B) = \dfrac{P(A \cap B)}{P(B)}$ provided $P(B) \neq 0$

Properties 2 and 3 $\begin{cases} \text{Let } P(A) \neq 0 \text{ and } P(B) \neq 0. \text{ Events } A \text{ and } B \text{ are} \\ \text{independent if and only if} \\ P(A|B) = P(A) \text{ and } P(B|A) = P(B) \end{cases}$

For any events

(4.22) $\quad P(A \cap B) = P(A)P(B|A) = P(B)P(A|B)$
$\qquad P(A \cap B \cap C) = P(A)P(B|A)P(C|A \cap B)$, one of six forms
(4.23) $\quad P(A \cup B) = P(A) + P(B) - P(A \cap B)$

4.14. EXERCISES

4.45. A person has graduated from college. Let I, J, and K denote the events "works for a pharmaceutical industry," "has a personnel position," and "lives in the suburbs," respectively. Give a concise verbal description of each of the following:
(a) $P(K|\bar{J})$
(b) $P(I \cap K|J)$
(c) $P(\bar{I} \cup K|J)$
(d) $P(\bar{J}|I \cap K)$
(e) $P(I|\overline{J \cup K})$

4.46. A person has graduated from high school. Let C, M, and O denote the events "go to college," "get married within one year," and "major in science," respectively. Give a concise verbal description of each of the following:
(a) $P(O|C)$
(b) $P(M|\bar{C})$
(c) $P(M \cap O|C)$
(d) $P(O|C \cap \bar{M})$
(e) $P(C \cap O|\bar{M})$
Express each of the following symbolically:
(f) the probability that a person graduated from high school will not go to college, given that he or she gets married within one year
(g) the probability that a person graduated from high school will not go to college or major in science, given that he or she does not get married within a year
(h) the probability that a person graduated from high school will go to college, given he or she does not major in science or does not get married within a year.

4.47. If $P(B) = 0.30$, $P(B|A) = 0.40$, and $P(A \cap B) = 0.20$, find $P(A|B)$, $P(A)$, and $P(A \cup B)$.

4.48. If $P(A) = 0.30$, $P(B) = 0.40$ and $P(A|B) = 0.50$, find $P(A \cap B)$, $P(B|A)$, and $P(A \cup B)$.

4.49. If A and B are mutually exclusive events, $P(A) = 0.30$, and $P(B) = 0.45$, find each of the following:
(a) $P(A \cap B)$
(b) $P(A|B)$
(c) $P(\bar{A} \cap \bar{B})$
(d) $P(\bar{A} \cup \bar{B})$

4.50. A well-balanced coin is tossed three times. Let A, B, C, D, E, and F denote the events "three heads," "two heads and one tail," "head on the first toss," "two or more heads," "tail on the third toss," and "head on first toss and tail on third toss," respectively.

Use the classical definition of probability to find each of the following:
(a) $P(B|D)$ (b) $P(D|B)$ (c) $P(E|A)$
(d) $P(B \cap C|E)$ (e) $P(E|B \cap C)$ (f) $P(B \cap C \cap D|E)$
(g) $P(\bar{B} \cup \bar{C}|E)$ (h) $P(D|C \cap E \cap F)$ (i) $P(E \cap F|B \cap C \cap D)$

4.51. Use the events defined in Exercise 4.50 and conditional probability to determine which of the following pairs of events are independent:
(a) C and D (b) C and E
(c) E and $B \cap C$ (d) $B \cap C$ and $E \cap F$
(e) E and $\bar{B} \cup \bar{C}$ (f) D and $C \cap E \cap F$
(g) $\bar{E}$ and $\overline{B \cap C \cap D}$

4.52. Use the events defined in Exercise 4.50 and a product of conditional probabilities to find each of the following probabilities:
(a) $P(B \cap C)$ (b) $P(B \cap C \cap D)$ (c) $P(B \cap \bar{C})$
(d) $P(\bar{B} \cap C \cap D)$ (e) $P(B \cap C \cap D \cap E)$ (f) $P(\overline{B \cap C \cap D \cap E})$

4.53. Two well-balanced dice are tossed.
(a) Find the probability that an even number appears on the first die if an odd number appeared on the second die.
(b) Find the probability that the sum on both dice is a number greater than five if it is known that the sum is an odd number.

4.54. A box contains 30 similar disks numbered from 1 to 30 inclusive. The disks are thoroughly mixed, and three are successively drawn without replacement. Find the probability that
(a) all three numbers are less than 11
(b) only the third number is less than 11
(c) two numbers are less than 11
(d) the sum of the three numbers is less than 11

E4.55. Use the events defined in Exercise 4.50. Find each of the following by two methods:
(a) $P(B \cup C|D)$ (b) $P(B \cup C|E \cap F)$
(c) $P[C \cap (B \cup E)|D]$ (d) $P[C \cap (B \cup E)|D \cap F]$

4.56. Each of the 20 dots in the Venn diagram of Figure 4.16 is equally likely.
(a) Using Equation (4.22) as a guide, write $P(A \cap B \cap C)$ in six ways as the product of conditional probabilities.
(b) Use Figure 4.16 and part (a) to evaluate $P(A \cap B \cap C)$ in three ways.

4.57. Prove that $P(A \cap B \cap C) = P(B)P(C|B)P(A|B \cap C)$.

E4.58. (a) Use events D and F of Exercise 4.50 to illustrate the four properties, 1, 2, 3, and 4, on page 108.

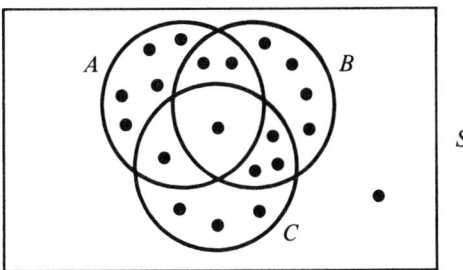

Figure 4.16. Venn Diagram

    (b) Prove Property 1
    (c) Prove Property 2
    (d) Prove Property 3
    (e) Prove Property 4

E4.59. Let E and F be any two events, and let $P(E)$, $P(F)$, and $P(E \cap F)$ be known numbers. Find the formula in terms of these numbers for each of the following:
    (a) $P(\bar{E} \cup \bar{F})$ (b) $P(\bar{E} \cap \bar{F})$
    (c) $P(\bar{E} \cup F)$ (d) $P(\bar{E} \cap F)$

E4.60. Prove: If A and B are any two independent events, then
    (a) A and $\bar{B}$ are independent events, and
    (b) $\bar{A}$ and $\bar{B}$ are independent events.

4.61. In three draws without replacement from a well-shuffled deck of playing cards, find the
    (a) probability of drawing an ace first, king second, and queen third
    (b) probability of drawing a king first, ace second, and queen third
    (c) probability of drawing ace, king, and queen, in any order

E4.62. Box A contains 4 white and 3 black balls. Box B contains 2 white and 4 black balls. One ball is drawn from box A and placed in box B, and then one ball is drawn from box B. What is the probability that the ball drawn from box B is white?

C4.63. In four draws without replacement from a well-shuffled deck of playing cards find the probability of drawing
    (a) four of a kind (same face, except for suit)
    (b) aces on the first two cards and kings on the last two cards
    (c) two aces and two kings
    (d) two pairs
    (e) ace, king, queen, and jack (there are 24 orders)
    (f) four of a suit

4.64. A scientist has to make a decision on each of two independent events. Suppose the probability of error on decision one is 0.05 and the probability of error on decision two is 0.02. Find the probability that the scientist will make the correct decision on
(a) both events
(b) only one event

4.65. A simple machine depends on five parts, a, b, c, d, e, for its most efficient operation, but it can operate satisfactorily if the only part which fails is either a or b or d. Suppose each part has a probability of 0.99 of successful performance. Assuming independence, what is the probability that
(a) the machine operates most efficiently
(b) the machine operates satisfactorily
(c) the machine does not operate satisfactorily or efficiently

E4.66. If A, B, and C are any events of a sample space S, then it can be shown that

(4.27)
$$P(A \cup B \cup C) = P(A) + P(B) + P(C) - P(A \cap B) - P(A \cap C)$$
$$- P(B \cap C) + P(A \cap B \cap C)$$

(a) Use events B, C, and D of Exercise 4.50 to verify Equation (4.27).
(b) Prove Equation (4.27).
(c) State a formula for $P(A \cup B \cup C \cup D)$ which is similar to that of Equation (4.27).

4.67. Over a ten-year period 1,250 students enrolled as freshmen at school M. Of these students 789 graduated. What is the probability that a person who entered M as a freshman will graduate?

4.68. Construct a dot diagram for each of the three parts of Example 4.27. Let the group number be located on the horizontal axis and the relative frequency of heads on the vertical axis. Write a brief report, comparing the variation of dots on the three diagrams.

C4.69. Using Figure 4.15 as a model construct a similar
(a) dot diagram for part (b) of Example 4.27
(b) dot diagram for part (c) of Example 4.27

4.70. (a) Toss a pair of dice ten times and count the number of times a total less than six appears. Repeat the experiment 20 times and record the results.
(b) Construct a dot diagram for the 200 tosses in part (a). Let the experiment number be located on the horizontal axis and the relative frequency of success on the vertical axis. Use your diagram to estimate the probability of a total less than six when your pair of dice is tossed one time.

4.71. (a) Find a book without formulas, graphs, etc., and use all full-length lines starting with Chapter 2. Count the number of lines in each sequence of 20 lines which contains no two-letter word. Record the counts for 20 sequences.

(b) Using part (b) of Exercise 4.70 as a guide, construct a dot diagram for the 400 lines of part (a). Use the diagram to estimate the probability that a line contains no two-letter word.

4.72. A coin is tossed three times. Assume the probability of each simple event is as follows:

$$P(HHH) = 0.14 \qquad P(TTT) = 0.11$$
$$P(HHT) = P(HTH) = P(THH) = 0.13$$
$$P(HTT) = P(THT) = P(TTH) = 0.12$$

Find the probability of each event defined in Exercise 4.50.

4.73. If $S = \{a, b, c, d\}$, $P(\{a,b\}) = 0.25$, $P(\{b,c\}) = 0.35$, and $P(\{a, b, c\}) = 0.40$, find the probability of each simple event.

4.74. If $S = \{a, b, c, d\}$, $P(\{a\}) = 0.25$, $P(\{b,c\}) = 0.35$, and $P(\{c,d\}) = 0.45$, find the probability of each simple event.

4.75. The possible outcomes to a question were "yes," "no," "don't know," and "refuse to answer." Assume the probability of a "no" is twice the probability of a "don't know" and half the probability of a "yes." The probabilities of "don't know" and "refuse to answer" were the same. Find the probability of each compound event with either two or three outcomes.

4.76. Let events of sample space $S = \{a, b, c\}$ be $A = \{a\}$, $B = \{b\}$, $C = \{c\}$, $D = \{a,b\}$, $E = \{a, c\}$, and $F = \{b, c\}$. Which of the eight probability assignments in Table 4.3 are acceptable? Explain your answer in each of the eight cases.

Table 4.3. Possible Probability Assignments for Events of Exercise 4.76

Event	I	II	III	IV	V	VI	VII	VIII
A	1/3	1/2	1/3	1/4	1/2	0	1/3	1/4
B	1/3	1/4	1/6	1/4	0	0	0	1/4
C	1/3	1/4	1/2	1/2	1/2	1	2/3	1/4
D	2/3	3/4	2/3	1	1/2	0	1/3	1/2
E	2/3	3/4	2/3	1	1	1	1	1/2
F	2/3	1/2	1/2	1	1/2	1	1/3	1/2

E4.77. Let p denote probability of "head" on a single toss of a coin, and let $q = 1 - p$ denote probability of "tails." In Example 4.6 the sample space is $S_7 = \{H, TH, TTH, TTTH, \ldots\}$. Show that Rule II'(b') is satisfied.

4.78. Use the probabilities of simple events in Exercise 4.72 to find the probabilities in parts (a) through (i) of Exercise 4.50.

4.79. Use the probabilities of simple events in Exercise 4.72 to determine which pairs of events are independent in parts (a) through (e) of Exercise 4.51.

C4.80. What is the probability that at least two of ten people at a party have the same birthday? Assume 365 days in the year, and assume no twins, triplets, or quadruplets are present. *Hint:* First, find the probability that no two of the ten people have the same birthday.

E4.81. Let E be any event of a finite sample space, and let $P(E)$ be the value of a set function P for event E. $P(E)$ is called "probability of event E" if

1. $P(A) \geq 0$ for each event A in S,
2. $P(S) = 1$, and
3. $P(A \cup B) = P(A) + P(B)$ provided A and B are mutually exclusive events in S.

Use set properties, algebra, and the above definition of probability to prove each of the following:
(a) $P(\bar{E}) = 1 - P(E)$
(b) $P(\emptyset) = 0$
(c) $0 \leq P(E) \leq 1$ for each event E in S
(d) $P(A \cup B \cup C) = P(A) + P(B) + P(C)$ provided A, B, and C are mutually exclusive events in S
(e) $P(A \cup B) = P(A) + P(B) - P(A \cap B)$

E4.82. Let $A_1, A_2, \cdots, A_n$ denote n mutually exclusive events whose union is the sample space S. Let E be any event in S for which $P(E) \neq 0$. Then

(4.28) $$P(E) = P(A_1)P(E|A_1) + \cdots + P(A_n)P(E|A_n)$$

and

(4.29) $$P(A_i|E) = \frac{P(A_i)P(E|A_i)}{P(A_1)P(E|A_1) + \cdots + P(A_n)P(E|A_n)} \quad (i = 1, \ldots, n)$$

Equation (4.28) is sometimes called the *formula on total probability* or the **rule of elimination**. Equation (4.29) is known as **Bayes' formula**.
(a) Prove Equation (4.28)
(b) Prove Equation (4.29)
(c) On a given day four machines, M_1, M_2, M_3, M_4, made 1,500; 1,000; 1,200; 1,800 similar components with 15, 5, 10, 20 percent defective, respectively. Suppose a machine is selected (at random) and then a component made by this machine is selected (at random). What is the probability that this component is defective?
(d) Suppose the component selected in (c) was found to be defective. What is the probability that it was taken from machine M_2?

5

MODEL DISTRIBUTIONS

Earlier we discussed procedures for describing distributions, and each procedure involved some numerical analysis. Now, we give some symbolic methods for describing *classes of distributions*. Such symbolic expressions may be used as models for experiments in much the same way that the rectangle, say, from Euclidean geometry is used as a model by architects, city planners, and paper manufacturers.

By knowing important properties of the model, we are able to make predictions in the real life situation, and the goodness of any prediction depends largely on (1) how well the "real problem" fits the theoretical model and (2) the extent of our knowledge of properties of the model. Usually there is no single model which is used to the exclusion of all others. But it is often possible to determine whether a particular model is satisfactory and to what degree. Such problems cannot be properly treated until some probability models have been studied.

The number of possible models for statistical study seems inexhaustible; but fortunately only a few models may adequately cope with a very large proportion of problems. We discuss the most typical from among those used with discrete variables or with continuous variables.

5.1. RANDOM VARIABLES AND PROBABILITY FUNCTIONS

The probability of events was the primary topic of Chapter 4. Now, we introduce *probability of numbers* associated with outcomes of experiments (i.e., with simple events). Two illustrations should make the basic ideas clear.

EXAMPLE 5.1. A balanced coin is fairly tossed three times. What is the probability of obtaining y heads where $y = 0, 1, 2, 3$?

Assuming each of the eight outcomes

$$HHH, HHT, HTH, HTT, THH, THT, TTH, TTT$$

to be equally likely, we find, by the methods of Chapter 4, the numbers shown in Table 5.1. In particular,

$$f(1) = P(y = 1) = P(\{HTT, THT, TTH\})$$
$$= P(\{HTT\}) + P(\{THT\}) + P(\{TTH\}) = \frac{1}{8} + \frac{1}{8} + \frac{1}{8} = \frac{3}{8}$$

Table 5.1. Probability Function for Number of Heads in Three Tosses of a Coin

Number of Heads, y	0	1	2	3
Probability of y, $f(y)$	$\frac{1}{8}$	$\frac{3}{8}$	$\frac{3}{8}$	$\frac{1}{8}$

There is only one probability value $f(y)$ associated with each number y. Since a value of y is a *number* resulting from outcomes of an experiment, we call y a **random variable**. Thus, the *set of ordered pairs*

$$\left\{\left(0, \frac{1}{8}\right), \left(1, \frac{3}{8}\right), \left(2, \frac{3}{8}\right), \left(3, \frac{1}{8}\right)\right\}$$

is the probability function of the random variable y where y is a function defined over the sample space of the experiment.

We have already expressed the distribution of y in two ways. However, it is often desirable that we find a formula, if possible, to give probabilities of values of the random variable. For Example 5.1 the formula is

$$(5.1) \qquad f(y) = \frac{3!}{y!(3-y)!}\left(\frac{1}{2}\right)^3 \quad \text{for } y = 0, 1, 2, 3$$

where $y!$, called "y-factorial," equals $1 \times 2 \times 3 \cdots (y-1) \times y$ when y is a positive integer, and $0! = 1$. By substitution, we verify that

$$f(0) = \frac{3!}{0!(3-0)!}\left(\frac{1}{2}\right)^3 = \frac{3!}{1(3)!}\left(\frac{1}{2}\right)^3 = \frac{1}{8},$$

$$f(1) = \frac{3!}{1!(3-1)!}\left(\frac{1}{2}\right)^3 = \frac{1 \times 2 \times 3}{1 \times 1 \times 2}\left(\frac{1}{8}\right) = \frac{3}{8}, \text{ etc.}$$

Sometimes, when there is no danger of confusion, we use a formula for $f(y)$, like Equation (5.1), to represent the *probability function of y*. Thus, $f(y)$ might represent (1) the probability function of a random variable y, or (2) a value of the probability function at a single y (in Example 5.1, $y = 0, 1, 2,$ or 3). The context of the statement should make clear which is intended.

EXAMPLE 5.2. Two well-balanced dice are tossed. What is the probability of obtaining a total of y where $y = 2, 3, \ldots, 12$?

Random Variables and Probability Functions

We assume the 36 outcomes of the form

$$(1,1), (1,2), (1,3), (1,4), (1,5), (1,6), (2,1), (2,2), \cdots, (6,5), (6,6)$$

to be equally likely. Thus, by the methods of Chapter 4, we obtain Table 5.2. Again, note that there is only one probability $f(y)$ associated with each total y. Also, y is a random variable since each of its values is the *numerical* value of outcomes of an experiment, and $f(y)$ is the value of a probability function at a specified y. We could say $f(y)$ represents a probability function of the random variable y which in turn is a function of the 36 outcomes of the experiment. That is, briefly, a *probability function is a function of a function of experimental outcomes*.

Table 5.2. Probability Function for Total on Two Dice

Total, y	2	3	4	5	6	7	8	9	10	11	12
Probability of y, $f(y)$	$\frac{1}{36}$	$\frac{2}{36}$	$\frac{3}{36}$	$\frac{4}{36}$	$\frac{5}{36}$	$\frac{6}{36}$	$\frac{5}{36}$	$\frac{4}{36}$	$\frac{3}{36}$	$\frac{2}{36}$	$\frac{1}{36}$

It is interesting to note that y is the sum of two other random variables u and v where u and v are numbers resulting from the toss of the first and second die, respectively. That is, $y = u + v$.

Graphs are often used to illustrate the nature of probability distributions. Figure 5.1 shows one type of graph of Example 5.2. The *points* with coordinates

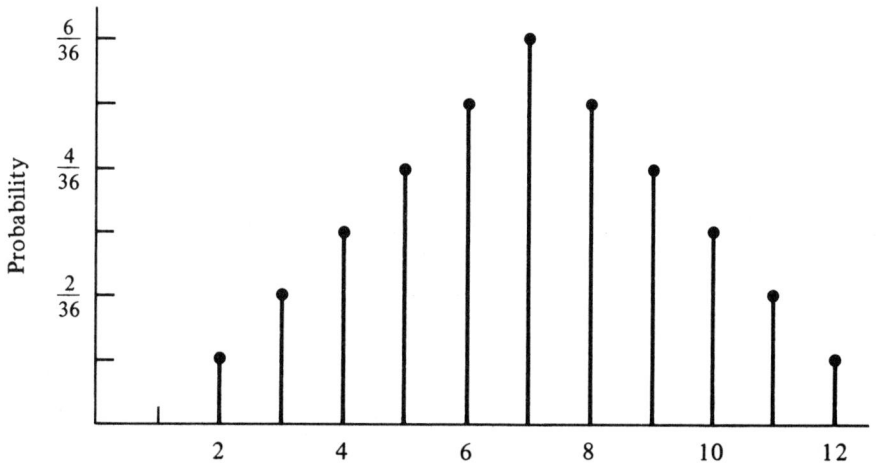

Figure 5.1. Probability Function of Example 5.2

y and $f(y)$ are plotted and vertical line segments (bars) are added to guide the eyes. The bars have lengths which are equal (sometimes proportional) to the probabilities of the corresponding y-values. Thus, the sum of all lengths of bars is equal to one. The histogram of Figure 5.2 is a more common and more useful graph. As explained in Chapter 2, each rectangle of the histogram is constructed so that its base is centered at a y-value and its height is the probability of that y-value. Note that the area of a rectangle is the same as the probability of the y-value of that rectangle, and that the total area is one.

These two illustrations have been useful as guides to the following general definitions:

I. A **random variable** y is a variable each of whose values is a *number* determined by the outcome of an experiment. (In other words, y is a real-valued function defined over the elements of a sample space.)

II. Let $y_1, y_2, y_3, \ldots$ be the possible values (either finite or countably infinite) of a discrete random variable y, and let $f(y_1), f(y_2), f(y_3), \ldots$ be the corresponding probabilities. Then the set of ordered pairs

$$\{(y_1, f(y_1)), (y_2, f(y_2)), (y_3, f(y_3)), \cdots\}$$

is called the **probability function** of y.

A probability function of a discrete random variable may also be expressed by a *formula* or by a *table* which lists all values y_i and the corresponding probabilities $f(y_i)$.

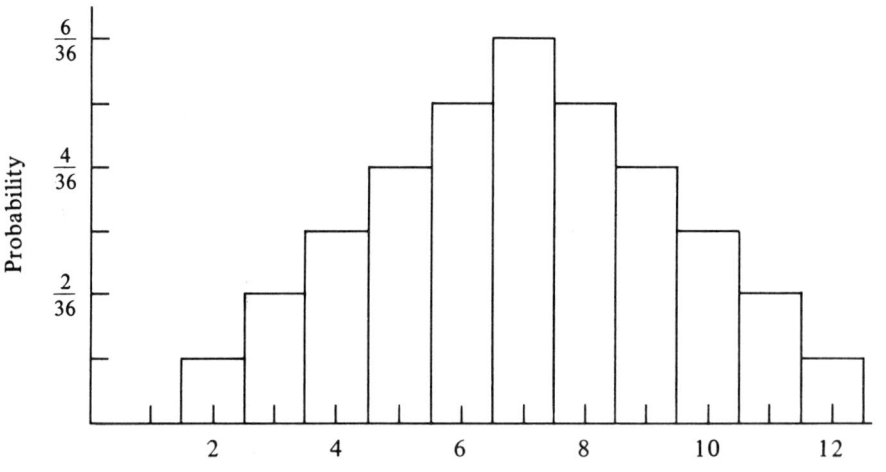

Figure 5.2. Histogram of Probability Function of Example 5.2

As a consequence of the definition of a probability function $f(y)$ we state the following three very important properties:

I. $f(y_i) \geq 0$ for each i where $i = 1, 2, 3, \ldots$.
II. $\sum_i f(y_i) = 1$ where summation is over all values of y.
III. The probability of any subset E of $Y = \{y_1, y_2, y_3, \ldots\}$ is the sum of the probabilities of those numbers which belong to E.

Two illustrations should indicate the usefulness of these properties.

EXAMPLE 5.3. Find the constant c so that

$$f(y) = cy \quad \text{for } y = 1, 2, 3, 4$$

is a probability function.

By Property I we know that the inequality $cy \geq 0$ must hold when $y = 1, 2, 3,$ and 4. Thus, c must be positive or zero. But, by Property II, we know that

$$c1 + c2 + c3 + c4 = 1$$

or

$$c = \frac{1}{10} = 0.1$$

Hence, the required probability function is

$$f(y) = 0.1y \quad \text{for } y = 1, 2, 3, 4$$

EXAMPLE 5.4. Use Example 5.2 to find the probability that y (a) is a multiple of 3, (b) is greater than 5, and (c) is between 5 and 8, inclusive.
From Table 5.2 and Property III it follows that

$$P(y = 3, 6, 9, \text{ or } 12) = f(3) + f(6) + f(9) + f(12)$$
$$= \frac{2}{36} + \frac{5}{36} + \frac{4}{36} + \frac{1}{36} = \frac{12}{36} = \frac{1}{3}$$

Thus, the required probability for (a) is $\frac{1}{3}$. It is possible to find many probabilities in terms of $P(y < \text{constant})$ or $P(y \leq \text{constant})$. In parts (b) and (c) we illustrate how this is done. Thus, for (b) we write

$$P(y > 5) = 1 - P(y \leq 4) = 1 - P(y = 2, 3, \text{ or } 4)$$
$$= 1 - [f(2) + f(3) + f(4)] = 1 - \left[\frac{1}{36} + \frac{2}{36} + \frac{3}{36}\right]$$
$$= 1 - \frac{6}{36} = \frac{5}{6}$$

For (c),

$$P(5 \leq y \leq 8) = P(y \leq 8) - P(y < 5)$$
$$= f(2) + f(3) + \cdots + f(8) - [f(1) + \cdots + f(4)]$$
$$= f(5) + \cdots + f(8) = \frac{4}{36} + \frac{5}{36} + \frac{6}{36} + \frac{5}{36}$$
$$= \frac{20}{36} = \frac{5}{9}$$

Any description (for example, verbal, tabular, graphic, or by formula) which gives each number of a random variable and the corresponding probability is called the *probability distribution* of that random variable. In describing the probability functions in Examples 5.1 and 5.2 we have at the same time described probability distributions. Actually, we use the two terms in the same sense.

5.2. FORMULAS FOR COUNTING

Some formulas for probability functions of a discrete random variable are expressed in terms of the *number of permutations* or the *number of combinations* of r objects selected from n. Before these probability functions, particularly the binomial and hypergeometric, are introduced we develop formulas for counting the number of points in a subset of the sample space.

Suppose we have three objects denoted by a, b, and c. We may select the following three groups of two objects

(5.2)   ab   ac   bc

and we may arrange them so as to have the six groups

(5.3)   ab   ba   ac   ca   bc   cb

Each group in Expression (5.2) is a *combination* of three objects taken two at a time, and each group in Expression (5.3) is a *permutation* of three objects taken two at a time. In general, for r objects taken from n we give the following definitions:

    I. Any collection of r objects selected, without regard to order, from a set of n distinct objects is called a **combination of n objects taken r at a time.**

    II. Any arrangement of r objects selected from n distinct objects is called a **permutation of n objects taken r at a time.**

From Expression (5.3) we know there are six permutations of three objects taken two at a time. In general, we want to find how many permutations there are of n objects taken r at a time.

Suppose we have r spaces to fill. The first space can be filled in n ways by any one of n distinct objects. After one of n objects fills the first space there are $(n-1)$ ways in which the second space can be filled by the remaining $(n-1)$ objects. Thus, the first two spaces can be filled in $n(n-1)$ different arrangements. For each of the $n(n-1)$ arrangements in the first two spaces we can fill the third space with any one of $(n-2)$ objects. So, the first three spaces can be filled in $n(n-1)(n-2)$ different arrangements. We continue this reasoning until we get to the rth space. At this point there are $n-(r-1)$ objects to fill the rth space, and the first $(r-1)$ spaces have been filled in $n(n-1) \cdots [n-(r-2)]$ different arrangements. Thus, the r spaces can be filled in

$$n(n-1) \cdots [n-(r-2)][n-(r-1)]$$

different arrangements.

I. If we let $P(n,r)$ denote the **number of permutations of n distinct objects taken r at a time**, then

(5.4) $$P(n,r) = n(n-1) \cdots [n-(r-1)]$$

II. In case $r = n$, then the **number of permutations of n distinct objects taken all together** is

(5.5) $$P(n,n) = n(n-1) \cdots 2 \times 1 = n!$$

EXAMPLE 5.5. A poker hand of five cards can be dealt in $52 \times 51 \times 50 \times 49 \times 48 = 311,875,200$ different ways (permutations) according to Equation (5.4). A particular hand of five cards can be arranged in $5 \times 4 \times 3 \times 2 \times 1 = 120$ permutations. Once a player has five cards, it is not an arrangement or the order in which they have been dealt that counts—instead, it is the particular combination of cards which is of interest. If we let C denote the total number of combinations which can be dealt, then clearly

$$120 \times C = P(52,5)$$

or

$$C = \frac{P(52,5)}{120} = \frac{311,875,200}{120} = 2,598,960$$

is the total number of different hands (combinations) which can be dealt.

We use Example 5.5 as a guide to the general development. Let $C(n,r)$ denote the **number of combinations** of n distinct objects taken r at a time. Each combination of r objects can be arranged in $r!$ ways according to Equation (5.5). Thus, $r!$ arrangements of each of $C(n,r)$ combinations gives $(r!)\,C(n,r)$ permutations, which is the same as the total number of permutations of r objects taken from n distinct objects. That is

$$(r!)\,C(n,r) = P(n,r)$$

or

(5.6) $$C(n,r) = \frac{P(n,r)}{r!}$$

Also, the number of combinations of n distinct objects taken r at a time is

(5.7) $$C(n,r) = \frac{n(n-1)\cdots[n-(r-1)]}{r!}$$

If both numerator and denominator of the right-hand side of Equation (5.7) are multiplied by $(n-r)!$ and simplified we see that

(5.8) $$C(n,r) = \frac{n!}{r!(n-r)!}$$

Sometimes we require the number of permutations of all n objects when some are the same. An illustration should make the ideas clear.

EXAMPLE 5.6. A coin is tossed ten times and six heads occur. In how many ways could this happen? That is, what is the number of permutations of ten objects, six being of one kind and four of a second kind?

If the ten outcomes were all different there would be 10! permutations. But the six heads are the same. That is, 6! arrangements which could have been distinguished if the objects were different now look alike. Also, since the four tails are the same, 4! arrangements which could have been distinguished if the objects were different now look the same. Let P denote the total number of permutations of ten objects, six being alike and of one kind and four alike and of a second kind. Thus, using an argument similar to that used in Example 5.5 and in the derivation of Equation (5.6), we find that

$$6!4!P = 10!$$

So

$$P = \frac{10!}{6!4!} = \frac{7 \times 8 \times 9 \times 10}{1 \times 2 \times 3 \times 4} = 210$$

In general, by a repeated application of the principle illustrated in Example 5.6, the following holds:

Let n_1 objects be alike of a first kind, n_2 objects be alike of a second kind, $\cdots$, and n_k objects be alike of a kth kind. Then the number of permutations P of all objects is

(5.9) $$P = \frac{n!}{n_1! n_2! \cdots n_k!}$$

where $n = n_1 + n_2 + \cdots + n_k$

Suppose $k = 2$. Let $n_1 = r$. Then $n_2 = n - n_1 = n - r$ and

(5.10) $$P = \frac{n!}{n_1! n_2!} = \frac{n!}{r!(n-r)!} = C(n,r)$$

Thus, the number of permutations of n objects, n_1 being of one kind and n_2 being of a second kind, is the same as the number of combinations of n_1 objects (or n_2 objects) selected from n distinct objects. There are many times when this dual interpretation seems useful. Thus, we introduce the symbol

$$\binom{n}{r}$$

read "$n\ C\ r$," to be used when either interpretation may be meaningful. The symbol $\binom{n}{r}$ is often termed the *binomial coefficient*, suggesting one place where the dual meaning applies.

5.3. EXERCISES

5.1. A well-balanced die is tossed and the number of dots counted.
  (a) Write the probability function in tabular form.
  (b) Use a graph like the one in Figure 5.1 to illustrate the probability function.
  (c) Give a formula for the probability function.
  (d) Use (c) to find the probability of the random variable being larger than 4.

5.2. Show that
$$f(y) = \frac{6 - |y - 7|}{36} \quad \text{for } y = 2, 3, \cdots, 12$$
is a formula for the probability function of Example 5.2.

5.3. Let the collection of students in your statistics class be a sample space S.
  (a) Define two random variables for the elements of S.
  (b) Using the equally likely assumption for each outcome in the sample space S, write in tabular form the probability function of each of the two random variables you defined.
  (c) Construct a histogram for each random variable defined in part (a).

5.4. If possible, find the constant c so that
$$f(y) = c \quad \text{for } y = 1, 2, \cdots, m$$
is a probability function. Assume m is a positive integer.

5.5. (a) Find c so that

$$f(y) = cy \quad \text{for } y = 2, 4, 6, 8$$

is a probability function.
(b) Find the probability that the random variable in part (a) has a value less than 5.
(c) Find the probability that the random variable in part (a) has a value not less than 6.

5.6. Find a constant c so that

$$f(y) = c \quad \text{for } y = -2, -1, 0, 1, 2$$

is a probability function.

*5.7. (a) Find the constant c so that the following function is a probability function:

$$f(y) = c \quad \text{for } y = a, a+1, a+2, \cdots, a+b$$

where a is an integer and b is a positive integer. The correct formula gives probabilities of the **discrete uniform distribution.** Often the formula is applied for the case where $a = 0$.
(b) Find the probability that the random variable in part (a) is greater than $a + 10$, assuming b is some positive integer greater than 10.
(c) Find the probability that the random variable in part (a) is within $b/3$ units of $(a + b)/2$ in case $a = 0$.

5.8. (a) Four well-balanced coins are tossed. Write the sample space for the 16 outcomes and find the probability function of the number of heads in tabular form.
(b) Toss four coins 100 times and make a relative frequency table of the number of heads. Compare the results with those of part (a).
(c) Show that

$$f(y) = \frac{4!}{y!(4-y)!} \left(\frac{1}{2}\right)^4 \quad \text{for } y = 0, 1, 2, 3, 4$$

gives the probabilities in part (a).
(d) Find $P(|y - 2| < 1.1)$.

5.9. Two well-balanced dice are tossed.
(a) Let y denote the number on one die plus twice the number on the second die. Make a table for the probability distribution of the random variable y.
(b) Let u denote the difference in the numbers on the two dice. Make a table for the probability distribution of the random variable u.
(c) Find a formula for the probability distribution of u.
(d) Define a third random variable w and make a table of the probability distribution of w.

5.10. If a college freshman must select five courses from 25 approved courses, how many selections can be made?

5.11. In how many ways can three different flavors of Jello be taken from a grocery shelf with 12 different flavors?

5.12. On a multiple-choice examination each of 20 questions allows 7 choices. How many ways can the examination be answered?

5.13. A homemaker bakes one cake each day for five consecutive days. If any of eight mixes could be used each day, how many choices does the homemaker have if
(a) a different mix is used each day;
(b) any of the eight mixes is used each day;
(c) a different mix is used on each of three days but a fourth mix is used on both of the other two days.

5.14. In how many ways may 7 people seat themselves around a circular table. Assume their relative positions are the only factors to be considered.

5.15. How many distinct arrangements can be made of the letters in "level"?

5.16. Suppose any integer from 1 through 20 is equally likely. Let y denote the number of divisors of an integer. For example, 12 has the following six divisors: 1, 2, 3, 4, 6, 12 Find the probability function of y.

5.17. Prove that $\binom{n}{r} = \binom{n}{n-r}$.

5.18. A committee of five is to be selected from 9 men and 6 women. How many committees can be selected with
(a) 3 men and 2 women
(b) at least 3 men

^C**5.19.** A box contains 6 black, 7 red, and 9 white balls.
(a) How many groups can be selected which contain 3 black, 2 red, and at most 2 white balls?
(b) How many groups can be selected which contain 6 balls, at most 2 of them being black?

^E**5.20.** If n and r are non-negative integers such that $n - 1 \geq r$, prove that

$$\binom{n}{r} = \binom{n-1}{r-1} + \binom{n-1}{r}$$

5.21. In how many ways can 2 clubs, 4 diamonds, 5 hearts, and 2 spades be dealt (or drawn without replacement) from a deck of 52 playing cards?

^C**5.22.** A room contains 20 seats on each side of a central aisle. In how many ways can 18 persons be seated if 3 people insist on taking seats to the left of the aisle and 2 people will not sit left of the aisle? *Hint:* Use Table 5.3.

5.23. A coin is tossed three times. Suppose the probability of "heads" on a single toss is π.
(a) Find the probability function of the number of heads y.
(b) Show that

$$f(y) = \frac{3!}{y!(3-y)!}\pi^y(1-\pi)^{3-y} \text{ for } y = 0, 1, 2, 3$$

gives the probabilities in part (a).
(c) Find the probabilities when $\pi = 0.7$.

5.24. On a test, a student is to match four *statements* with four *answers*. There is only one correct answer to each statement. Suppose the student has no knowledge of the correct answers. If the student guesses, what is the probability function of the number of correct matches?

5.25. On a multiple-choice test each of four questions has five possible answers listed and only one is correct.
(a) If the student guesses on each question, what is the probability function of the number of correct answers?
(b) If the student can correctly eliminate two wrong choices on question one, one wrong choice on question two, and three wrong choices on question four, but guesses on all other choices, what is the probability function of the number of correct answers?

5.26. In an experiment with triads a pair-wise preference rating is to be made on four similar items (for example, four types of soap). If the items are denoted by a, b, c, d, then two possible triads are

$$a\text{—}b \qquad b\text{—}c$$
$$\diagdown\diagup \qquad \diagdown\diagup$$
$$c \qquad d$$

In the first triad, the pair-wise comparisons are between

$$a \text{ and } b \qquad a \text{ and } c \qquad b \text{ and } c$$

(a) If every pair of items is to be represented the same number of times, what is the smallest number of triads to be used in the experiment? List these triads.
(b) Answer part (a) if there are five items.

5.4. BINOMIAL DISTRIBUTIONS

Many simple experiments (or trials) are dichotomous in nature. A few examples follow. A tossed coin shows a "tail" or "head." A manufactured part can be "defective" or "non-defective." The response to a question might be "yes" or "no." A person selected is "male" or "female." An egg has "hatched" or "not hatched." A rat in a maze turns "left" or "right." A candidate is "elected" or "not elected." The decision is "yes" or "no." *In a dichotomous experiment it is customary to call one of the outcomes "success" and the other "failure."*

Our interest in such experiments normally centers in finding the *probability of a specified number of successes in n trials.* As examples, a production manager wants to know the probability of 2 defective parts in a box of 100 parts, a pollster asks for the probability of getting 300 or more "yes" responses in a group of 500 persons, an animal husbandman might want to know the probability that at most 3 animals in a herd of 200 will be deformed, or a traffic supervisor might wish to know the probability that two cars will make a left-hand turn during a week at an intersection where no left-hand turn is permitted.

Several important probability distributions are derived from experiments whose trials are dichotomous. The most important is the binomial distribution, which is discussed in this section. Other important probability distributions are introduced in Exercises of Section 5.6.

Now, we find the probability of exactly y successes in n trials of a dichotomous experiment in which it is assumed that

1. The n repeated trials are independent
2. The probability of success is the same on each trial

Let π denote the probability of success on a single trial. Then $1 - \pi$ is the probability of failure on a single trial. First, for the particular sequence of y consecutive successes followed by $n - y$ failures the probability is

$$\text{(5.11)} \qquad \underbrace{\pi \times \pi \cdots \pi}_{y \text{ times}} \times \underbrace{(1 - \pi)(1 - \pi) \cdots (1 - \pi)}_{(n-y) \text{ times}} = \pi^y(1 - \pi)^{n-y}$$

since the n trials are independent and the probabilities do not change with the trial. The probability of obtaining y successes and $n - y$ failures in any sequence with y successes is also

$$\text{(5.12)} \qquad \pi^y(1 - \pi)^{n-y}$$

since the factors are identically those of Expression (5.11) except for order in which they occur. The number of sequences with y successes and $n - y$ failures (that is, sample points with y successes and $n - y$ failures) is the same as the number of permutations of n objects, y being successes and $n - y$ being failures; it is

$$\text{(5.13)} \qquad \frac{n!}{y!(n - y)!} = \binom{n}{y}$$

These $\binom{n}{y}$ sequences are mutually exclusive simple events of an experiment, each with a probability of $\pi^y(1 - \pi)^{n-y}$. Thus, by Equation (4.14), the probability of exactly y successes is the sum of $\binom{n}{y}$ probabilities $\pi^y(1-\pi)^{n-y}$. That is,

The probability of y successes in n independent performances of a dichotomous experiment is

(5.14) $$f(y) = \binom{n}{y} \pi^y (1-\pi)^{n-y} \quad \text{for } y = 0, 1, \cdots, n$$

where π is the probability of success on each performance of the dichotomous experiment.

Equation (5.14) is known as the formula for the **binomial probability function**. The notation $b(y; n, \pi)$, or $b(y)$, is often used to represent *binomial probability function with parameters* n *and* π. The term "binomial" is applied because the right-hand side of Equation (5.14) is a term in the binomial expansion of

$$[(1-\pi) + \pi]^n$$

In Exercise 5.30 the reader will have an opportunity to justify the last statement.

Equation (5.14) is actually the formula for many binomial distributions. For a fixed probability π there are many possible value of n, and for a fixed n there are infinitely many possible values of π. Thus, when both n and π are specified we have one member of a very large family of binomial distributions. In Example 5.1 and in Exercises 5.8 and 5.23 the reader has been introduced to particular simple binomial distributions. Now, we look at a slightly more cumbersome example.

EXAMPLE 5.7. Suppose 0.6 is the probability of head on a single toss of a particular coin. In 15 independent tosses what is the probability of getting more than 10 heads?

We wish to find

$$P(y > 10) = b(11) + b(12) + b(13) + b(14) + b(15)$$

when $n = 15$ and $\pi = 0.6$. First, we find, after some lengthy calculations that

$$b(11) = \frac{15!}{11!4!} (0.6)^{11}(0.4)^4$$
$$= 1{,}365(0.00362797056)(0.0256)$$
$$\doteq 0.1268$$

Each of the other four probabilities could be found by the same method. However, we prefer to use the following **recursion formula** to shorten the computations:

(5.15) $$b(y+1) = \frac{\pi}{1-\pi} \times \frac{n-y}{y+1} \times b(y)$$

It will be left to the reader to derive this formula (see Exercise 5.34). By the recursion formula and some simple calculations we find

$$b(12) = \frac{0.6}{0.4} \times \frac{4}{12}(0.1268) \doteq 0.0634$$

$$b(13) = \frac{0.6}{0.4} \times \frac{3}{13}(0.0634) \doteq 0.0219$$

$$b(14) = \frac{0.6}{0.4} \times \frac{2}{14}(0.0219) \doteq 0.0047$$

$$b(15) = \frac{0.6}{0.4} \times \frac{1}{15}(0.0047) \doteq 0.0005$$

Thus, by addition

$$P(y > 10) \doteq 0.2173$$

In Example 5.7 we indicated one way to shorten calculations involving $b(y)$. In other relatively simple problems it is also helpful to refer to a table of binomial coefficients (see Table 5.3). Clearly, direct application of the binomial formula would not be desirable if one wanted to compute something like the

Table 5.3. Binomial Coefficients $\binom{n}{y}$

n \ y	0	1	2	3	4	5	6	7	8	9	10*
0	1										
1	1	1									
2	1	2	1								
3	1	3	3	1							
4	1	4	6	4	1						
5	1	5	10	10	5	1					
6	1	6	15	20	15	6	1				
7	1	7	21	35	35	21	7	1			
8	1	8	28	56	70	56	28	8	1		
9	1	9	36	84	126	126	84	36	9	1	
10	1	10	45	120	210	252	210	120	45	10	1
11	1	11	55	165	330	462	462	330	165	55	11
12	1	12	66	220	495	792	924	792	495	220	66
13	1	13	78	286	715	1287	1716	1716	1287	715	286
14	1	14	91	364	1001	2002	3003	3432	3003	2002	1001
15	1	15	105	455	1365	3003	5005	6435	6435	5005	3003
16	1	16	120	560	1820	4368	8008	11440	12870	11440	8008
17	1	17	136	680	2380	6188	12376	19448	24310	24310	19448
18	1	18	153	816	3060	8568	18564	31824	43758	48620	43758
19	1	19	171	969	3876	11628	27132	50388	75582	92378	92378
20	1	20	190	1140	4845	15504	38760	77520	125970	167960	184756

*If $y > 10$, use the identity $\binom{n}{y} = \binom{n}{n-y}$

probability of fewer than 400 responses to 2,000 mailed questionnaires. There are other procedures and approximations designed to make the job as easy as possible, but the computations still remain cumbersome. In practice, we normally use one of many published tables of binomial probabilities or other distributions which approximate binomial distributions. However, abbreviated Tables 5.4, 5.5, and 5.6 [computed with the aid of Equation (5.14) and appropriate recursion formulas] are used to illustrate how such extensive published tables apply; in Section 5.11 we show how to apply the normal approximation to the calculation of binomial probabilities.

Table 5.4. Probabilities of Binomial Distributions for $n = 10$

$$b(y; 10, \pi) = \binom{10}{y} \pi^y (1 - \pi)^{10-y}$$

Number of Successes y	Probability for					
	$\pi = .05$	$\pi = .1$	$\pi = .2$	$\pi = .3$	$\pi = .4$	$\pi = .5$
0	.5987	.3487	.1074	.0282	.0060	.0010
1	.3151	.3874	.2684	.1211	.0403	.0098
2	.0746	.1937	.3020	.2335	.1209	.0439
3	.0105	.0574	.2013	.2668	.2150	.1172
4	.0010	.0112	.0881	.2001	.2508	.2051
5	.0001	.0015	.0264	.1029	.2007	.2461
6	.0000	.0001	.0055	.0368	.1115	.2051
7	.0000	.0000	.0008	.0090	.0425	.1172
8	.0000	.0000	.0001	.0014	.0106	.0439
9	.0000	.0000	.0000	.0001	.0016	.0098
10	.0000	.0000	.0000	.0000	.0001	.0010

Tables 5.4 and 5.5 show probabilities for the selected values $n = 10$ and $\pi = 0.5$, respectively. Table 5.6 is derived from Table 5.4 and gives the summed (cumulative) probabilities less than or equal to a given number of successes.

The reader should observe in Table 5.4 that the distributions become more symmetric as π approaches 0.5, and at $\pi = 0.5$ the distributions are always symmetric (see Table 5.5). Actually, it was possible to abbreviate Table 5.5 because of the symmetry property. Further, both tables were abbreviated to include only columns for $\pi \leq 0.5$. This was possible since the choice of "success" is quite arbitrary, and, therefore, we can always term that outcome a success which has smaller probability. (When $\pi = \frac{1}{2}$, $1 - \pi = \frac{1}{2}$ and either outcome could be termed a success.) However, binomial tables can best be utilized if the reader notes that

(5.16) $$b(y; n, \pi) = b(n - y; n, 1 - \pi)$$

since

$$\frac{n!}{y!(n-y)!}\pi^y(1-\pi)^{n-y} = \frac{n!}{(n-y)!y!}(1-\pi)^{n-y}\pi^y$$

Table 5.5. Probabilities of Binomial Distributions for $\pi = 0.5$

$$b(y; n, 0.5) = \binom{n}{y}(.5)^n = \binom{n}{n-y}(.5)^n = b(n-y; n, 0.5)$$

Number of Successes y	$n = 5$	$n = 10$	Probability for $n = 15$	$n = 20$*	$n = 25$*	$n = 30$*
0	.0312	.0010	.0000	.0000	.0000	.0000
1	.1562	.0098	.0005	.0000	.0000	.0000
2	.3125	.0439	.0032	.0002	.0000	.0000
3	.3125	.1172	.0139	.0011	.0001	.0000
4	.1562	.2051	.0417	.0046	.0004	.0000
5	.0312	.2461	.0916	.0148	.0016	.0001
6		.2051	.1527	.0370	.0053	.0006
7		.1172	.1964	.0739	.0143	.0019
8		.0439	.1964	.1201	.0322	.0055
9		.0098	.1527	.1602	.0609	.0133
10		.0010	.0916	.1762	.0974	.0280
11			.0417		.1328	.0509
12			.0139		.1550	.0806
13			.0032			.1115
14			.0005			.1354
15			.0000			.1445

*Due to symmetry, all probabilities are not listed.

Table 5.6. Summed Probabilities of Binomial Distributions for $n = 10$

$$P(y \leq y') = \sum_{y=0}^{y'} b(y; 10, \pi)$$

Largest Number of Successes y'	$\pi = 0.05$	0.10	Single Trial Probability 0.20	0.30	0.40	0.50
0	.5987	.3487	.1074	.0282	.0060	.0010
1	.9139	.7361	.3758	.1493	.0464	.0107
2	.9885	.9298	.6778	.3828	.1673	.0547
3	.9990	.9872	.8791	.6496	.3823	.1719
4	.9999	.9984	.9672	.8497	.6331	.3770
5	1.0000	.9999	.9936	.9527	.8338	.6230
6	1.0000	1.0000	.9991	.9894	.9452	.8281
7	1.0000	1.0000	.9999	.9984	.9877	.9453
8	1.0000	1.0000	1.0000	.9999	.9983	.9893
9	1.0000	1.0000	1.0000	1.0000	.9999	.9990
10	1.0000	1.0000	1.0000	1.0000	1.0000	1.0000

As an example of Equation (5.16), to obtain the probability of $y = 2$ successes in $n = 10$ independent trials when $\pi = 0.6$ is the probability of success on a single trial, we may equivalently find the probability of $10 - 2 = 8$ failures where the probability of failure on a single trial is $1 - .6 = .4$. That is

$$b(2;10,.6) = b(8;10,.4) = .0106, \text{ by Table 5.4.}$$

Table 5.6 is similar to the cumulative "less than" tables of Chapters 2 and 3. Now we accumulate probabilities instead of relative frequencies. If y' denotes a fixed number of successes, then Table 5.6 gives the probability that the number of successes is equal to or less than y'.

EXAMPLE 5.8. A binomial distribution has $n = 10$ and $\pi = 0.6$. (a) Construct a histogram, and (b) find $P(3 \le y < 6)$.

Figure 5.3 shows the required probability histogram where the probabilities were obtained from Table 5.4. Since the base of each rectangle is one, probability can be measured in terms of either height or area of a rectangle. Thus, for a probability histogram the sum of the areas of all rectangles is 1.

For (b) of Example 5.8 we must find

$$b(3; 10, 0.6) + b(4; 10, 0.6) + b(5; 10, 0.6)$$

Since $\pi = 0.6$, we use Equation (5.16) and replace the last sum by the following equivalent sum

$$b(7; 10, 0.4) + b(6; 10, 0.4) + b(5; 10, 0.4)$$

From Table 5.4 we find that the required probability is

$$0.0425 + 0.1115 + 0.2007 = 0.3547$$

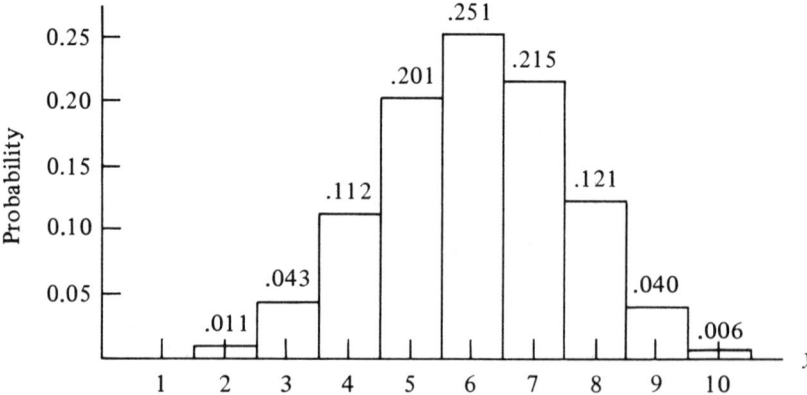

Figure 5.3. Probability Histogram for $n = 10$ and $\pi = 0.6$

Also, entering Table 5.6 with $\pi = 0.40$ we find the required probability is

$$P(y \leq 7) - P(y \leq 4) = 0.9877 - 0.6331 = 0.3546$$

where the difference in the last decimal place is due to rounding-off errors.

5.5. THE MEAN AND THE STANDARD DEVIATION OF PROBABILITY MODELS

Our primary objective is to find the mean and standard deviation for the binomial probability function. We must first introduce new formulas since probabilities are used in place of the frequencies of Chapter 3.

If a random variable y has values $y_1, y_2, y_3, \cdots$ which occur with probabilities $f(y_1), f(y_2), f(y_3), \cdots$ respectively, then the *mean μ of the random variable y* is defined by

(5.17) $$\mu = \sum_i y_i f(y_i)$$

This new formula does not require (1) that each observation be equally likely, or (2) that y have a finite number of values. However, for the purpose of comparison we do consider such a distribution. For N objects, suppose each measurement y_i has frequency f_i ($i = 1, 2, \cdots, h$). Then, by the classical definition, the probability of y_i is $f(y_i) = f_i/N$ for each i. Thus

$$\mu = \sum_{i=1}^{h} y_i f(y_i) = \sum_{i=1}^{h} y_i \left(\frac{f_i}{N}\right) = \frac{\sum_{i=1}^{h} y_i f_i}{N}$$

Hence our first definition of μ [see Equation (3.7)] is a special case of the above definition [Equation (5.17)]. [Also, note that when the weights w_i in Equation (3.10) are the probabilities $f(y_i)$ of the values y_i, then $\sum_i f(y_i) = 1$ and Equation (3.10) becomes Equation (5.17).]

The mean of a random variable y may also be written as μ_y and $E(y)$, and read as "mu sub y" and "expected value of y," respectively. The symbol μ_y is used when it is necessary to distinguish between two or more means. The term "expected value of" has its origin in gambling and is typically associated with probability models.

EXAMPLE 5.9. Find the mean of the random variable in Example 5.1. Letting $y_1 = 0$, $y_2 = 1$, $y_3 = 2$, and $y_4 = 3$, it follows from the definition that

$$E(y) = \mu = 0\left(\frac{1}{8}\right) + 1\left(\frac{3}{8}\right) + 2\left(\frac{3}{8}\right) + 3\left(\frac{1}{8}\right) = \frac{12}{8} = 1.5$$

EXAMPLE 5.10. Find the mean of the random variable of the histogram in Example 5.8(a).

Since the y values are integers we may write

$$E(y) = \mu = \sum_{i=1}^{11} y_i f(y_i) = \sum_{y=0}^{10} y f(y)$$

Then by substitution from Table 5.4

$$\mu = 0(0.0001) + 1(0.0016) + 2(0.0106) + \cdots + 10(0.0060)$$
$$= 5.9995 \doteq 6.0$$

If the random variable y has only a few values it is easy, as in Example 5.10, to find the mean by Equation (5.17). If y has a large number of values, the calculation of μ may be quite time-consuming, as in an experiment involving 200 tosses of a balanced coin.

Fortunately, one of the advantages of a model is that we can often find the mean by mathematical methods. In fact, it can be shown that

> For the binomial probability distribution of Statement (5.14) the mean of the random variable y is

(5.18)
$$\mu = n\pi$$

That is, the mean of a random variable which has a binomial probability distribution is the product of the number of trials and the probability of success on a single trial. It is customary to abbreviate the statement "mean of a random variable which has a binomial probability distribution" to "mean of a binomial distribution."

In Examples 5.9 and 5.10, binomial distributions were used. Thus, by Equation (5.18), we easily verify that the mean in Example 5.9 is $3\left(\frac{1}{2}\right) = 1.5$ and in Example 5.10 it is $10(0.6) = 6.0$.

In Chapter 3 the standard deviation was defined in terms of frequencies. Now, we generalize the definition as follows:

> If a random variable y has values $y_1, y_2, y_3, \cdots$ which occur with probabilities $f(y_1), f(y_2), f(y_3), \cdots$ respectively, then the *variance* σ^2 of the random variable y is defined by

(5.19)
$$\sigma^2 = \sum_i (y_i - \mu)^2 f(y_i)$$

and the *standard deviation is the positive square root* of the variance.

Just as with the mean, it can be shown that the Chapter 3 definitions of variance and standard deviation are special cases of the above definitions. See Exercise 5.44 for a computing formula, analogous to Equation (3.32), for the variance defined by Equation (5.19).

The variance of a random variable y may also be written as $E(y - \mu)^2$, and read as "expected value of $(y - \mu)^2$." That is, the variance is the expected value of a function of a random variable.

EXAMPLE 5.11. Find the variance and standard deviation of the random variable y which has a binomial probability distribution with $n = 5$ and $\pi = \frac{1}{2}$. The binomial probabilities are given by the formula

$$b(y) = \binom{5}{y}\left(\frac{1}{2}\right)^5 \quad \text{for } y = 0, 1, \cdots, 5$$

and are shown in Table 5.7 along with other calculations required to obtain the variance. By Equation (5.18) the mean is $5(\frac{1}{2}) = 2.5$. Table 5.7 shows a variance of 1.2500. Thus, the standard deviation is $\sqrt{1.2500} \doteq 1.12$.

Table 5.7. Calculations for Variance of Example 5.11

y	$f(y)$	$y - 2.5$	$(y - 2.5)^2$	$(y - 2.5)^2 f(y)$
0	$\frac{1}{32}$	-2.5	6.25	0.1953
1	$\frac{5}{32}$	-1.5	2.25	0.3516
2	$\frac{10}{32}$	-0.5	0.25	0.0781
3	$\frac{10}{32}$	0.5	0.25	0.0781
4	$\frac{5}{32}$	1.5	2.25	0.3516
5	$\frac{1}{32}$	2.5	6.25	0.1953
				$\sigma^2 = 1.2500$

As we have seen in Example 5.11, the formula [Equation (5.19)] for the variance of a random variable of a general distribution leads to lengthy calculations even for simple distributions. When a specific probability distribution is known, a much simpler formula can often be derived. For example, it can be shown that

> For the binomial probability distribution of Statement (5.14) the variance of the random variable y is

(5.20) $$\sigma^2 = n\pi(1 - \pi)$$

> and the standard deviation of the random variable y is

(5.21) $$\sigma = \sqrt{n\pi(1 - \pi)}$$

That is, the variance of a random variable which has a binomial probability distribution is the product of the number of trials, probability of success on a single trial, and probability of failure on a single trial. As with the mean, it is customary to abbreviate the statement "variance of a random variable which has a binomial probability distribution" to "variance of a binomial distribution." Similar statements hold for the standard deviation.

For the binomial distribution of Example 5.11, $n = 5$ and $\pi = \frac{1}{2}$. Thus, the variance is

$$\sigma^2 = 5\left(\frac{1}{2}\right)\left(\frac{1}{2}\right) = 1.25$$

which agrees with the result already obtained in Table 5.7. Further, in Example 5.8, $n = 10$ and $\pi = 0.6$. Hence

$$\sigma^2 = 10(0.6)(0.4) = 2.4$$

and

$$\sigma \doteq 1.55$$

EXAMPLE 5.12. Use the binomial distribution with $n = 30$ and $\pi = \frac{1}{2}$ to make probability statements about the concentration of successes about the mean.

By Equations (5.18), (5.20), and (5.21) we find the mean, variance, and standard deviation for the binomial distribution to be

$$\mu = 30\left(\frac{1}{2}\right) = 15,$$

$$\sigma^2 = 30\left(\frac{1}{2}\right)\left(\frac{1}{2}\right) = 7.5, \text{ and}$$

$$\sigma = \sqrt{7.5} \doteq 2.74$$

Thus, using properties of symmetry and Table 5.5, we find the following approximate probabilities (that y is within 3, 2, and 1 standard deviations of the mean):

$$P[\mu - 3\sigma \leq y \leq \mu + 3\sigma] = P[15 - 3(2.74) \leq y \leq 15 + 3(2.74)]$$
$$= P[6.78 \leq y \leq 23.22]$$
$$= 1 - 2P[y < 6.78]$$
$$\doteq 1 - 2[b(0) + b(1) + \cdots + b(6)]$$
$$= 1 - 2[0.0000 + 0.0000 + \cdots + 0.0006]$$
$$= 1 - 2[0.0007] = 0.9986$$

$$P[\mu - 2\sigma \leq y \leq \mu + 2\sigma] = P[9.52 \leq y \leq 20.48]$$
$$\doteq 1 - 2[b(0) + b(1) + \cdots + b(9)]$$
$$= 1 - 2[0.0214] = 0.9572$$

$$P[\mu - \sigma \leq y \leq \mu + \sigma] = P[12.26 \leq y \leq 17.74]$$
$$\doteq b(13) + b(14) + \cdots + b(17)$$
$$= 0.1115 + 0.1354 + \cdots + 0.1115 = 0.6383$$

For each computation, each probability was determined over a restricted interval. For example, in the last case we actually found $P[12.5 \leq y \leq 17.5]$ instead of $P[12.26 \leq y \leq 17.74]$. By visualizing a histogram, we observe that the probability over the complete interval may be computed as follows:

$P[12.26 \leq y \leq 17.74]$
$$= P[12.26 \leq y < 12.5] + P[12.5 \leq y \leq 17.5] + P[17.5 < y \leq 17.74]$$
$$= 0.24b(12) + [b(13) + \cdots + b(17)] + 0.24b(18)$$
$$= 0.24(.0806) + 0.6383 + 0.24(.0806)$$
$$\doteq 0.6770$$

According to Chebyshev's theorem (see page 60) the corresponding probabilities for $k = 3, 2$, and 1 are $1 - (1/3^2) = 8/9 \doteq 0.8889$, 0.7500, and 0.0000, respectively. Clearly, knowing the probability distribution allows us to improve on the estimates given by Chebyshev's theorem, especially for small values of k.

Binomial distributions are likely the most important discrete probability distributions—at least they play a central role in the development of other distributions and they are probably used more in applications. Also, the binomial probability function is relatively easy to derive and to relate to the following underlying assumptions:

1. The number of trials in the experiment is fixed.
2. Each trial has exactly two possible outcomes, called success and failure.
3. The probability of success (or failure) is the same on each trial.
4. The trials are mutually independent.

Finally, computations are relatively simple since extensive tables are available and easy to read.

If this were not an introduction to statistics, we would study five other families of important discrete probability distributions. However, we do not ignore them completely, for they are introduced in the exercises. The reader has already seen *uniform distributions* in Exercises 5.4, 5.6, and 5.7. In the exercises of Section 5.6 we define and illustrate *hypergeometric* (5.53 through 5.59), *geometric* (5.60), *negative binomial* (5.60 through 5.63), and *Poisson* (5.64 through 5.66) distributions. More information on uniform distributions can be found in Exercises 5.49, 5.50, and 5.51.

5.6. EXERCISES

5.27. Let S denote sucess and F denote failure. List the $\binom{5}{2}$ sequences of two successes and three failure in five dichotomous trials.

5.28. (a) Do the computations and prepare a table of binomial probabilities (correct to four decimal places) for $b(y; 6, 0.3)$ when $y = 0, 1, \ldots, 6$.
(b) Construct a histogram for part (a).

5.29. The binomial coefficients of Table 5.3 can be obtained from the following pattern known as **Pascal's triangle**:

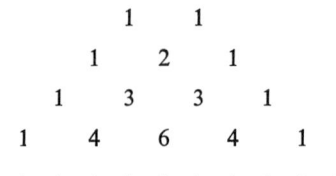

where each row starts with a 1 and ends with a 1, and every other entry is the sum of the nearest two numbers in the preceding row.
(a) Verify that the numbers in the fourth row of Pascal's triangle are the binomial coefficients for $y = 0, 1, 2, 3, 4$.
(b) Complete the fifth, sixth, and seventh rows of Pascal's triangle and verify that the numbers of the seventh row are the binomial coefficients for $y = 0, 1, \cdots, 7$.

E5.30. Apply the binomial expansion to $[(1 - \pi) + \pi]^n$ and show that the $(r + 1)$st term in the expansion is the same as the binomial when $y = r$. Illustrate the equality for the cases where $r = 0, 1,$ and n.

C5.31. Construct a sequence of histograms for the binomial distributions with
(a) $n = 4$ and $\pi = 0.10, 0.25, 0.50, 0.75,$ or 0.90
(b) $\pi = 0.25$ and $n = 3, 5,$ or 7

5.32. In 10 tosses of a well-balanced coin find
(a) the probability of tossing 3 heads and 7 tails
(b) the probability of tossing at most 3 heads
(c) the probability of tossing at least 7 heads

5.33. In a toss of a die the occurrence of a 1 or 6 is considered a success. Suppose a balanced die is tossed 7 times. Find the probability of obtaining at least 5 failures.

E5.34. (a) Prove the following recursion formula, Equation (5.15), for binomial probabilities:

$$b(y + 1) = \frac{\pi}{1 - \pi} \times \frac{n - y}{y + 1} \times b(y) \qquad \text{for } y = 0, 1, \cdots, n - 1$$

(b) Derive a recursion formula for binomial probabilities which expresses $b(y - 1)$ in terms of $b(y)$ for $y = 1, 2, \cdots, n$.

5.35. (a) If $n = 10$, $\pi = 0.3$, and $b(3) = 0.2668$, use the recursion formulas in Exercise 5.34 to find $b(1), b(2), b(4),$ and $b(5)$.

(b) If $n = 30$, $\pi = 0.5$, and $b(14) = 0.1354$, use the recursion formulas for a binomial distribution to find $b(13)$, $b(15)$, $b(16)$, and $b(17)$.

5.36. Suppose rats have been trained to run a certain maze and it is claimed that there is an 80 percent chance that each rat will run the maze in an allotted time. If 15 rats run, find
(a) the expected number of rats that will run the maze in the allotted time
(b) the probability that at least 12 will run the maze in the allotted time

5.37. The batting average of a big-league baseball player over the past five years is 0.340. In five times at bat what is the probability that he gets a hit? How many hits would he be expected to get?

5.38. A manager of a resort hotel estimates that guests in 40 percent of the occupied rooms have breakfast in their rooms. What is the probability that breakfast will be served in 5 of 20 occupied rooms?

5.39. A forest ranger has found that one person in ten (in the long run) misses a well-marked trail. Of 15 lone hikers what is the probability that at most two will miss the trail? Find the expected number of hikers who would not miss the trail.

c5.40. Ted Wagner went to Lake Owen to fish. Based on past experience he believed that a fish would bite in one cast out of four (on the average).
(a) What is the probability that Ted will get at most three bites in twelve casts?
(b) If a fish bites, the probability of catching the fish is 0.40. What is the probability of a fish being caught? What is the probability that Ted will catch at most three fish in twelve casts?
(c) Find the expected number of fish to be caught in 100 casts.
(d) Suppose Ted throws two out of three fish back in the lake. What is the probability that he will catch and keep at most three fish in twelve casts?

5.41. Find the mean of the random variable y in Example 5.2.

5.42. For a binomial distribution with $n = 10$ and $\pi = 0.4$, find
(a) the probability that the total number of successes will not deviate from the mean by more than two standard deviations
(b) the probability that the total number of successes will deviate from the mean by more than three standard deviations

5.43. If a binomial distribution has $n = 20$ and $\pi = 0.5$, find the answers to parts (a) and (b) of Exercise 5.42.

5.44. Prove the following for any discrete random variable y:

$$\sigma^2 = \sum_{i=1}^{} y_i^2 f(y_i) - \mu^2$$

5.45. Compute the variance for Example 5.2 using the formula of Exercise 5.44.

*5.46. Prove that $\mu = n\pi$ for a binomial distribution.

Hint: Substitute the binomial probabilities in Equation (5.17), note that the first of $(n+1)$ terms is zero, factor $n\pi$ out of the remaining n terms, and show that the other long factor can be reduced to $[(1-\pi)+\pi]^{n-1}$.

*5.47. If the random variable y has a binomial probability distribution $b(y)$ where $y = 0, 1, \ldots, n$, show that

$$\sum_{y=0}^{n} y^2 f(y) = n\pi - n\pi^2 + n^2\pi^2$$

Hint: Let $y^2 = y(y-1) + y$ and use the method of proof of Exercise 5.46.

E5.48. If the random variable y has a binomial probability distribution $b(y)$, show that

$$\sigma^2 = n\pi(1-\pi)$$

Hint: Use Exercises (5.44) and (5.47).

E5.49. Suppose the random variable y has a *discrete uniform distribution* with probability function

$$u(y;m) = \frac{1}{m} \quad \text{for } y = 1, 2, \ldots, m$$

where m is any positive integer. Show that the mean and variance of y are

$$\mu = \frac{m+1}{2} \quad \text{and} \quad \sigma^2 = \frac{m^2-1}{12}$$

Hint: Use Exercise 5.44 and the identities

$$1 + 2 + 3 + \cdots + m = \frac{m(m+1)}{2}$$

and

$$1^2 + 2^2 + 3^2 + \cdots + m^2 = \frac{m(m+1)(2m+1)}{6}$$

(By now the reader should have observed from Exercises 5.4, 5.6, and 5.7 that there are many different uniform distributions. Actually, the values of y do not have to be integers.)

E5.50. A well-balanced die is tossed. Let the random variable y be the number of spots on the top side of the die.
(a) Use Exercise 5.49 to find the mean, variance, and standard deviation of y.
(b) What is the probability that y will be within one standard deviation of the mean? Within two standard deviations of the mean?

E5.51. An outcome of an experiment is an integer from 1 through 100. Assume each outcome is equally likely. Since an outcome is a number, it is also the value of a random variable, say y.

(a) If a single number is drawn, what is the probability that it will be a number from 30 to 71 inclusive?

(b) The numbers 30 and 71 are how many standard deviations apart?

(The reader should note that a uniform distribution is the model for numbers drawn from a random number table.)

5.52. Five successes were obtained for a binomial distribution with $n = 10$.
(a) Find π if $P(y > 5)$ is approximately 0.05.
(b) Find π if $P(y < 5)$ is approximately 0.38.

E5.53. A finite dichotomous population has N objects, A being of the first kind and B of the second kind. Suppose n objects are drawn without replacement.
(a) Write an expression for the number of ways in which y objects of the first kind *and* $n - y$ objects of the second kind can be selected from the collection of N objects.
(b) The probability that y objects of the first kind *and* $n - y$ objects of the second kind can be selected (i.e., randomly drawn without replacement) from the collection of N objects is given by

$$h(y; A, B, n) = \frac{\binom{A}{y}\binom{B}{n-y}}{\binom{A+B}{n}} \quad \text{for } y = \max(0, n-B), 1, \cdots, \min(A, n)$$

where max $(0, n - B)$ means take the larger number of 0 and $n - B$ and min (A, n) means take the smaller number of A and n. Use part (a) to prove this. We call this probability function the *hypergeometric probability function*, or simply, *hypergeometric distribution*. The notation $h(y)$ is used in place of $h(y; A, B, n)$ if there is no danger of confusion. *Note:* Of the four assumptions (see page 147) underlying the binomial distribution, only the first two hold. The last two fail since the trials are *dependent* and the *probability of success changes* from trial to trial.

(c) Four children are selected from a group of 3 girls and 5 boys. Make a table showing the probability (to three decimal places) of selecting 0, 1, 2, 3 girls.

E5.54. In a box of 88 fancy apples 3 are defective. If an inspector randomly selects 5, what is the probability that none will be defective? That 3 will be defective? (See Exercise 5.53.)

E5.55. Five cards are dealt from a well-shuffled deck of 52 playing cards. Make a table showing the probability of getting 0, 1, 2, 3, 4 aces.

E5.56. A hypergeometric distribution may be useful in acceptance sampling. To illustrate, suppose items are regularly shipped in lots (for example, cartons, boxes, packages) of 24. Four items from each lot are inspected, and the lot is accepted if no item is defective. What is the probability of accepting a lot if it contains (a) 2 defective items, (b) 7 defective items, and (c) 12 defective items? *Note:* In practice, tables are used to aid in the computation of such probabilities.

152 *Model Distributions*

E5.57. Eight pennies and two dimes are in box A, and 15 pennies are in box B. Six coins are taken from box A and placed in box B, and then the 21 coins of box B are thoroughly mixed. Then seven coins are taken from box B and placed in box A. What is the probability that one dime is in box A?

E5.58. Derive a recursion formula which expresses the hypergeometric probability $h(y+1)$ in terms of $h(y)$.

E5.59. Seven objects are drawn without replacement from 15 thoroughly mixed objects, 6 being of the first kind and 9 of the second kind.
 (a) Make a table showing the probability of selecting 0, 1, 2, 3, 4, 5, or 6 objects of the first kind.
 (b) Compute the mean and variance of the random variable of part (a).
 (c) It can be shown that the mean and variance of a hypergeometric distribution are given by

$$\mu = n\left(\frac{A}{N}\right) \quad \text{and} \quad \sigma^2 = n\left(\frac{A}{N}\right)\left(1 - \frac{A}{N}\right)\left(\frac{N-n}{N-1}\right)$$

Use these formulas to verify your answers of part (b). *Note:* If the relative frequency A/N is set equal to π, the mean of the hypergeometric distribution is the same as for the binomial distribution, but the variance is less. However, if n is small relative to N, the variances of the two distributions are approximately the same. In this case, the binomial is often used as an approximation to the hypergeometric.

E5.60. If all the assumptions (see page 147) underlying the binomial distribution are satisfied except that the number of trials is not fixed, and if the random variable y is the *number of the trial on which the first success occurs*, then the resulting probability function of y is given by

$$g(y; \pi) = \pi(1 - \pi)^{y-1} \quad \text{for } y = 1, 2, 3, \cdots$$

This probability function is called the *geometric probability function*, or simply, *geometric distribution*. Special aspects of the distribution have already been presented in Examples 4.6 and 4.32 and in Exercise 4.77.
 (a) Suppose a well-balanced coin is tossed until a head appears. What is the probability that a head appears on the sixth toss for the first time?
 (b) In the game of billiards a player continues until he misses a shot. Suppose 0.1 is the probability that player X misses a shot, what is the probability that player X misses on the sixth shot? Assume the shots are independent.
 (c) It can be shown that the mean and variance of the geometric distribution are

$$\mu = \frac{1}{\pi} \quad \text{and} \quad \sigma^2 = \frac{1-\pi}{\pi^2}$$

For part (b), find the mean number of shots.
 (d) For part (b), find the median number of shots.

E5.61. If all the assumptions underlying the binomial distribution (see page 147) are satisfied except that the number of trials is not fixed, and if the random variable y is the *number of the trial on which the kth success occurs*, then the resulting probability function of y is given by

$$nb(y; k, \pi) = \binom{y-1}{k-1} \pi^k (1-\pi)^{y-k} \quad \text{for } y = k, k+1, k+2, \cdots$$

This function is called the *negative binomial probability function*, or simply *negative binomial distribution*, since the probabilities are successive terms of the expansion of

$$\left[\frac{1}{\pi} - \frac{1-\pi}{\pi}\right]^{-k}$$

(a) Show that the geometric distribution (see Exercise 5.60) is a special case of the negative binomial distribution in case $k = 1$.

(b) A well-balanced coin is tossed until a head appears the third time. Find the probability of this happening on the fifth trial; eighth trial; eleventh trial.

(c) It can be shown that the mean and variance of the negative binomial distributions are

$$\mu = \frac{k}{\pi} \quad \text{and} \quad \sigma^2 = \frac{k(1-\pi)}{\pi^2}$$

Find the mean for each distribution in part (b).

E5.62. A well-balanced die is tossed until an even number appears the third time. What is the probability that this will happen on the sixth trial? *Hint:* See Exercise 5.61.

E5.63. Four balanced coins are tossed until all heads or all tails occur the fifth time.
(a) What is the probability that this will happen on trial 5, 6, 7, 8, 9, or 10?
(b) Use Exercise 5.61(c) to find the mean and variance of this distribution.

E5.64. If the random variable y denotes the number of rare events which occur independently in a specified time interval (or specified region in space), then the probability function is given by

$$p(y; \mu) = \frac{e^{-\mu} \mu^y}{y!} \quad \text{for } y = 0, 1, 2, 3, \cdots$$

where μ is the mean of the random variable y and e is the irrational number used in connection with natural logarithms—its value being approximately 2.71828. We may define e^a to be equal to the infinite sum

$$1 + \frac{a}{1!} + \frac{a^2}{2!} + \frac{a^3}{3!} + \cdots$$

where a is any number. However, rather than approximate $e^{-\mu}$ by this infinite sum we use Table I in the Appendix.

The above probability function is called the *Poisson probability function* of the family of *Poisson distributions*. (A Poisson distribution is often useful as an approximation to that binomial distribution with large n and small π. The mean μ of the approximating Poisson distribution is equated to $n\pi$.) The following problems illustrate some applications:

(a) The mean number of accidents per month on a dangerous curve is 1.3. What is the probability of at most one accident on this curve during a specified month?

(b) At a large publishing company the mean number of typesetter errors per page is 0.8. What is the probability that a specific page will have more than one error?

(c) The mean number of bacteria per square area in a given culture is 2.3. What is the probability that the number of bacteria on a specified square will not deviate more than 1.5 from the mean?

(d) The mean number of telephone calls per minute at a switchboard is 3.8. Suppose the switchboard is overloaded with more than nine calls per minute. What is the probability that the switchboard will be overloaded in a specified minute?

E5.65. A pollster finds that one percent of the people interviewed responded favorably to a certain question. If 500 people are asked the question, what is the probability that at most three will give a favorable answer? *Hint:* Use a Poisson approximation.

*5.66. If a random variable y has the Poisson distribution, prove that $\sigma^2 = \mu$. *Hint:* Use the infinite series of Exercise 5.64 and the methods of Exercises 5.46, 5.47, and 5.48.

5.7. CONTINUOUS DISTRIBUTIONS OF ONE VARIABLE

In passing from the discrete to the continuous distribution we encounter several new problems. Many problems can be traced to the fact that there is always an uncountable number of possible outcomes in the sample space of a continuous random variable. Thus, we do not build our probability measure on individual outcomes, but rather on collections as in an interval. (For a continuous random variable, the probability of a single value is always zero.)

We have already learned that a histogram is useful in finding the relative frequencies (Chapter 2), proportions (Section 3.3), or probabilities (Sections 5.1 and 5.4) of a random variable in an interval (or in intervals). We have seen that when the histogram is constructed so that the total area is one, then the probability of a value of the random variable falling in an interval (or in intervals) is the area above the interval (or intervals). So we use histograms to relate the discrete to the continuous random variable.

By considering a sequence of histograms each with area one we may visualize what happens in passing from the discrete to the continuous case. Suppose y is a continuous random variable with values in some interval on the line of real numbers. If many observations are made, we may group the data in classes and construct a relative frequency histogram with area one. Now suppose the number of observations is increased indefinitely. As the number of observations increases suppose we subdivide each interval into two, then four, then eight, ... smaller intervals. The sequence of histograms would look smoother and smoother until conceptually, for an infinite collection of observations, we would expect the limiting graph to be a smooth curve with area one. Further, we would expect the area under the curve and above any subinterval, say $a < y < b$, in the domain of the function (whose graph is the curve) to be the probability of the random variable y falling between a and b. Formally, we say

(5.22)   The continuous random variable y has a **probability density function** with values $f(y)$ if it satisfies the following properties:

I. $f(y)$ is a non-negative real number for all real values of y.
II. The area under the curve $w = f(y)$, and above the y-axis is one.
III. The area under the curve $w = f(y)$, and above the interval $a < y < b$, denoted by $A(a, b)$, is the probability that y has a value between a and b; that is, $A(a, b) = P(a < y < b)$.

The reader should note that $f(y)$ is not the probability of a particular value of a random variable y. Instead $f(y)$ is simply the vertical distance between a point on the y-axis and the corresponding point on the curve of the graph of $w = f(y)$. Thus, for a continuous random variable y, it is customary to refer to $f(y)$ as probability density function instead of probability function. Also, since probability of a continuous random variable is defined in terms of area, it should be observed that

$$P(a < y \leq b) = P(a \leq y < b) = P(a \leq y \leq b) = A(a, b)$$

(Recall that for the discrete case this is not necessarily so.)

EXAMPLE 5.13.   Find the constant c so that

$$f(y) = cy \quad \text{for } 0 \leq y \leq \frac{3}{2}$$

is a probability density function of the continuous random variable y, and find the probability that y has a value between 0.3 and 0.6.

We know, by Property I, that cy must be non-negative. Since $y \geq 0$, it follows that $c > 0$. The graph of $f(y) = cy$ is a straight line segment with end points at $(0, 0)$ and $(\frac{3}{2}, \frac{3c}{2})$ Since the figure under the line segment and above the interval $0 \leq y \leq \frac{3}{2}$ is a triangle, and since its area must be one (by Property

II), it follows that

$$\frac{1}{2}\left(\frac{3}{2}\right)\left(\frac{3c}{2}\right) = 1$$

or

$$c = \frac{8}{9}$$

Thus, the required probability density function is

(5.23) $$f(y) = \frac{8y}{9} \quad \text{for } 0 \leq y \leq \frac{3}{2}$$

and its graph is shown in Figure 5.4. The reader should observe that for any value of y in the interval $\frac{9}{8} < y \leq \frac{3}{2}$ the probability density function value $f(y)$ is greater than one. Thus, we note again that $f(y)$ does not give probability at a specific value of y.

From that part of the triangle shaded with slant lines we find the area to be

$$A(0.3, 0.6) = A(0, 0.6) - A(0, 0.3)$$

$$= \frac{1}{2}(0.6)\left[\frac{8}{9} \times (0.6)\right] - \frac{1}{2}(0.3)\left[\frac{8}{9} \times (0.3)\right]$$

$$= \frac{1.08}{9} = 0.1200 = P(0.3 < y < 0.6)$$

The area is the probability that y falls between .3 and .6.

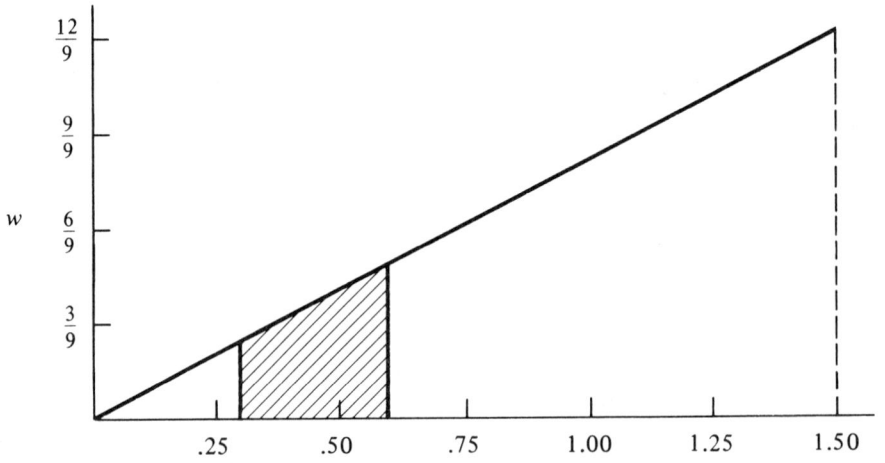

Figure 5.4. Graph of the Continuous Distribution of Example 5.13

5.8. CONTINUOUS UNIFORM DISTRIBUTIONS

If a probability density function is constant for all possible values of a continuous random variable in some interval, the variable is said to be **uniformly distributed**. We consider two illustrations. First, suppose a clock-like face has a perfectly balanced hand that can, after put in motion, come to rest at any point on the circumference of the face. The probability that it stops between 1 and 4 is $\frac{3}{12} = 0.25$. (In fact, any circular object would do if, as a result of an experiment, the occurrence of the points on the circumference were assumed to be equally likely—flywheel, oval track, and roulette wheel are illustrations.) Second, suppose a thin wire 20 inches long is randomly cut; that is, each point on the wire is equally likely to be cut. The probability of cutting the wire between the third and seventh inch marks (from the left, say) is $\frac{4}{20} = 0.20$. (Any experiment in which an object randomly comes to rest along a line would do—readers should think of their own experiments.)

In general, we write a *continuous uniform probability distribution* as

$$(5.24) \quad u(y; \alpha, \beta) = \begin{cases} \dfrac{1}{\beta - \alpha} & \text{for } \alpha \leq y \leq \beta \\ 0 & \text{elsewhere} \end{cases}$$

where α and β are any real numbers with $\alpha < \beta$. This distribution is also referred to as a rectangular probability distribution. When y is a continuous random variable the custom is to add the condition "$f(y) = 0$ elsewhere" [in this case "$u(y; \alpha, \beta) = 0$ elsewhere"] so that $f(y)$ is defined for all real values of y.

A graph of the continuous uniform probability density function of Equation (5.24) is shown in Figure 5.5, and the shaded area gives the probability that the random variable y has a value between a and b, that is

$$A(a, b) = P(a < y < b) = \frac{b - a}{\beta - \alpha}$$

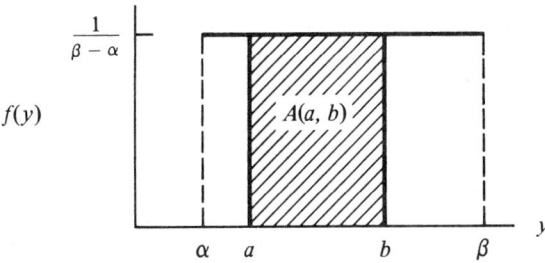

Figure 5.5. A Typical Uniform Density Curve

The simplicity of the continuous uniform distribution is a primary reason for using this distribution in illustrating basic properties of all continuous distributions.

It can be shown that the mean and variance for the continuous random variable with uniform distribution are

(5.25) $$\mu = \frac{\beta + \alpha}{2}$$

and

(5.26) $$\sigma^2 = \frac{(\beta - \alpha)^2}{12}$$

respectively. The standard deviation is

$$\sigma = \frac{\beta - \alpha}{\sqrt{12}} = \frac{\sqrt{3}}{6}(\beta - \alpha)$$

Since $u(y) > 0$ is an interval of length $\beta - \alpha$, the end points of the interval are $\sqrt{3}\sigma$ units distant from the mean. Thus, for the uniform distribution

$$P(\mu - \sqrt{3}\sigma \leq y \leq \mu + \sqrt{3}\sigma) = 1$$

The uniform distribution with $\alpha = 0$ and $\beta = 1$ is one of special interest. For example, it is used in studies with transformations and in constructing probability graph paper which has many applications in practical problems.

The student should note that the uniform distribution is a model when significant figures are used. That is, when a measurement is recorded as a number c to the closest unit, say, we understand that the true measurement y is in the interval $c - \frac{1}{2} < y < c + \frac{1}{2}$ and the density function is

$$u(y; c) = \begin{cases} 1 & \text{for } c - \frac{1}{2} < y < c + \frac{1}{2} \\ 0 & \text{elsewhere} \end{cases}$$

For density functions with simple graphs the areas (i.e., probabilities) can be found by methods of geometry, but for most curves areas must be found by the calculus or other fairly sophisticated numerical methods. Fortunately, for the most important curves, areas may be found in tables.

5.9. NORMAL DISTRIBUTIONS

Normal distributions are central to the study of statistics and have been for a long time. In 1733 De Moivre (1667–1754), a French refugee living in London, derived the formula for normal curves and presented it privately to some friends.

He recognized that when $\pi = 0.5$, binomial distributions approach a definite form as n becomes larger, and this limiting form is what we know as a normal distribution. Normal distributions were also later discovered by Laplace (1749–1827) in France and were applied in science and in other practical affairs. Properties of normal distributions were extensively developed by Gauss (1777–1855) in Germany. Many types of applications of normal distributions were first made to social statistics by Quetelet (1796–1874) in Belgium and to the biological sciences by Galton (1822–1911) in England.

The graph of a typical normal distribution is shown in Figure 5.6. The curve is bell-shaped and symmetrical about the line $y = \mu$. The maximum point of the curve is where $y = \mu$, and its inflection points are located above the values $y = \mu - \sigma$ and $y = \mu + \sigma$. The curve gets closer and closer to the y-axis as we move away from the mean in either direction but it never reaches the y-axis. The fact that the model curve goes to infinity in two directions is not troublesome in practice since the area under the curve more than three (or four) standard deviations away from the mean is usually considered negligible.

The normal probability density function, denoted by $n(y; \mu, \sigma)$ or $n(y)$, is given by

$$(5.27) \qquad n(y; \mu, \sigma) = \frac{k}{\sigma} e^{-1/2[(y-\mu)/\sigma]^2} \qquad \text{for } -\infty < y < \infty$$

where the parameters μ and σ are the mean and standard deviation of the distribution, $e \doteq 2.71828$ is the base to the natural logarithm, and $k = (1/\sqrt{2\pi})$ is approximately equal to 0.39894. (The normal distribution is also called the *error distribution* and the *Gaussian distribution*.) Since extensive tables of the normal function and of normal probabilities are readily available we do not need Equation (5.27) for computational purposes—in fact, we never use the formula for such purposes. We included Equation (5.27) so that the reader could notice how much it differs from the binomial probability function,

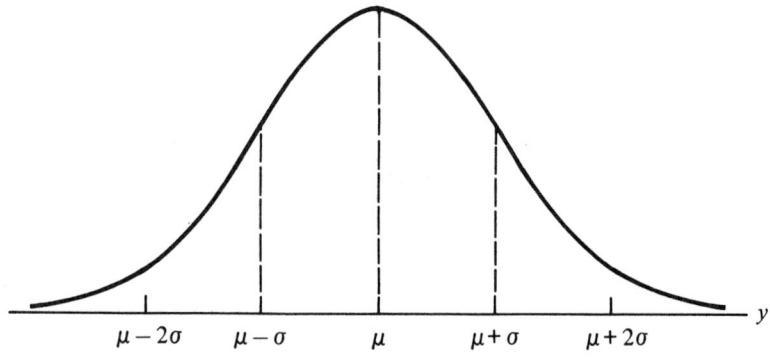

Figure 5.6. Typical Normal Probability Curve

for the sake of completeness, and for those who like to "play around" with such things.

Equation (5.27) defines a two parameter family of curves. If σ is held constant and μ allowed to vary we obtain an infinite family of curves as illustrated in Figure 5.7. If μ is held constant and σ allowed to vary we obtain another infinite family of curves as illustrated in Figure 5.8.

One curve of special interest, shown in Figure 5.9, is obtained by letting $\mu = 0$, $\sigma = 1$ and $y = z$. The probability density function becomes

$$(5.28) \qquad n(z; 0, 1) \doteq .39894\, e^{-z^2/2} \quad \text{for } -\infty < z < \infty$$

and is known as the **standard normal probability density function**, or as the **standard normal distribution**. The area under the curve is 1. At $z = 0$, the maximum is $f(0) \doteq 0.40$. The inflection points are approximately $.39894\, e^{-.5} = .39894(.6065) \doteq 0.24$ above $z = -1$ and $z = 1$ [using Table I and Equation (5.28)].

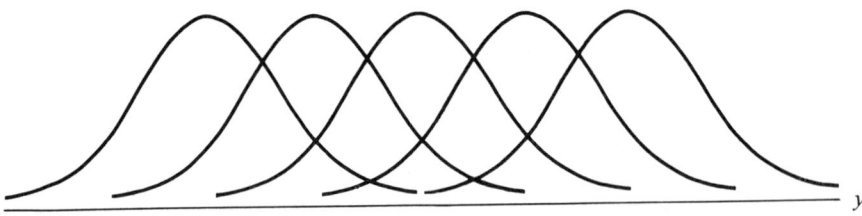

Figure 5.7. Family of Normal Curves with Fixed σ

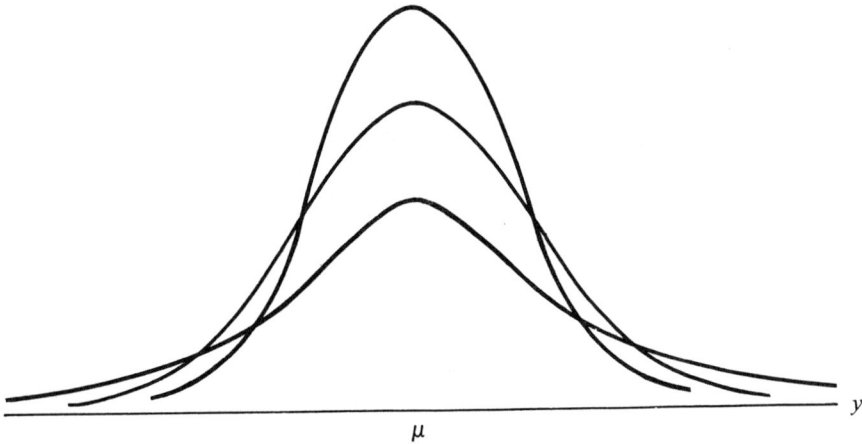

Figure 5.8. Family of Normal Curves with Fixed μ

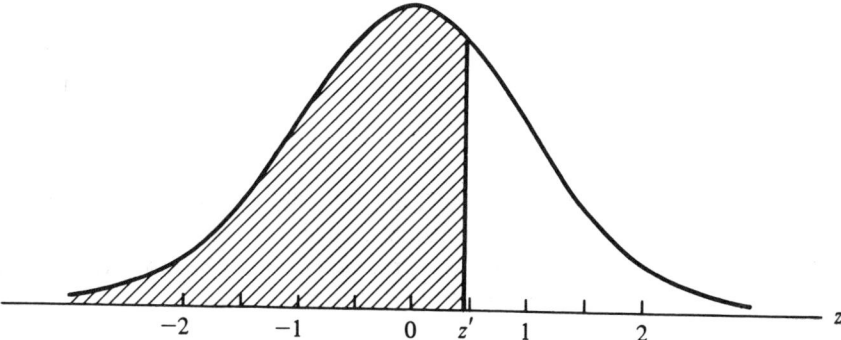

Figure 5.9. The Standard Normal Curve

The probability that z is less than some specified z' is given in Table II of the Appendix and is illustrated by the shaded area in Figure 5.9. That is

(5.29) $$P(z < z') = P(-\infty < z < z') = A(-\infty, z')$$

Due to symmetry we observe that

(5.30a) $$P(z > z') = P(z < -z')$$

or

(5.30b) $$P(z > z') = 1 - P(z < z')$$

EXAMPLE 5.14. Use Table II for the standard normal distribution to find each of the following:

(a) $P(z > 2.0)$ (b) $P(|z| > 1.5)$ (c) $P(-3.0 < z < 2.0)$
(d) z' such that $P(0 < z < z') = 0.4000$
(e) z' such that $P(|z| < z') = 0.9500$
(f) z' such that $P(|z| > z') = 0.0010$

With the aid of some basic relations of events, a graph, and Table II we find the following solutions:

(a) $P(z > 2.0) = P(z < -2.0) = 0.0228$
(b) $P(|z| > 1.5) = P(z < -1.5 \text{ or } z > 1.5)$
$= P(z < -1.5) + P(z > 1.5) = 2P(z < -1.5)$
$= 2(0.0668) = 0.1336$
(c) $P(-3.0 < z < 2.0) = P(z < 2.0) - P(z < -3.0)$
$= 0.9772 - 0.0013 = 0.9759$

For part (d), adding $P(z \leq 0) = 0.5000$ to $P(0 < z < z') = 0.4000$ gives

$$P(z \leq 0) + P(0 < z < z') = 0.5000 + 0.4000$$

or

$$P(z < z') = 0.9000$$

From the second part of Table II we find the required value of z to be $z' = 1.282$. In case a specified probability (such as 0.9000) cannot be found in the bottom part of Table II, it may be necessary to use the top part of Table II. Illustrating the procedure for part (d) we first note that

$$P(z < 1.2) = 0.8849 \quad \text{and} \quad P(z < 1.3) = 0.9032$$

If only one decimal place is required we select $z' = 1.3$ since 0.9000 is closer to 0.9032 than 0.8849. For two decimal places we use linear interpolation to find

$$z' = 1.2 + (0.10)\left(\frac{0.9000 - 0.8849}{0.9032 - 0.8849}\right)$$

$$= 1.2 + 0.08 = 1.28$$

More decimal places cannot be justified with the use of the first part of Table II. (Since our primary purpose is to illustrate principles, linear interpolation will seldom be required with prepared tables of standard distributions.)

For (e), we must find z' such that $P(-z' < z < z') = 0.9500$. With the aid of a graph we find that z' is a value such that $P(z < -z') = 0.0250$. By interpolation, we find that $z' = 1.96$. Also, from the second part of Table II, $z' = 1.960$.

For (f), z' is a value such that $P(z < -z') = 0.0005$. Thus, from Table II, $z' = 3.3$.

It would be impossible to construct normal tables for every pair of values of μ and σ since there are infinitely many values of both μ and σ. Fortunately, we can use the standard normal distribution to find areas for any member of the family of normal distributions. For example, suppose we wanted to find $P(y < y')$. We would convert the units of measurement of y into standard units by the formula

(5.31) $$z = \frac{y - \mu}{\sigma}$$

Then

$$z' = \frac{y' - \mu}{\sigma}$$

and

(5.32) $$P(y < y') = P(z < z')$$

since $y < y'$ implies that $y - \mu < y' - \mu$ implies that $(y - \mu)/\sigma < (y' - \mu)/\sigma$ means that $z < z'$. Also, for any two real numbers y'_1 and y'_2 such that $y'_1 < y'_2$ we find that

(5.33) $$P(y'_1 < y < y'_2) = P(z'_1 < z < z'_2)$$

where

$$z'_1 = \frac{y'_1 - \mu}{\sigma} \quad \text{and} \quad z'_2 = \frac{y'_2 - \mu}{\sigma}$$

An illustration should clarify the method.

EXAMPLE 5.15. Suppose the height y of adult males is normally distributed with mean 68 inches and standard deviation 2.5 inches. Find

(a) $P(y > 72 \text{ inches})$
(b) $P(63 \text{ inches} < y < 73 \text{ inches})$
(c) y' such that $P(y > y') = 0.05$
(d) k such that $P(|y - \mu| < k) = 0.80$

It is desirable that we work from a graph (see Figure 5.10) showing the change in scale in random variables. (It is not necessary to show the vertical change in scale.) Recall that the mean on any scale is always directly below the highest point on the normal curve and that points one standard deviation to the left and right of the mean are always directly below the points of inflection of the curve. The shaded area is for part (a) only.

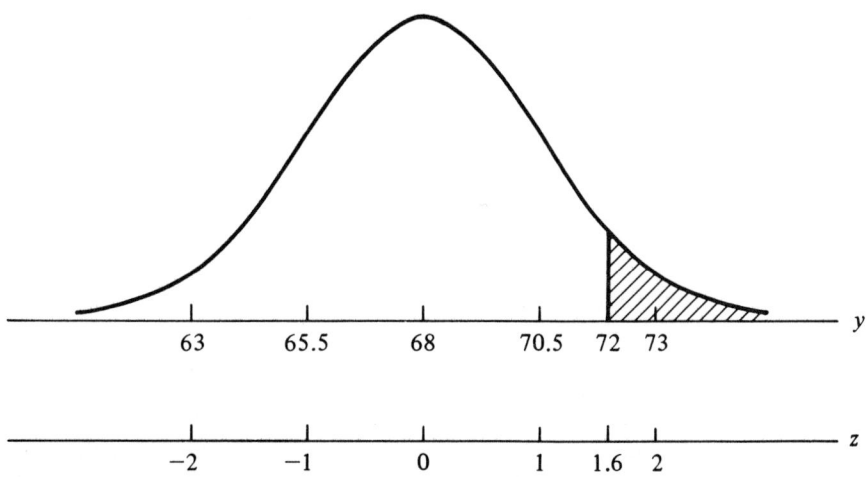

Figure 5.10. Normal Curve for Example 5.15

When $y = 72$ inches, $z = \dfrac{72 - 68}{2.5} = 1.6$ and

$$P(y > 72 \text{ inches}) = P(z > 1.6) = P(z < -1.6) = 0.0548$$

which is the probability for part (a). For part (b), when $y'_1 = 63$ inches and $y'_2 = 73$ inches, then

$$z'_1 = \dfrac{63 - 68}{2.5} = -2.00, \quad z'_2 = \dfrac{73 - 68}{2.5} = 2.00, \text{ and}$$

$$\begin{aligned} P(63 \text{ inches} < y < 73 \text{ inches}) &= P(-2.00 < z < 2.00) \\ &= P(z < 2.00) - P(z < -2.00) \\ &= 0.9772 - 0.0228 = 0.9544 \end{aligned}$$

For part (c), $P(y > y') = 0.05$ when $P(z > z') = 0.05$ where

$$z' = \dfrac{y' - 68}{2.5} \quad \text{or} \quad y' = 2.5\, z' + 68$$

From the bottom part of Table II we find $P(z > z') = 0.05$ or $P(z < z') = 0.95$ when $z' = 1.645$. Thus

$$y' = 2.5(1.645) + 68 \doteq 72.1 \text{ inches}$$

For part (d), k must be a value which satisfies

$$P(-k < y - \mu < k) = 0.80$$

or

$$P\left(-\dfrac{k}{\sigma} < z < \dfrac{k}{\sigma}\right) = 0.80$$

or

$$P\left(z < -\dfrac{k}{\sigma}\right) + P\left(z > \dfrac{k}{\sigma}\right) = 1 - 0.80 = 0.20$$

or

$$P\left(z < -\dfrac{k}{\sigma}\right) = 0.10$$

From Table II we find $P(z < -1.282) = 0.10$ and thus $-\dfrac{k}{\sigma} = -1.282$. So

$$k = 1.282\sigma = 1.282(2.5) \doteq 3.2$$

This means that the central 80 percent of the observations are within 3.2 inches of the mean of 68 inches. Similarly, it is easy to show that the central 68.3 percent of all values are within one standard deviation of the mean; 95.4 percent within two standard deviations; and 99.7 percent within three standard deviations.

5.10. EXERCISES

5.67. Find the constant c so that

$$f(y; m) = \begin{cases} cy & \text{for } 0 \le y \le m \\ 0 & \text{elsewhere} \end{cases}$$

is a probability density function of the continuous random variable y. Assume m is any positive real number. Also, make a graph when $m = 2$.

5.68. A continuous uniform distribution has density function

$$u(y; m) = \begin{cases} \dfrac{1}{m} & \text{for } 0 < y < m \\ 0 & \text{elsewhere} \end{cases}$$

(a) Find the mean and variance of the random variable y.
(b) Find $P(|y - \mu| \le k\sigma)$ when $k = 1, 1.5, 2$.
(c) Find $P(1 \le y \le 3)$, $P(1 \le y < 3)$, $P(1 < y \le 3)$, and $P(1 < y < 3)$ for both the discrete and continuous uniform distributions when $m = 6$. *Hint:* See Exercise 5.49 for the discrete distribution.
(d) Construct a histogram for the discrete uniform distribution and compare it with the graph of the continuous uniform distribution when $m = 6$.

E5.69. (a) Graph

$$f(y) = \begin{cases} \dfrac{1}{2}\left[1 - \dfrac{|y-3|}{2}\right] & \text{for } 1 \le y \le 5 \\ 0 & \text{elsewhere} \end{cases}$$

(b) Use geometry to show that the function in part (a) is a probability density function.
(c) Let

$$f(y; \alpha, \beta) = \begin{cases} \dfrac{1}{\alpha}\left[1 - \dfrac{|y-\beta|}{\alpha}\right] & \text{for } \beta - \alpha < y < \beta + \alpha \\ 0 & \text{elsewhere} \end{cases}$$

where $\alpha > 0$ and $\beta > 0$. Show that part (a) is a special case of this function.
(d) Graph the function in part (c) when $\beta = 1$ and $\alpha = 2$.
(e) What is the mean of the random variable y in part (c)? Explain.

E5.70. A random variable y with probability density function

$$f(y; \theta) = \begin{cases} \theta e^{-\theta y} & \text{for } y > 0 \\ 0 & \text{elsewhere} \end{cases}$$

is said to have the **exponential distribution.**
(a) Use Table I to graph the exponential distribution when $\theta = 2$.

(b) It can be shown that
$$P(0 < y < y') = 1 - e^{-\theta y'}$$
where y' is a specific value of a random variable y having an exponential distribution. For an exponential distribution with $\theta = 2$, find the probability that the random variable y falls between 0 and 1. *Hint:* Use Table I.

(c) It can be shown that the mean and variance of the exponential distribution are
$$\mu = \frac{1}{\theta} \text{ and } \sigma^2 = \frac{1}{\theta^2}$$
Find the probability that a value of the random variable y falls within two standard deviations of the mean of an exponential distribution with $\theta = 2$.

(d) Find the probability that a random variable y having an exponential distribution with mean 5 has a value greater than 6.

(e) Find an illustration of the exponential distribution in the scientific or technological literature.

5.71. Use a table of the standard normal distribution to find
(a) the probability that z is less than 1.3
(b) the probability that z is between -1.2 and 0.7
(c) the probability that z is less than -0.8 or greater than 1.3
(d) the probability that z is greater than -0.8 and less than 1.3
(e) $P(|z - 1| < 2)$
(f) z' such that $P(z > z') = 0.0968$
(g) z' such that $P(z > z') = 0.05$
(h) z' such that $P(|z| > z') = 0.02$

5.72. Find the probability that the standard normal random variable z is within k units of the mean if $k = 1; 2; 3$.

5.73. Find z' and $-z'$ so that the area symmetric about the mean of a standard normal distribution is $0.99; 0.98; 0.95; 0.90; 0.80$.

5.74. The random variable y is normally distributed with mean 50 and standard deviation 10. Find the
(a) probability that y is less than 42
(b) probability that y is between 42 and 65
(c) probability that y is greater than 38 and less than 70
(d) probability that y is less than 38 or greater than 70
(e) value y' for which $P(y > y') = 0.7580$
(f) value y' for which $P(y > y') = 0.01$

5.75. Find the standard deviation of a normal distribution if its mean is 60, and 0.0446 is the probability that a value of the random variable y is less than 50.

5.76. Find the mean of a normal distribution if its standard deviation is 15, and 0.1151 is the probability that a value of the random variable y is greater than 72.

5.77. The lifetime of a certain type of battery is normally distributed with mean 1,000 days and standard deviation 65 days. What percentage of this type of battery can be expected to last anywhere from 900 to 1,200 days?

5.78. Suppose the diameters (at eye level) of 1,090 teak trees are approximately normally distributed with mean 21.3 inches and standard deviation 5.5 inches.
(a) What proportions of the trees have diameters between 10 and 30 inches?
(b) How many of the trees have diameters greater than 25 inches?

5.79. Suppose the *excess* yardage of 100 denier acetate yarn over the specified minimum (100,000 yards per bobbin) is normally distributed with mean 19 yards and standard deviation 4.2 yards.
(a) What proportion of the bobbins have more than 10 yards excess yarn?
(b) How many of 1,000 bobbins have more than 25 yards of excess yarn?
(c) What proportion of the bobbins have less than 0.03 percent excess yarn?

5.80. A large group of public school children took a spelling test. The scores were approximately normally distributed with mean 62 and standard deviation 4.5. Find the numerical grades (rounded-off to the closest whole number and based on 100 points) required to achieve the letter grades A, B, C, and D if the highest 7 percent received A, the next 18 percent received B, the middle 50 percent received C, the next 18 percent received D, and the lowest 7 percent received F.

5.81. At a large university three departmental tests were given to nearly a thousand students in first semester mathematics. Tests T_1, T_2, and T_3 had mean scores 68, 80, and 50 and standard deviations 15, 10, and 5, respectively. On these tests a student made scores of 80, 91, and 62, respectively. Arrange the grades in order of decreasing excellence and give reasons.

5.82. On an examination, the mean grade was 72.0 and the standard deviation was 8.3. The instructor gave all students with grades from 80.0 through 89.9 the grade of B, and 10 students received a B grade. If the grades were approximately normally distributed, how many students took the examination?

*5.83. For a normal distribution of standardized scores with mean 500 and standard deviation 100 there are 775 students with scores between 450 and 600. How many students have scores between 350 and 650?

*5.84. Suppose the grades on an examination are normally distributed with mean 66.0. Further, suppose the instructor assigns 15 percent A's, 30 percent B's, 35 percent C's, 15 percent D's, and 5 percent F's.
(a) If the border grade between C and B is 70.0, what is the standard deviation of the class?
(b) If the standard deviation is 12.3, find the four border grades.

*5.85. Suppose an instructor assigns 0.15 A's, 0.30 B's, 0.40 C's, 0.10 D's, and 0.05 F's, and

C ranges from 65.5 to 79.5. If the grades are normally distributed, what are the mean and standard deviation of the grades?

5.86. A random variable y is known to be normally distributed with variance 36 and mean unknown. Find the mean μ if
(a) 0.025 is the probability that y is greater than 50
(b) 0.025 is the probability that y is less than 50

E5.87. The area $A(-\infty, z)$ under a standard normal curve from $-\infty$ to z may be found by the series expansion

$$A(-\infty, z) = \frac{1}{2} + \frac{1}{\sqrt{2\pi}}\left(z - \frac{z^3}{2 \times 3} + \frac{1}{2!} \times \frac{z^5}{2^2 \times 5} - \cdots\right)$$

By taking enough terms in the expansion, we can compute the area to any specified accuracy. For values of z in the interval from -1 to 1 the expansion above (to the term in z^5) gives satisfactory approximations for most purposes.
(a) Compute $A(-\infty, 0.1)$ and $A(-\infty, -0.2)$ and compare with values obtained from the normal table.
(b) Write the next three terms in the expansion. What is a maximum value for the sum of all terms after the term containing z^3?

E5.88. For large values of z, the expansion in Exercise 5.87 converges too slowly to be of practical use. For large positive values of z the following series may be used:

$$A(-\infty, z) = 1 - n(z) \times \frac{1}{z}\left(1 - \frac{1}{z^2} + \frac{3}{z^4} - \frac{3 \times 5}{z^6} + \cdots \pm R\right)$$

where $n(z)$ is defined by Equation (5.28) and R is numerically less than the last term considered.
(a) Compute $A(-\infty, 2)$ and $A(-\infty, -3)$ using the first four terms in the parentheses and then compare with values found from the normal table.
(b) What is the maximum error in $A(-\infty, 2)$ as computed in (a)? In $A(-\infty, -3)$?

5.11. NORMAL APPROXIMATIONS FOR BINOMIAL DISTRIBUTIONS

In Sections 5.4 and 5.5 we observed that in many problems the calculations associated with binomial distributions are formidable, and we suggested that normal distribution can often be used as approximations. Later (in Section 5.9) we noted that De Moivre derived the normal distribution from the binomial distribution with $\pi = 0.5$ as n increased indefinitely. Actually, when $\pi \neq 0.5$ it is possible to find a satisfactory normal approximation to a binomial provided n is large enough—as π moves further away from 0.5 the size of n must increase.

The effect on the normal approximation of varying π and n is illustrated by the four histograms of Figure 5.11. When π is fixed at 0.4 (or at 0.2) it is clear

169 Normal Approximations for Binomial Distributions

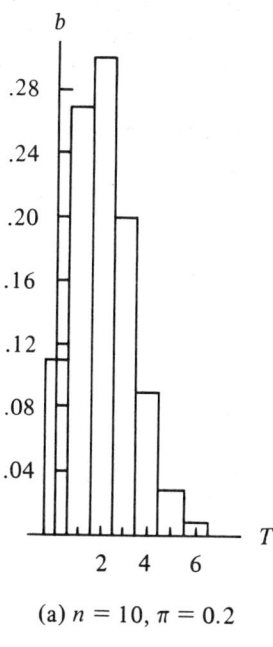

(a) $n = 10, \pi = 0.2$

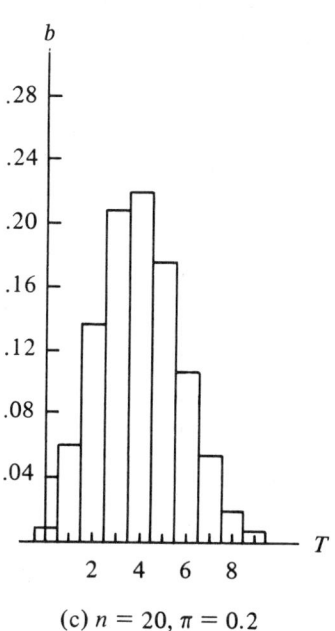

(c) $n = 20, \pi = 0.2$

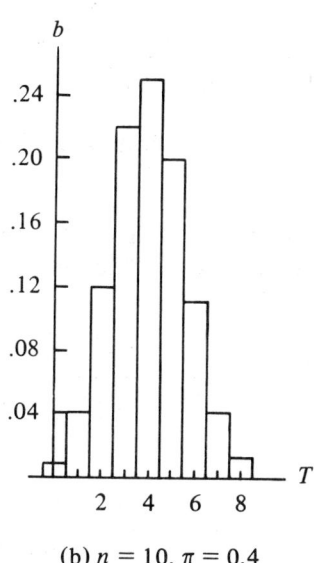

(b) $n = 10, \pi = 0.4$

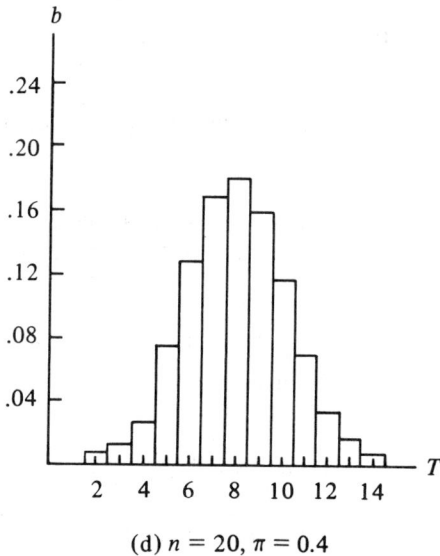

(d) $n = 20, \pi = 0.4$

Figure 5.11. Histograms of Binomial Distributions

that the histogram with larger n more nearly approximates a normal curve. Also, when n is fixed at 20 (or at 10) the histogram with π closer to .5 fits a normal curve better. In Figure 5.12 we show a normal curve approximation to the histogram of Figure 5.11(d). (Both the area of the histogram and the area under the normal curve are equal to one.) From the graph in Figure 5.12 it appears that the normal approximation is adequate; Table 5.8 gives a numerical comparison of the area of a rectangle above an interval with the area under that part of the normal curve above the same interval (see Example 5.16 for an illustration). As a practical rule

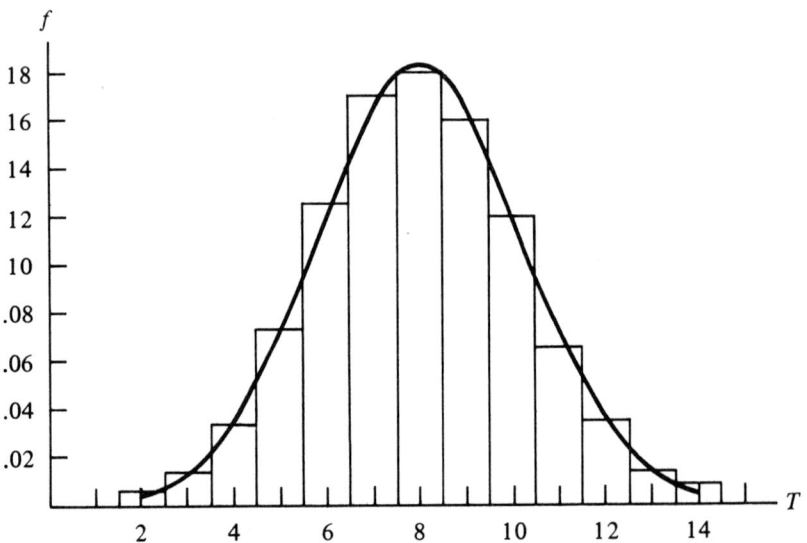

Figure 5.12. Normal Curve Approximation to the Binomial of Figure 5.11(d)

Table 5.8. Comparison of Binomial Probabilities (with $n = 20$ and $\pi = 0.4$) and Corresponding Normal Approximations (with $\mu = 8$ and $\sigma = \sqrt{4.8}$)

No. of Successes	0	1	2	3	4	5	6	7	8
Binomial Prob.	.0000	.0005	.0031	.0123	.0350	.0746	.1244	.1659	.1797
Normal Prob.	.0003	.0012	.0045	.0140	.0351	.0718	.1197	.1632	.1803

No. of Successes	9	10	11	12	13	14	15	16	17*
Binomial Prob.	.1597	.1171	.0710	.0355	.0146	.0049	.0013	.0003	.0000
Normal Prob.	.1632	.1197	.0718	.0351	.0140	.0045	.0012	.0003	.0001

* All other probabilities are zero to four decimal places.

we use the normal distribution to approximate binomial probabilities when $n\pi \geq 5$ and $n(1 - \pi) \geq 5$.

In this section and other places where the binomial probabilities are approximated by normal probabilities, we let the random variable of the binomial distribution be denoted by T (the total number of successes) and the random variable of the normal distribution be denoted by y. When the binomial distribution has parameters n and π the approximating normal distribution has parameters

(5.34) $$\mu_y = n\pi = \mu_T$$

and

(5.35) $$\sigma_y = \sqrt{n\pi(1 - \pi)} = \sigma_T$$

That is, the mean and standard deviation of the approximating normal distribution are set equal to the mean and standard deviation of the corresponding binomial distribution. Furthermore, the probability of a particular value T' of the random variable T is computed by

(5.36) $$b(T') = P(T = T') \doteq P\left(T' - \frac{1}{2} < y < T' + \frac{1}{2}\right)$$

except at the extremes of T. Since the normal curve reaches from $-\infty$ to ∞, it is customary to let $T = 0$ correspond to the interval from $-\infty$ to $\frac{1}{2}$ and to let $T = n$ correspond to the interval from $n - \frac{1}{2}$ to ∞.

EXAMPLE 5.16. A well-balanced die is tossed 30 times. Use the normal distribution to find approximations to the

(a) probability of at most six 5's
(b) probability of at least ten 5's
(c) probability of exactly five 5's

Suppose the probability of a success, occurrence of a 5, on a single toss of the die is $\pi = \frac{1}{6}$. Then for 30 independent trials we find

$$\mu = n\pi = 30\left(\frac{1}{6}\right) = 5$$

$$n(1 - \pi) = 30\left(\frac{5}{6}\right) = 25, \text{ and}$$

$$\sigma = \sqrt{n\pi(1 - \pi)} = \sqrt{30\left(\frac{1}{6}\right)\left(\frac{5}{6}\right)} \doteq 2.04$$

Since $n\pi = 5$ and $n(1 - \pi) = 25$, we feel the normal approximation is justified.

172 Model Distributions

By the normal approximation and linear interpolation we find the following:

(a) $P(\text{at most six 5's}) = P(T = 0 \text{ or } 1 \text{ or } \cdots \text{ or } 6)$

$$\doteq P\left(y \leq 6 + \tfrac{1}{2}\right)$$

$$= P\left(\frac{y-5}{2.04} \leq \frac{6.5-5}{2.04}\right)$$

$$= P(z \leq 0.74)$$

$$\doteq 0.7700 \doteq 0.77$$

(b) $P(\text{at least ten 5's}) = P(T = 10 \text{ or } 11 \text{ or } \cdots \text{ or } 30)$

$$\doteq P\left(y \geq 10 - \tfrac{1}{2}\right)$$

$$= P\left(\frac{y-\mu}{\sigma} \geq \frac{9.5-5}{2.04}\right)$$

$$= P(z \geq 2.21) = P(z \leq -2.21)$$

$$= 0.0136 \doteq 0.014$$

(c) $P(\text{exactly five 5's}) = P(T = 5)$

$$\doteq P\left(5 - \tfrac{1}{2} < y < 5 + \tfrac{1}{2}\right)$$

$$= P\left(\frac{4.5-5}{2.04} < \frac{y-\mu}{\sigma} < \frac{5.5-5}{2.04}\right)$$

$$= P(-0.25 < z < 0.25) = P(z < 0.25) - P(z < -0.25)$$

$$\doteq 0.5986 - 0.4014$$

$$= 0.1972 \doteq 0.20$$

It is informative to compare this last probability with the exact probability obtained by computing $b(5; 30, \tfrac{1}{6})$ directly. We have

$$b(5) = \binom{30}{5}\left(\tfrac{1}{6}\right)^5\left(\tfrac{5}{6}\right)^{25} = \frac{(174)(819)\,5^{25}}{6^{30}}$$

$$\log b(5) = \log 174 + \log 819 + 25 \log 5 - 30 \log 6$$

$$= 2.2405 + 2.9133 + 25(0.6990) - 30(0.7782)$$

$$= 9.2828 - 10$$

and, therefore

$$b(5) \doteq 0.1918 \doteq 0.19$$

In this case the answers disagree one unit in the second decimal place, and a large part of this difference could be because the normal values of z had only two significant digits. However, generally, the normal approximation is better for a "less than" probability such as in part (a) or a "greater than" probability such as in part (b) than for a probability such as in part (c).

5.12. EXERCISES

5.89. A well-balanced coin is tossed 20 times. Use the normal curve approximation to find the probability of obtaining
(a) exactly 7 heads
(b) 5 or fewer heads
(c) Use Table 5.5 to find the actual values for parts (a) and (b). Compare binomial and normal values.

5.90. Use the normal approximation to find the probability of obtaining
(a) exactly ten 2's in 100 tosses of a well-balanced die
(b) at most twenty 1's or 2's in 50 tosses of a well-balanced die

5.91. A manufacturer of electronic tubes finds that the mean proportion of defectives is 1.2 percent. What is the probability of obtaining 16 or more defectives in a random sample of 1,000 tubes.

*5.92. How many times must two dice be tossed in order that there is at least a 0.75 chance of getting at least six 7's?

5.93. Suppose that in response to a certain question 0.4 people say "yes." What is the probability that at least 1,000 of 2,400 people say "yes"? (In problems where n is greater than some large number, say 200, the correction factor of 0.5 is usually ignored.)

*5.94. The value of π is unknown for a certain binomial distribution. In a random sample of 1,000 it was found that 400 were successes. Use the normal approximation to estimate π (ignore the 0.5 factor) if
(a) the probability of 400 or fewer successes is 0.05
(b) the probability of 400 or more successes is 0.05

6

SAMPLING DISTRIBUTIONS

6.1. INTRODUCTION

All procedures of statistical inference are based on samples and variation among samples. It is through information gained from a sample that we make statements (usually in terms of probability) about the population from which the sample was selected. That is, we use the sample to generalize or predict the nature of some characteristic of the larger group from which the sample was drawn [see statement (7.1) for a definition of statistical inference]. For example, a lumberman walked through a forest of timber and estimated that 60 percent of the trees large enough to be harvested were Englemann spruce. If he observed a representative sample, we might expect the estimate to be close to the actual percent Englemann spruce. As a second illustration, a member of the personnel office of a large steel company found on examination of the files of 55 employees that their median term of employment was 5 years and 7 months. On the basis of the sample information what can we conclude about the median term of employment of all employees? As a third illustration, consider the variation in running time of rats trained to run a maze. Suppose it was found that the standard deviation of 10 rats was 5.2 seconds. An experimenter might like to know how often one could expect the standard deviation to be less than 6.0 seconds.

In each of the above illustrations the value of a statistic (i.e., number computed from a sample) is used to make a statement about a population. We recognize that the value of a statistic may change from sample to sample, and that the magnitude of such fluctuations would depend on such things as the *type of population*, *size of sample*, and *method of sampling*. Further, we observe that it would be impossible to use the value of a statistic from a particular sample to make a meaningful statement about a population unless we knew what other values were possible. Thus, we are led to consider the distribution of values of a statistic obtained under the same conditions.

6.2. SAMPLING DISTRIBUTIONS OF THE MEAN

Let

(6.1) $$y_1, y_2, \ldots, y_n$$

denote n values of a random variable y which may be either discrete or continuous. The values of Expression (6.1) are the values of a *sample of size n* drawn from a **univariate population**. The sample mean, denoted by $\bar{y}$ and defined by

(6.2) $$\bar{y} = \frac{\sum_{i=1}^{n} y_i}{n}$$

is a statistic.

In this section we discuss distributions of sample means, called **sampling distributions of the mean**, as we illustrate general principles of sampling distributions of any statistic. We use very small finite populations and two of the simplest sampling plans to introduce the basic concepts. The sampling plans are called **sampling with replacement** and **sampling without replacement**. (For finite populations the two plans lead to different results; for infinite populations the results are the same.) Each sampling plan may be carried out *experimentally* or *theoretically*—both are illustrated.

EXAMPLE 6.1. Let four disks of the same kind, marked 1, 3, 5, and 7, respectively, be placed in a bag. After the disks have been thoroughly mixed, two are drawn one after another and the numbers recorded. Then the disks are returned to the bag and the experiment is repeated. In such a procedure each sample of size two is *obtained experimentally* and is said to be a *random sample drawn without replacement*.

Suppose 120 such samples are drawn and the mean of each computed. The distribution of these 120 means is shown in Table 6.1. If another 120 samples

Table 6.1. Experimental Sampling Distribution of the Mean of Random Samples of Size Two Drawn without Replacement from the Population of Example 6.1

$\bar{y}$	Frequency	Relative Frequency
2	19	.158
3	16	.133
4	43	.358
5	20	.167
6	22	.183

were drawn and a relative frequency table of their means constructed, we would expect to get a different experimental sampling distribution of the mean. If two such distributions were noticeably different, we could obtain a more stable experimental sampling distribution by increasing the number of samples. The larger the number of samples, the less variation we would expect from the model sampling distribution.

The *theoretical sampling distribution*, or briefly *sampling distribution*, obtained by mathematical methods is the model for an experimental sampling distribution, and it might be thought of as the true sampling distribution. In any case, the theoretical sampling distribution does not vary. This method of sampling is illustrated in Example 6.2, and properties of the sampling distribution of the mean are illustrated in other examples.

EXAMPLE 6.2. Find the theoretical sampling distribution of the mean of samples of size two drawn without replacement from the finite population given in Example 6.1.

Note: When sampling without replacement, we may (1) draw one value after another, or (2) obtain all values simultaneously. For (1) order of draw may be taken into account; for (2) order of draw is meaningless. For example, for a sample of size three we could list 1 1 2, 1 2 1, and 2 1 1 for case (1), and only one of the three for case (2). But both methods of drawing values lead to the same relative frequency sampling distribution (see Exercise 6.1). Thus, when constructing theoretical relative frequency tables one may ignore order of draw for sampling without replacement.

Returning to Example 6.2, all possible samples which can be drawn, without regard for order, are

$$1, 3 \quad 1, 5 \quad 1, 7 \quad 3, 5 \quad 3, 7 \quad 5, 7$$

and the corresponding means are

$$2 \quad 3 \quad 4 \quad 4 \quad 5 \quad 6$$

respectively. Assuming the six samples to be equally likely (i.e., each sample has probability $\frac{1}{6}$) we obtain the required sampling distribution of the mean shown in Table 6.2. Also, we point out that the probability of each sample may be obtained in terms of a product of probabilities of individual outcomes (see Exercise 6.3).

Readers should observe that the relative frequencies of Table 6.1 follow the same pattern as the probabilities of Table 6.2, and, from their knowledge of Chapter 4, should realize that by taking enough samples the relative frequencies get sufficiently close to their respective probabilities. (Members of the class might draw another 180 samples to add to the 120 of Table 6.1 and compare the new set of relative frequencies with the probabilities of Table 6.2.)

Table 6.2. Sampling Distribution of the Mean of Random Samples of Size Two Drawn without Replacement from the Population of Example 6.1

$\bar{y}$	Frequency	Probability
2	1	$\frac{1}{6} \doteq 0.167$
3	1	$\frac{1}{6} \doteq 0.167$
4	2	$\frac{2}{6} \doteq 0.333$
5	1	$\frac{1}{6} \doteq 0.167$
6	1	$\frac{1}{6} \doteq 0.167$

Whenever a sample is chosen in such a way that each possible sample has the same probability of being selected, we say we have a simple random sample, or briefly **random sample**. Thus, each of the six samples of Example 6.2 is a random sample (or if we assigned probability $\frac{1}{12}$ to each of the 12 possible ordered samples, then each such sample would be a random sample). To be specific, when we sample without replacement, a sample of size n chosen from a population of size N is a simple random sample provided each of the $\binom{N}{n}$ possible samples, without regard to order, has the same probability $1 / \binom{N}{n}$ of being selected.

Since all possible samples of a fixed size are represented in a theoretical sampling distribution of the mean, such a distribution is a **population of means** $\bar{y}$. The original population from which the samples were drawn may be called the **parent population of** y. The mean and standard deviation of the theoretical sampling distribution are denoted by $\mu_{\bar{y}}$ and $\sigma_{\bar{y}}$, respectively, so that they may be distinguished from the mean μ and standard deviation σ of the parent population. (It should be observed that an experimental distribution of the mean is usually different from the corresponding theoretical sampling distribution of the mean. Thus, some symbolism other than $\mu_{\bar{y}}$ and $\sigma_{\bar{y}}$ should be used to denote the mean and standard deviation. For example, the mean might be designated by $\hat{\mu}_{\bar{y}}$ or $\bar{y}_{\bar{y}}$, to indicate that the value is subject to change.)

EXAMPLE 6.3. Compute the mean and variance of the sampling distribution of Example 6.2 and compare with the corresponding parent population parameters.

For the parent population

$$\mu = \mu_y = \frac{1 + 3 + 5 + 7}{4} = \frac{16}{4} = 4$$

and

$$\sigma^2 = \sigma_y^2 = \frac{(1-4)^2 + (3-4)^2 + (5-4)^2 + (7-4)^2}{4} = \frac{20}{4} = 5$$

For the sampling distribution of the mean

$$\mu_{\bar{y}} = 2\left(\frac{1}{6}\right) + 3\left(\frac{1}{6}\right) + 4\left(\frac{2}{6}\right) + 5\left(\frac{1}{6}\right) + 6\left(\frac{1}{6}\right) = \frac{24}{6} = 4$$

and

$$\sigma_{\bar{y}}^2 = (2-4)^2\left(\frac{1}{6}\right) + (3-4)^2\left(\frac{1}{6}\right) + (4-4)^2\left(\frac{2}{6}\right) + (5-4)^2\left(\frac{1}{6}\right) + (6-4)^2\left(\frac{1}{6}\right)$$

$$= \frac{10}{6} = \frac{5}{3}$$

(Note that the variance of $\bar{y}$ could have been found by the computing formula of Exercise 5.44.) Thus, for this particular example we find that

(6.3) $$\mu_{\bar{y}} = \mu$$

and

(6.4) $$\sigma_{\bar{y}}^2 < \sigma^2$$

Indeed, the last two relations are quite general when we sample without replacement from a finite population of size N. Equation (6.3) holds for any sample size from 1 through N. Further, it can be shown that

(6.5) $$\sigma_{\bar{y}}^2 = \frac{\sigma^2}{n}\left(\frac{N-n}{N-1}\right)$$

So we see that the inequality of (6.4) is satisfied whenever n is greater than 1. We verify Equation (6.5) for Example 6.3 as follows:

$$\frac{\sigma^2}{n}\left(\frac{N-n}{N-1}\right) = \frac{5}{2}\left(\frac{4-2}{4-1}\right) = \frac{5}{3} = \sigma_{\bar{y}}^2$$

It is interesting to note that for any specific n the factor $(N-n)/(N-1)$ approaches 1 as N becomes very large. This is easy to see when we write

$$\frac{N-n}{N-1} = \frac{1 - \frac{n}{N}}{1 - \frac{1}{N}}$$

Thus, as N approaches infinity and n remains fixed $\sigma_{\bar{y}}^2$ approaches σ^2/n. In practice, σ^2/n is used as an approximation of $\sigma_{\bar{y}}^2$ when n is small relative to N. It is important to note that $\sigma_{\bar{y}}^2$ becomes smaller as the sample size increases.

EXAMPLE 6.4. Use histograms to compare the sampling distribution of the mean in Example 6.2 with its parent population.

Probability histograms are shown in Figure 6.1. Both histograms are symmetric about a vertical line through 4. Thus, it is clear that $\mu_{\bar{y}} = \mu = 4$. The

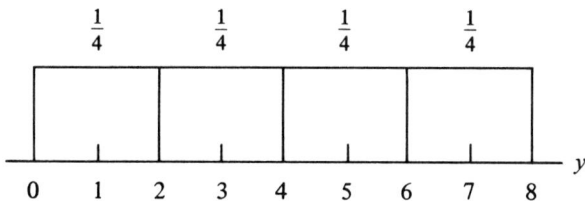

(a) Parent Population of Example 6.1

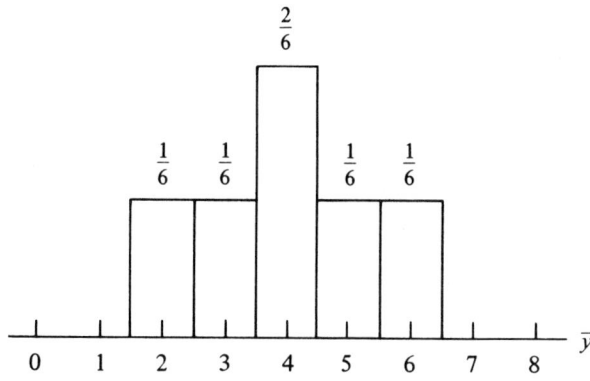

(b) Sampling Distribution of Example 6.2

Figure 6.1. Probability Histograms (total area of one)

histogram of the sampling distribution is not as spread out as the parent population. Thus, it is obvious that $\sigma_{\bar{y}} < \sigma$; i.e., the fluctuations in the random variable $\bar{y}$ are less than for the random variable y. (Actually, we know that $\sigma = \sqrt{3}\,\sigma_{\bar{y}} \doteq 1.73\sigma_{\bar{y}}$.) However, a primary reason for obtaining a sample mean is to estimate a fixed unknown population mean. For example, we observe that the probability is $\frac{2}{3}$ that a random sample of size two has a mean within one unit of the true mean. Also, the probability is $\frac{1}{3}$ that a sample mean is equal to the population mean.

We now turn our attention to sampling with replacement. This allows us to make a fairly smooth transition from the finite to the infinite population. We illustrate the properties with theoretical sampling, leaving experimental sampling for the reader (see Exercise 6.5).

EXAMPLE 6.5. Find the sampling distribution of the mean of samples of size two drawn with replacement from the finite population with observations

1, 3, 5, and 7. Also, find the mean and variance of the sampling distribution of the mean.

All possible samples are in Table 6.3. (The third and fourth columns of Table 6.3 are listed for later references.) Assuming the 16 samples to be equally likely, we find the sampling distribution of the mean shown in Table 6.4. The last two columns of Table 6.4 are to be used in the computation of $\mu_{\bar{y}}$ and $\sigma_{\bar{y}}^2$. We find that

$$\mu_{\bar{y}} = \frac{64}{16} = 4 \quad \text{and} \quad \sigma_{\bar{y}}^2 = \frac{296 - \frac{(64)^2}{16}}{16} = \frac{296 - 256}{16} = \frac{5}{2}$$

Table 6.3. Samples of Size Two Drawn with Replacement from the Population of Example 6.5

Sample	$\bar{y}$	s^2	s
1,1	1	0	0.00
1,3	2	2	1.41
1,5	3	8	2.83
1,7	4	18	4.24
3,1	2	2	1.41
3,3	3	0	0.00
3,5	4	2	1.41
3,7	5	8	2.83
5,1	3	8	2.83
5,3	4	2	1.41
5,5	5	0	0.00
5,7	6	2	1.41
7,1	4	18	4.24
7,3	5	8	2.83
7,5	6	2	1.41
7,7	7	0	0.00

Table 6.4. Sampling Distribution of the Mean of Samples of Size Two Drawn with Replacement from the Population of Example 6.5

$\bar{y}$	Frequency (f)	Probability [$f(\bar{y})$]	$\bar{y}f$	$\bar{y}^2 f$
1	1	.0625	1	1
2	2	.1250	4	8
3	3	.1875	9	27
4	4	.2500	16	64
5	3	.1875	15	75
6	2	.1250	12	72
7	1	.0625	7	49
Totals	16		64	296

(Note that the mean and variance of the theoretical distribution of the mean were computed by frequency formulas instead of by probability formulas.) Since $\mu = 4$ and $\sigma^2 = 5$ (from page 177) we see that

$$\mu_{\bar{y}} = \mu$$

and

$$\sigma_{\bar{y}}^2 = \frac{\sigma^2}{2}$$

These last two relations are very similar to those obtained when we sampled without replacement and are illustrative of two important general properties. For it can be shown that when sampling with replacement

(6.6) $$\mu_{\bar{y}} = \mu$$

and

(6.7) $$\sigma_{\bar{y}}^2 = \frac{\sigma^2}{n}$$

for both finite and infinite populations. Thus,

(6.8) $$\sigma_{\bar{y}} = \frac{\sigma}{\sqrt{n}}$$

Any sample in Example 6.5 is a simple random sample since they all have the same probability. More generally, when we sample with replacement, a sample of size n chosen from a population of size N is a simple random sample provided each of the N^n possible samples has the same probability $1/N^n$ of being selected. So in sampling with and without replacement from a finite population we have used simple random samples to obtain the sampling distributions of the mean and the resulting sampling properties.

It will be instructive to look at an alternative way of defining a random sample. Perhaps it is more usual to define a random sample in terms of the probabilities of the individual observations which make up the sample. According to such a definition, *a simple random sample is one in which each obervation of the population that has not been previously drawn has an equal probability of being selected.* Thus, the probability of an ordered sample is the product of probabilities of the individual observations. For example, suppose we consider the ordered sample, 7 and 3, drawn from the population with observations 1, 3, 5, 7. If the sample is drawn with replacement, then the probability is

$$P(\text{ordered sample 7 and 3}) = P(\text{observation 7}) \times P(\text{observation 3})$$
$$= \frac{1}{4} \times \frac{1}{4} = \frac{1}{16}$$

If the sample is drawn without replacement, then the probability is

$$P(\text{ordered sample 7 and 3}) = P(7) \times P(3|7) = \frac{1}{4} \times \frac{1}{3} = \frac{1}{12}$$

There are 2! arrangements of a sample of size 2. Thus, the probability of sample 7 and 3, regardless of order, is

$$P(\text{7 and 3 in any order}) = P(\text{ordered sample 7 and 3})$$
$$+ P(\text{ordered sample 3 and 7})$$
$$= \frac{1}{12} + \frac{1}{12} = \frac{1}{6}$$

All these results agree with our definitions of simple random sample.

Now we make the concepts of the last example more general. Suppose an ordered random sample of size n is drawn from a population of size N. If the random sample is drawn with replacement, then the probability is

$$\left(\frac{1}{N}\right)\left(\frac{1}{N}\right)\cdots\left(\frac{1}{N}\right) = \left(\frac{1}{N}\right)^n = \frac{1}{N^n}$$

which is consistent with our definition of *random sample drawn with replacement*. Furthermore, *we observe that the individual observations are mutually independent.* If an ordered random sample is drawn without replacement, then the probability of that sample is

$$\left(\frac{1}{N}\right)\left(\frac{1}{N-1}\right)\cdots\left(\frac{1}{N-(n-1)}\right)$$

There are $n!$ arrangements of any sample of size n. Thus, the probability of a sample of n specific observations, regardless of order, is

$$n!\left(\frac{1}{N}\right)\left(\frac{1}{N-1}\right)\cdots\left(\frac{1}{N-(n-1)}\right) = \frac{1}{\dfrac{N(N-1)\cdots(N-n+1)}{n!}}$$

$$= \frac{1}{\binom{N}{n}}$$

which is consistent with our definition of *random sample drawn without replacement*. Furthermore, *we observe that the individual observations are dependent.*

We mentioned that Equations (6.6), (6.7), and (6.8) hold for infinite populations, but we did not derive them. (As a matter of fact, we have not shown that these equations hold for every finite population. However, we hope the reader understands what they mean and will be able to apply them to problems in the future.) Since an infinite population is a useful model in so many problems, it might be informative to observe that *in a sense selecting a random sample from an infinite population is like drawing with replacement a random sample of size n from a finite population—the population is the same on the next draw and the sample observations are independent.*

In practice, if we wanted to make an inference about a population mean, we would usually draw only one sample and apply any useful known properties. Fortunately, in working with the mean the key sampling distribution properties can be summarized in two statements. The one which brings together the concepts of this section follows (the other is the Central Limit Theorem of Section 6.3):

(6.9)   The sampling distribution of the mean $\bar{y}$ of random samples of size n from a population with mean μ and standard deviation σ has mean

$$\mu_{\bar{y}} = \mu$$

and standard deviation

$$\sigma_{\bar{y}} = \frac{\sigma}{\sqrt{n}} \sqrt{\frac{N-n}{N-1}} \quad \text{or} \quad \sigma_{\bar{y}} = \frac{\sigma}{\sqrt{n}}$$

depending on whether the samples were drawn without replacement from a population of size N or with replacement from any population (or without replacement from an infinite population).

For random samples, the equation $\mu_{\bar{y}} = \mu$ holds regardless of the size of a sample or of the type of population. This points up a desirable property of random sampling.

The standard deviation of a sampling distribution of the mean is also called the **standard error of the mean** to distinguish it from the standard deviation of the parent population. For random samples drawn with replacement, the magnitude of the standard error of the mean is influenced by two factors. First, for a fixed sample size, a large variation in the parent population causes a relatively large variation in the sample means. Second, for a fixed σ, the larger the sample size the smaller the variation of the sample means about the true mean. For example, a change in sample size from 3 to 12 changes the standard error of the mean from $\sigma/\sqrt{3}$ to $\sigma/\sqrt{12} = \frac{1}{2}(\sigma/\sqrt{3})$. That is, by multiplying the sample size by four we decrease the standard error of the mean by one half—which means that fluctuations in sample means are about half as great. For random samples drawn without replacement from a finite population, the factor $\sqrt{\frac{N-n}{N-1}}$ is usually ignored when n is small compared to N; otherwise, its presence tends to decrease the standard error of the mean.

When the sampling distribution of the mean is known we can find the probability that a sample mean will have one of a set of values of $\bar{y}$ (usually in an interval about μ). Without the sampling distribution of the mean we can use

Chebyshev's Theorem (see page 60) to determine a minimum probability of obtaining a sample mean within k standard deviations of the true mean. This is,

(6.10) $$P(|\bar{y} - \mu| \leq k\sigma_{\bar{y}}) \geq 1 - \frac{1}{k^2}$$

where k is a real number equal to or greater than one. (For applications see Exercises 6.10, 6.11, and 6.12.) Fortunately, due to the Central Limit Theorem, we can do much better.

6.3. CENTRAL LIMIT THEOREM

To introduce this section we take as our parent population the simple uniform distribution with values 1, 7, 13, and 19. Using the method of Example 6.5, we find sampling distributions of the mean for samples of sizes 1, 2, 3, and 6 drawn with replacement. Assuming the 4 samples of size 1 to be equally likely, the 16 samples of size 2 to be equally likely, the 64 samples of size 3 to be equally likely, and the $4^6 = 4{,}096$ samples of size 6 to be equally likely, we obtain, in terms of probability, the four sampling distributions shown in Table 6.5. Note that for

Table 6.5. Sampling Distributions of the Mean of Random Samples of Sizes 2, 3, and 6 Drawn with Replacement from the Uniform Distribution with Values 1, 7, 13, and 19

Mean $\bar{y}$	Probability when			
	$n = 1$ (parent pop.)	$n = 2$	$n = 3$	$n = 6$
1	.2500	.0625	.0156	.0002
2				.0015
3			.0469	.0051
4		.1250		.0137
5			.0938	.0293
6				.0527
7	.2500	.1875	.1562	.0820
8				.1113
9			.1875	.1333
10		.2500		.1416
11			.1875	.1333
12				.1113
13	.2500	.1875	.1562	.0820
14				.0527
15			.0938	.0293
16		.1250		.0137
17			.0469	.0051
18				.0015
19	.2500	.0625	.0156	.0002

samples of size $n = 1$, the sampling distribution of the mean is identical to the distribution of the parent population, except that $\bar{y}$ replaces y. It is left as an exercise (See Exercise 6.13) for the reader to show that Table 6.5 is correct.

We observe from Table 6.5 that as n increases the sampling distributions of the mean have a larger proportion of the sample means near the true mean and the shape gets more bell-shaped. For some it might be useful to construct a sequence of histograms (see Exercise 6.14). As a matter of fact, these properties hold when sampling from any parent population with a finite variance provided the sample size is sufficiently large.

This leads to the following formal statement.

(6.11) **Central Limit Theorem:** Let the random variable y have a distribution with mean μ and standard deviation σ. Then the sampling distribution of the mean $\bar{y}$ of random samples from the distribution of y approaches the distribution of a normal variable with mean μ and standard deviation $\sigma/\sqrt{n}$ as n gets large.

This is the most important theorem in statistics. Due to this theorem the normal distribution has a central place in both theory and application and the mean becomes a favorite statistic.

If the *parent population is normal,* then the sampling distribution of the mean is normal. For other parent populations the "goodness of the approximation" depends on the nature of the parent population *and* the size of the sample. If the parent population is unimodal and symmetric, the sampling distribution of the mean may be approximately normal for small sample sizes, say $n = 5$. For almost any distribution found in practice it appears that a sample size of around 30 gives a satisfactory approximation.

When n is large enough for the Central Limit Theorem to apply, the sampling distribution of the statistic

$$\frac{\bar{y} - \mu_{\bar{y}}}{\sigma_{\bar{y}}} = \frac{\bar{y} - \mu}{\sigma/\sqrt{n}}$$

can be approximated by the standard normal distribution with statistic z. Of course, if y is normally distributed then

$$\frac{\bar{y} - \mu}{\sigma/\sqrt{n}} = z$$

EXAMPLE 6.6. A random sample of 25 adult males is drawn from a normal population of heights for which the mean and standard deviation are assumed to be 69 and 2.5 inches, respectively. (a) What is the probability that the sample mean will be greater than 70 inches? (b) Find a symmetric interval about 69 inches which can be expected to contain 99 percent of all sample means.

First, we note that $\bar{y}$ has a normal distribution with mean

$$\mu_{\bar{y}} = 69$$

and standard error

$$\sigma_{\bar{y}} = \frac{2.5}{\sqrt{25}} = 0.5$$

Thus

$$\frac{\bar{y} - \mu_{\bar{y}}}{\sigma_{\bar{y}}} = \frac{\bar{y} - 69}{0.5}$$

has the standard normal distribution. Hence, for (a)

$$P(\bar{y} > 70'') = P\left(z > \frac{70 - 69}{0.5}\right)$$
$$= P(z > 2) = P(z < -2)$$
$$= .0228$$

For (b) we find that

$$P(-2.576 < z < 2.576) = .99$$

since

$$P(z < -2.576) = P(z > 2.576) = 0.005$$

Therefore since $\bar{y} = 0.5z + 69$ the two end points of the interval are

$$\bar{y}_1 = (0.5)(-2.576) + 69 = -1.29 + 69 = 67.71''$$

and

$$\bar{y}_2 = (0.5)(2.576) + 69 = 1.29 + 69 = 70.29''$$

Thus, for samples of size 25, about 99 percent of the sample means fall in the interval

$$67.7 \text{ inches} < \bar{y} < 70.3 \text{ inches}$$

Note: By the method of part (b) it is easy to show that for samples of size 100 the corresponding 99 percent interval is

$$68.36 \text{ inches} < \bar{y} < 69.64 \text{ inches}$$

and for samples of size 400 the interval is

$$68.68 \text{ inches} < \bar{y} < 69.32 \text{ inches}$$

Thus, by making n large enough (see Exercise 6.17) we could be 99 percent sure that $\bar{y}$ falls in an interval like

$$68.9 \text{ inches} < \bar{y} < 69.1 \text{ inches}$$

say. So even when we do not know μ we can be reasonably sure (with high probability) that the sample mean $\bar{y}$ is very close to the true mean μ provided n is large enough.

We have used sample means to introduce the concept of sampling distributions; we could have used any other statistic. In the exercises of Section 6.4 we take a look at sampling distributions of some other measures of location. Sampling distributions of other statistics, important in the application of statistics, are considered in Sections 6.6 and 6.7 and in Chapters 8, 9, and 10.

6.4. EXERCISES

6.1. Find the sampling distribution of the mean of random samples of size two drawn without replacement, one after another, from the population with values 1, 3, 5, and 7. Compare your probability sampling distribution of 12 sample means with the one shown in Table 6.2.

6.2. Five disks of a finite population are marked 2, 5, 8, 11, and 14.
  (a) Find the mean and standard deviation of this population.
  (b) Random samples of size three are drawn without replacement from this population. List the 10 possible samples, without regard for order, and find the probability sampling distribution of the mean.
  (c) Calculate the mean and standard deviation of the probability sampling distribution of the mean found in part (b) and compare with the corresponding values obtained by using Equations (6.3) and (6.5).

6.3. (a) Use probabilities of individual observations to find the probability of any random sample of Exercise 6.1.
  (b) Use part (a) to find the probability of the six samples listed in the solution of Example 6.2 and verify that they are the same as those shown in Table 6.2.

6.4. Four disks of a finite population are marked 1, 4, 7, and 10.
  (a) Find the mean and standard deviation of this population.
  (b) Random samples of size three are drawn with replacement from this population. List all possible samples (and keep them for Exercises 6.7, 6.8, and 6.9), find the mean of each, and make the probability sampling distribution of the mean. *Hint:* List all selections and count number of arrangements of each selection.
  (c) Calculate the mean and standard deviation of the probability sampling distribution of the mean found in part (b) and compare with the corresponding values obtained by using Equations (6.6) and (6.8).
  (d) Use probabilities of individual observations to find the probability of any random sample in part (b).
  (e) Construct and compare probability histograms of the parent population and the

188 *Sampling Distributions*

sampling distribution of the mean of part (b). What is the probability that a sample mean will be within 3 units of the true mean? Within 2 standard errors of the true mean?

^{C}6.5. (a) Experimentally draw 100 random samples of size three with replacement from the parent population of Exercise 6.4 and find the relative frequency sampling distribution of the mean. Compare your distribution with the one obtained in Exercise 6.4(b).

(b) Combine the relative frequency distributions of the mean of all members of the class and compare the resulting distribution with the sampling distribution of Exercise 6.4(b).

*6.6. Six disks of a finite population are marked 2, 2, 2, 4, 4, and 6.
(a) Find the mean and standard deviation of this population.
(b) Random samples of size two are drawn without replacement from this population. Find the probability sampling distribution of the mean. *Hint:* Remember your sample objects.
(c) Calculate the mean and standard deviation of the sampling distribution found in part (b) and compare with the corresponding values obtained by using Statement (6.9).
(d) Construct and compare probability histograms of the parent population and the sampling distribution of the mean of part (b). What is the probability that a sample mean will be within one unit of the population mean?
(e) Random samples of size two are drawn with replacement from the parent population. Find the probability sampling distribution of the mean.
(f) Calculate the mean and standard deviation of the sampling distribution found in part (e) and compare with the corresponding values obtained by using Statement (6.9).

*6.7. (a) Use the sample means $\bar{y}$ found in Exercise 6.4(b) and the values of $\mu_{\bar{y}}$ and $\sigma_{\bar{y}}$ found in Exercise 6.4(c) to find the probability sampling distribution of

$$u = \frac{\bar{y} - \mu_{\bar{y}}}{\sigma_{\bar{y}}}$$

(b) Construct a histogram for the sampling distribution of u.
(c) Construct a cumulative "less than" probability polygon of u.
(d) Find $P(|u| \geq 2)$, and find u' such that $P(-u' < u < u') = 0.90$.

^{E}6.8. (a) Use the samples found in Exercise 6.4(b) to find the sampling distribution of the median.
(b) Calculate the mean and standard deviation of the sampling distribution of the median found in part (a) and compare with the mean and standard deviation of the sampling distribution of the mean found in Exercise 6.4(c).
(c) It can be shown that when the parent population is approximately normal that the

standard error of the median (i.e., the standard deviation of the median) is approximately $1.25\sigma/\sqrt{n}$ for random samples of size n. Compare the value of the standard error of the median found in part (b) with the value computed from $1.25\sigma/\sqrt{n}$.

^{E}6.9. (a) Use the samples found in Exercise 6.4(b) to find the sampling distribution of the mid-range (i.e., the mean of the largest and smallest sample value).
(b) Calculate the mean and standard deviation of the sampling distribution of the mid-range found in part (a) and compare with the mean and standard deviation of the sampling distribution of the mean found in Exercise 6.4(c).

6.10. A random sample of size 16 is drawn from a population with standard deviation 4.8. If the mean of the sample is used to estimate the population mean, what is the least probability that the error in the estimate will be less than 3?

6.11. The mean of a random sample of size 49 is used to estimate the mean of a population with standard deviation 28. What statement can you make about the probability that the error in the estimate will be less than 10 if
(a) Chebyshev's Theorem is used?
(b) the Central Limit Theorem is used?

6.12. The probability is at least 0.90 that a sample mean differs from the population mean by at most 4 units. If it is known that the variance of the parent population is 250, find the minimum sample size so that the above statement is satisfied.

^{C}6.13. (a) Verify the probabilities in Table 6.5 when $n = 2$ and $n = 3$.
(b) Verify the probabilities in Table 6.5 for $n = 6$.
Hint: List all selections and count the number of arrangements of each selection.

6.14. (a) Construct probability histograms for the parent population and the three sampling distributions of Table 6.5.
(b) Find the probability that a sample mean is within three units of the true mean for each of the three sampling distributions of Table 6.5.

6.15. A random sample of size 16 is drawn from a normal population with mean 50 and variance 100.
(a) Find the probability that the sample mean is less than 48.
(b) Find a symmetric interval about 50 which contains 0.95 of the sample means.
(c) Find $\bar{y}_1$ and $\bar{y}_2$ such that $P(\bar{y} < \bar{y}_1) = 0.02$ and $P(\bar{y}_1 < \bar{y} < \bar{y}_2) = 0.95$.

6.16. The verbal scores of 324 freshmen in a college are approximately normally distributed with mean 610 and standard deviation 60. If a random sample of 36 students were selected, what is the probability that the mean score would be
(a) greater than 630?
(b) between 600 and 640?

6.17. A random sample of size n is drawn from a normal population with mean 69 and standard deviation 2.5. Find the smallest n such that $P(68.9 < \bar{y} < 69.1) = 0.99$.

6.18. A random sample of size n is drawn from a normal population with mean μ and variance 100. What is the smallest sample size one can use in order that at least 0.95 of the sample means $\bar{y}$ fall within 4 units of μ?

6.19. The sample size changes from 30 to 120.
(a) What happens to the standard error of the mean?
(b) What happens to the standard error of the median [see Exercise 6.8(c)]?

6.20. The standard error of the mean changes from 10 to 5.
(a) What happens to the size of the sample?
(b) What are the sample sizes if one standard error is three times the size of the other?

6.21. Suppose the hourly wages of all workers in factory A have a distribution with mean $5.25 and standard deviation $0.60. If a random sample of 49 workers were selected, what is the probability that their mean hourly wage would be
(a) greater than $5.50?
(b) between $5.00 and $5.40?

6.22. Suppose the daily wages in a certain industry are normally distributed with mean $25.65. If 10 percent of the mean daily wages of samples of 16 workers fall below $25.00, what is the standard deviation of the daily wages of this industry?

6.5. EXPECTED VALUE

We know that a statistic is a numerical characteristic of a sample, and that the number of possible statistics is unlimited. We have already indicated that each statistic has a sampling distribution. For a sample of size one the sampling distribution of the value of the sample is the same as the distribution of the parent population. Thus, we may think of every distribution of a random variable as a sampling distribution of some statistic.

Statistical inference deals with the sampling distributions of all types of statistics. Now we introduce a few useful definitions that are common to all sampling distributions.

Already we have given the standard deviation of the sampling distribution of the mean a special name—the *standard error of the mean*. In general, if t is any statistic and σ_t denotes the standard deviation of the sampling distribution of the statistic t, then σ_t is also termed the **standard error of the statistic** t. For example, if t is the sample median, m, then σ_m is the *standard error of the median;* if t is the sample variance, s^2, then σ_{s^2} is the *standard error of the variance*.

Also, the mean of a sampling distribution of a statistic t is given a special name; it is called the **expected value of** t and denoted by μ_t or $E(t)$. For example, the expected value of a random sample mean is the population mean, that is $E(\bar{y}) = \mu$.

A parameter is a numerical characteristic of a population. Let θ be any parameter of a population P, and t a statistic for a random sample of size n from P. Whenever

$$E(t) = \theta$$

we say t is an **unbiased estimator of** θ, and a particular value of t is called an **unbiased estimate of** θ. Thus, $\bar{y}$ is an unbiased estimator of μ, and any value of $\bar{y}$ is an unbiased estimate of μ. In Example 6.5, $\bar{y} = 3$ is an unbiased estimate of $\mu = 4$, because it is the value of an unbiased estimator. As a matter of fact, each value of $\bar{y}$ in Table 6.4 is an unbiased estimate of $\mu = 4$, but only $\frac{4}{16} = \frac{1}{4}$ of the sample means are actually the same as the parameter mean which they estimate.

An advantage of the mean of a random sample is that it is an unbiased estimator of the mean of the parent population, regardless of the type of population or of the size of the sample. This is not true of every statistic. For example, the median of a random sample may or may not be an unbiased estimator of the median of a parent population (see Exercise 6.23).

6.6. SAMPLING DISTRIBUTIONS OF THE VARIANCE AND OF THE STANDARD DEVIATION

Sample variances could be defined in many ways. We consider the three following candidates of random samples of size n:

$$v_1 = \frac{\sum_i (y_i - \mu)^2}{n}$$

$$v_2 = \frac{\sum_i (y_i - \bar{y})^2}{n}$$

(6.12) $$v_3 = \frac{\sum_i (y_i - \bar{y})^2}{n - 1}$$

The positive square root of each of these variances is, by definition, a standard deviation.

Variance v_1 has the obvious disadvantage that it is defined in terms of the population mean μ which is typically unknown. Thus, v_1 cannot usually be computed for a sample. The variance v_2 would appear to be the most likely candidate since it is defined like a population variance. However, v_2 can be shown (see Exercise 6.24) to be a biased estimator of σ^2.

The third variance, v_3, is the most desirable since it turns out to be an unbiased estimator of σ^2. The corresponding standard deviation is denoted by s and defined by $s = \sqrt{v_3}$. Thus, the variance is sometimes denoted by s^2.

EXAMPLE 6.7. Find the sampling distributions of the variance s^2 and standard deviation s of random samples of size two drawn with replacement from the finite population with values 1, 3, 5, and 7 and construct their probability histograms.

The samples of size two and the corresponding variances and standard deviations are shown in Table 6.3. This information is brought together in the form of sampling distributions shown in Tables 6.6 and 6.7. (The last column in

Table 6.6. Sampling Distribution of the Variance s^2 of Random Samples of Size Two Drawn with Replacement from the Population of Values 1, 3, 5, and 7

s^2	Frequency (f)	Probability [$f(s^2)$]	$s^2 f$
0	4	.2500	0
2	6	.3750	12
8	4	.2500	32
18	2	.1250	36
Totals	16		80

each table is to be used for later computations.) The frequencies (or probabilities) for corresponding values of s^2 and s are the same. But the distributions are actually different since the values fall at different intervals along the horizontal axes. For example, successive values in Table 6.7 are $\sqrt{2}$ units apart, but successive values in Table 6.6 are 2, 6, and 10 units apart. Thus, in the construction of a probability histogram we take the successive boundaries for values of s to be

$$-\frac{\sqrt{2}}{2}, \frac{\sqrt{2}}{2}, \frac{3\sqrt{2}}{2}, \frac{5\sqrt{2}}{2}, \frac{7\sqrt{2}}{2}$$

Table 6.7. Sampling Distribution of the Standard Deviation s of Random Samples of Size Two Drawn with Replacement from the Population of Values 1, 3, 5, and 7

s	f	$f(s)$	sf
0	4	.2500	0
$\sqrt{2}$	6	.3750	$6\sqrt{2}$
$2\sqrt{2}$	4	.2500	$8\sqrt{2}$
$3\sqrt{2}$	2	.1250	$6\sqrt{2}$
Totals	16		$20\sqrt{2}$

and the corresponding successive boundaries for values of s^2 are

$$-\frac{1}{2}, \frac{1}{2}, \frac{9}{2}, \frac{25}{2}, \frac{49}{2}$$

The probability histograms are shown in Figures 6.2 and 6.3. The areas of the rectangles are made equal to the probabilities. Thus, their heights are equal to the ratio

$$\frac{\text{Probability of a value}}{\text{Length of base}}$$

For s the heights are then

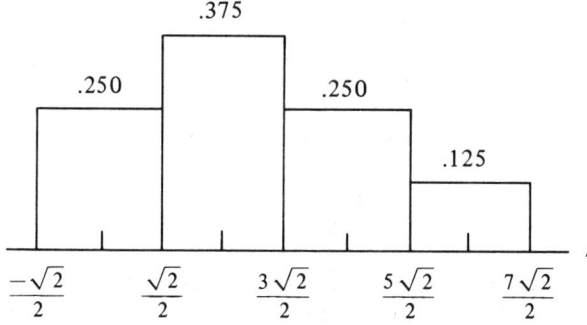

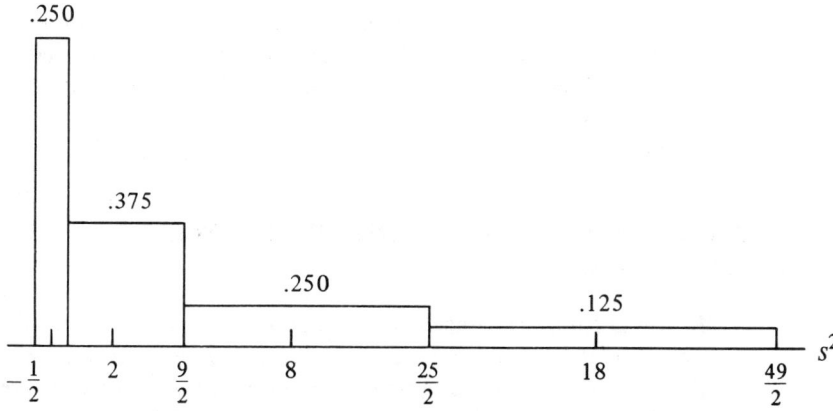

Figure 6.2. Probability Histogram for Table 6.7

Figure 6.3. Probability Histogram for Table 6.6

For s^2 the heights are

$$\frac{.2500}{1} = .250, \quad \frac{.3750}{4} = .094, \quad \frac{.2500}{8} = .031, \quad \frac{.1250}{12} = .010.$$

Note that the vertical scales are the same. If the horizontal scales had been made the same the areas under the two histograms would look the same. (The student should note that the histograms show negative values of s and s^2. Actually, neither s nor s^2 can be negative—the constructions are made in such a way as to preserve the "shape" and "area" of the histograms.) Clearly, the distributions of s^2 and s are quite different, and their "shapes" change with the parent populations or the size of the sample. (See Exercises 6.24, 6.25, and 6.26 for further insights into the nature of sampling distributions of the variance and standard deviation.)

EXAMPLE 6.8. Find the expected values of s^2 and s of Example 6.7. From Table 6.6 we find that

$$E(s^2) = \frac{80}{16} = 5 = \sigma^2$$

Thus, s^2 is an unbiased estimate of σ^2. From Table 6.7 we find that

$$E(s) = \frac{20\sqrt{2}}{16} = \frac{5\sqrt{2}}{4} < \sqrt{5} = \sigma$$

Hence, s is a biased estimator of σ which actually underestimates σ.

In Example 6.8 we have illustrated general properties. That is, it can be shown that if *a random sample of any size n is drawn from any population, the sample variance s^2 is an unbiased estimator of σ^2*. But it is somewhat surprising to find that *the standard deviation s is a biased estimate of σ and that it is too small on the average*. Furthermore, the amount of bias in s depends on the parent population and the size of the sample. This explains why the sample variance, instead of the standard deviation, is defined so as to be an unbiased estimator of the corresponding parameter of the parent population. (See Exercise 6.31 for further information.)

6.7. SAMPLING DISTRIBUTION OF THE DIFFERENCE IN SAMPLE MEANS

In this section, we consider the properties of one more sampling distribution. Let populations 1 and 2 have means μ_1 and μ_2 and variances σ_1^2 and σ_2^2, respectively. Suppose random samples of sizes n_1 and n_2 drawn with replacement (or, if infinite, without replacement) from populations 1 and 2 have means $\bar{y}_1$ and $\bar{y}_2$,

respectively. Let
$$\bar{d} = \bar{y}_1 - \bar{y}_2$$
denote the difference in any two means. Then it can be shown that

(6.13) $$E(\bar{d}) = \mu_{\bar{y}_1} - \mu_{\bar{y}_2} = \mu_1 - \mu_2$$

and

(6.14) $$\sigma_{\bar{d}} = \sqrt{\sigma_{\bar{y}_1}^2 + \sigma_{\bar{y}_2}^2} = \sqrt{\frac{\sigma_1^2}{n_1} + \frac{\sigma_2^2}{n_2}}$$

In summary, we can be assured that *the statistic $\bar{y}_1 - \bar{y}_2$ has the properties of Equations (6.13) and (6.14) provided the two samples are randomly and independently drawn from two populations;* i.e., provided a random sample is drawn from population 1, a random sample is drawn from population 2, and the means $\bar{y}_1$ and $\bar{y}_2$ are independently distributed.

In particular, it the two populations are normally distributed, then $\bar{d}$ is also normally distributed with mean $E(\bar{d})$ and standard deviation $\sigma_{\bar{d}}$. Thus, it follows that

(6.15) $$\frac{\bar{d} - E(\bar{d})}{\sigma_{\bar{d}}} = \frac{\bar{y}_1 - \bar{y}_2 - (\mu_1 - \mu_2)}{\sqrt{\frac{\sigma_1^2}{n_1} + \frac{\sigma_2^2}{n_2}}}$$

is distributed as the standard normal random variable z. If the parent populations fail to be normally distributed, we may say, applying the Central Limit Theorem, that the statistic in Expression (6.15) approaches the distribution of the standard normal variable z as the sample sizes get large.

EXAMPLE 6.9. Suppose the heights of adult males are normally distributed with mean 69 inches and standard deviation 2.5 inches and the heights of adult females are normally distributed with mean 65 inches and standard deviation 2.0 inches. If a random sample of size 8 is drawn from each population, what is the probability that the mean height of the females will exceed the mean height of the males?

Let y_1 and y_2 denote the random variables for the male and female populations. If $\bar{d} = \bar{y}_1 - \bar{y}_2$, we wish to find

$$P(\bar{d} < 0)$$

Since

$$E(\bar{d}) = \mu_1 - \mu_2 = 69 - 65 = 4$$

and

$$\sigma_{\bar{d}} = \sqrt{\frac{(2.5)^2}{8} + \frac{(2.0)^2}{8}} = \sqrt{1.281} \doteq 1.13$$

it follows that

$$P(\bar{d} < 0) = P\left[\frac{\bar{d} - E(\bar{d})}{\sigma_{\bar{d}}} < \frac{0-4}{1.13}\right]$$
$$= P(z < -3.54) \doteq 0.0002$$

Thus, for samples of size eight, we would expect the mean height of females to exceed the mean height of males only two times out of 10,000.

EXAMPLE 6.10. It is informative to verify the statements we made about $\bar{d}$. Suppose the population of y_1 values is 3, 5, and 7 and the population of y_2 values is 1, 4, 7, and 10. Let $\bar{y}_1$ be the mean of random samples of size two drawn with replacement from the population of y_1 values. Let $\bar{y}_2$ be the mean of random samples of size three drawn with replacement from the population of y_2 values. Let $\bar{d} = \bar{y}_1 - \bar{y}_2$. We wish to find the sampling distribution of $\bar{d}$, the mean $E(\bar{d})$, and the standard deviation $\sigma_{\bar{d}}$.

The reader should have no difficulty in finding the sampling distributions of $\bar{y}_1$ and $\bar{y}_2$ shown in Table 6.8. These two sampling distributions are put in one table for easy reference and rapid computations.

The differences $\bar{d}$ may be obtained by subtracting each value of $\bar{y}_2$ from every value of $\bar{y}_1$. To be specific, if $\bar{y}_1 = 5$ and $\bar{y}_2 = 3$, then $\bar{y}_1 - \bar{y}_2 = 5 - 3 = 2$ with frequency $3 \times 6 = 18$ since 3 is subtracted six times from each of the three 5's. Also, $\bar{y}_1 - \bar{y}_2 = 2$ when $\bar{y}_1 = 6$ and $\bar{y}_2 = 4$ or when $\bar{y}_1 = 7$ and $\bar{y}_2 = 5$. The sampling distribution of all $\bar{d}$ is shown in Table 6.9. The last two columns are to be used in the computation of $E(\bar{d})$ and $\sigma_{\bar{d}}$.

From the totals in Table 6.9 we find that

$$E(\bar{d}) = \frac{-288}{576} = -\frac{1}{2}$$

Table 6.8. Sampling Distributions of the Mean of Random Samples of Sizes Two and Three Drawn with Replacement from the Two Parent Populations of Example 6.10

$\bar{y}_1$	f	$\bar{y}_2$	f
3	1	1	1
4	2	2	3
5	3	3	6
6	2	4	10
7	1	5	12
Total	9	6	12
		7	10
		8	6
		9	3
		10	1
		Total	64

Table 6.9. Sampling Distribution of $\bar{d} = \bar{y}_1 - \bar{y}_2$ Obtained from Table 6.8

$\bar{d}$	f	$\bar{d}f$	$\bar{d}^2 f$
−7	1	−7	49
−6	5	−30	180
−5	15	−75	375
−4	33	−132	528
−3	57	−171	513
−2	81	−162	324
−1	96	−96	96
0	96	0	0
1	81	81	81
2	57	114	228
3	33	99	297
4	15	60	240
5	5	25	125
6	1	6	36
Totals	576	−288	3072

and

$$\sigma_{\bar{d}}^2 = \frac{3{,}072 - \frac{(-288)^2}{576}}{576} = \frac{3{,}072 - 144}{576} = \frac{61}{12}$$

From the parent populations we find that

$$\mu_1 = 5,\ \mu_2 = 5.5,\ \sigma_1^2 = \frac{8}{3},\ \text{and}\ \sigma_2^2 = \frac{45}{4}$$

Hence

$$\mu_1 - \mu_2 = 5 - 5.5 = -0.5 = E(\bar{d})$$

and

$$\frac{\sigma_1^2}{n_1} + \frac{\sigma_2^2}{n_2} = \frac{\frac{8}{3}}{2} + \frac{\frac{45}{4}}{3} = \frac{4}{3} + \frac{15}{4} = \frac{61}{12} = \sigma_{\bar{d}}^2$$

Thus, the properties of Equations (6.13) and (6.14) are verified.

6.8. EXERCISES

6.23. (a) If random samples of size four are drawn without replacement from the uniform population with values 1, 3, 5, 7, and 9, show that the sample median, m, is an unbiased estimate of the population median, M.

(b) If random samples of size four are drawn without replacement from the uniform populations with values 1, 3, 7, 7, 9, show that the sample median is a biased estimate of the population median.

(c) If random samples of size three are drawn without replacement from the population of part (b), what can you say about the expected value of the sampling distribution of medians? Compare $E(m)$ with both M and μ.

6.24. Random samples of size two are drawn with replacement from the finite population with values 1, 3, 5, and 7.
(a) Find the sampling distribution of the variance v_1.
(b) Find the sampling distribution of the variance v_2.
(c) Show that v_1 is an unbiased estimator of σ^2 while v_2 is a biased estimate of σ^2.
(d) Show that both $\sqrt{v_1}$ and $\sqrt{v_2}$ are biased estimators of σ.

6.25. Five disks of a finite population are marked 2, 5, 8, 11, and 14. Random samples of size three are drawn without replacement (see Exercise 6.2).
(a) Find the sampling distribution of the variance s^2.
(b) Find the sampling distribution of the standard deviation s.
(c) Find the sampling distribution of the range.
(d) Which of the statistics in parts (a), (b), and (c) are unbiased estimators of their corresponding population parameters?
(e) Find the sampling distribution of $2s^2/\sigma^2$ and find the mean of the distribution.
(f) Construct probability histograms of the distributions in parts (a) and (b).

6.26. Random samples of size three are drawn with replacement from the uniform population with values 1, 4, 7, and 10 (see Exercise 6.4 for list of samples).
(a) Find the sampling distribution of the variance s^2.
(b) Find the sampling distribution of the variance v_2.
(c) Which, if either, of the statistics in parts (a) and (b) is an unbiased estimator of σ^2?
(d) Find the sampling distribution of $2s^2/\sigma^2$.
(e) Find the expected value of $2s^2/\sigma^2$.
(f) Construct a probability histogram for the distribution in part (d).

6.27. Suppose the population of y_1 values is 2, 4, 6, and 8 and the population of y_2 values is 4, 7, and 10. Let $\bar{y}_1$ be the mean of random samples of size two drawn with replacement from the population of y_1 values. Let $\bar{y}_2$ be the mean of random samples of size three drawn with replacement from the population of y_2 values. Let $\bar{d} = \bar{y}_1 - \bar{y}_2$. Find the sampling distribution of $\bar{d}$, the mean $E(\bar{d})$, and the standard deviation $\sigma_{\bar{d}}$. Check your values for $E(\bar{d})$ and $\sigma_{\bar{d}}$ by using Equations (6.13) and (6.14).

C6.28. (a) Use the sample means $\bar{d}$ found in Exercise 6.27 to find the sampling distribution of
$$u = [\bar{d} - E(\bar{d})]/\sigma_{\bar{d}}$$
(b) Find $E(u)$ and σ_u^2.
(c) Construct a probability histogram of the sampling distribution of u and find $P(|u| \geq 2)$.

6.29. Normal population one has mean 50 and variance 100. Normal population two has mean 40 and variance 200. If random and independent samples of size 10 are drawn from each population, what is the probability that the mean of sample one differs from the mean of sample two by less than five?

6.30. The mean gasoline mileage for cars of make A is 18 miles per gallon with standard deviation of 2.2 miles per gallon; comparable figures for cars of make B are 27 and 3.4 miles per gallon. Nine cars of make A and twenty cars of make B are randomly and independently drawn. What is the probability that the mean gasoline mileage for make A is more than 7 miles per gallon less than the mean gasoline mileage of make B?

C6.31. Random samples are drawn with replacement from the population with values 1, 7, and 13.
(a) Find the sampling distribution of the standard deviation of samples of size two.
(b) Find the sampling distribution of the standard deviation of samples of size three.
(c) Find the means of the two sampling distributions in parts (a) and (b). Compare the biases of the standard deviations for the two sample sizes.
(d) Find the sampling distribution of the standard deviation of random samples of size two drawn with replacement from the population with values 1, 7, and 19.
(e) Compare the biases of the standard deviations in parts (a) and (d).

6.32. A random sample of size 25 and a second random sample of size n_2 were independently drawn from the same population. What is n_2 if the ratio of the standard error of the mean of the first sample to the standard error of the mean of the second sample is 3:4?

E6.33. Suppose $\sigma_1^2 = \sigma_2^2$ and $n_1 + n_2 = 20$. Find n_1 and n_2 such that $\sigma_{\bar{y}_1 - \bar{y}_2}$ is a minimum.

E6.34. Suppose $\sigma_1 = 3$, $\sigma_2^2 = 27$, and $n_1 + n_2 = 20$. Find n_1 and n_2 such that $\sigma_{\bar{y}_1 - \bar{y}_2}^2$ is a minimum.

6.9. RANDOM NUMBERS AND SIMULATED EXPERIMENTS

We did theoretical and experimental sampling to illustrate principles. In practice we take one sample (sometimes, under very special circumstances, more than one sample) from one population and make inferences. Usually, the populations are very large, so getting a single random sample by the experimental method is not always easy. As examples, consider the problem of drawing a random sample of size 138 from a student body of 917; of sampling 3 percent of the trees in an orchard, grove, or forest; or of getting opinions of 5 percent of the farm households in 10 agricultural counties. In each case we could assign a number to each object of the population, put the numbers on similar pieces of paper (or disks or tags), mix them well in a large box, and draw the desired sample, one object at a time. Actually, going through such a routine before each experiment could become very time-consuming. Fortunately, the Interstate Commerce Commission, The Rand Corporation, and others have already selected and published tens of thousands of random digits to be used in selecting a random sample.

Prepared tables of random digits not only save an experimenter time in drawing random digits by devices such as shuffling cards, tossing dice, or drawing pieces of paper out of a box, but they are likely to be "more random." Any method of selecting random numbers is likely to depart slightly from randomness, but the best published tables minimize such biases. Table III in the Appendix contains 7,500 random digits and illustrates a typical arrangement of digits for easy reading. The digits can be read in groups of two, three, or more to give numbers with two, three, or more digits. The starting point in the table should be selected by chance, and the direction of reading decided on in advance. (The object is to avoid getting the same sequence of random numbers in several different experiments.) Remember that every digit is a random digit, and that any predetermined sequential (but nonoverlapping) selection is permissible.

EXAMPLE 6.11. Use a table of random digits to draw a random sample of 138 (15 percent) students from a student body of size 917. Also, show that the sample is a simple random sample.

First, obtain an accurate list of the student body and assign them consecutive three digit numbers 001, 002, . . . , 917. If a directory is available, simply give the first name the number 001, the second name the number 002, . . . , and the last name the number 917. Care should be taken to be sure that the names in the listed population satisfy the definition of being a member of the student body. Decisions may need to be made about such items as to whether a student is a part-time student, absent for the term, or away on foreign study.

Next, 138 three-digit random numbers must be selected from the table of random digits. To be specific, suppose our starting point is the 11th digit in the fourth line of page 406, and we read from left to right across a line and move down the page. The first 35 digits are

06243   61680   07856   16376   39440   53537   71341

From these digits we obtain the following sequence of 11 three-digit numbers.

062   436   168   007   856   163   763   944   053   537   713

We continue in this way until 138 different permissible three-digit numbers are obtained. *Any number exceeding 917 along with 000 is discarded and any number which has already been selected is ignored*—number 944 in this case.

Now, we show that the sample selected is a simple random sample. First, by the classical definition, the probability of each digit from 0 through 9 is $\frac{1}{10}$.

Thus, the probability of any sequence of three digits is

$$\left(\frac{1}{10}\right)\left(\frac{1}{10}\right)\left(\frac{1}{10}\right) = \frac{1}{1,000}$$

Since each student is assigned one three-digit number, the probability of any three-digit number being first is $\frac{1}{1,000}$. So the probability that some student number will be selected in the first draw is $\frac{917}{1,000}$, and the probability is $\frac{83}{1,000}$ that a nonstudent number will be selected. However, by the classical definition, the conditional probability that *if a student is selected* it will be a particular one is $\frac{1}{917}$ for each student. In particular, the probability of selecting the student number 740 on the first draw is $\frac{1}{917}$.

Once the first student number is selected there are only 916 permissible student numbers left. So, by the classical definition, the probability of any remaining student number being selected on the second draw is $\frac{1}{916}$. Similarly, after two numbers have been selected the probability of any remaining student number being selected is $\frac{1}{915}$. Continuing in this way, we see that after 137 numbers have been selected the probability of any remaining student number being selected is $1/(917 - 137) = \frac{1}{780}$.

Finally, the probability of selecting any particular sequence of 138 student numbers is

$$\left(\frac{1}{917}\right)\left(\frac{1}{916}\right)\left(\frac{1}{915}\right) \cdots \left(\frac{1}{780}\right)$$

Thus, any sample selected is a simple random sample since each sample has the same probability of being selected.

We observe that action is required to select a sample of n objects from a population of N objects. *Arbitrary or haphazard selections of* n *objects do not constitute a randon sample.* In the above example, it is not enough to select the first 138 students we see or to select those in the same dormitory. It might be interesting for the reader to write a sequence of 200 digits in an arbitrary manner and let someone else check for evidence of regularity in the digits. All sorts of subtle biases tend to influence people to write (or draw) random sequences with patterns.

Another common misconception about a random sample is that it is necessarily representative of the parent population. If readers look at all the samples in Table 6.3, they should soon be convinced that not all random samples appear to be representative of the population. Furthermore, our opinion of what is representative is likely to be influenced by the characteristic of the population we wish to examine.

A third common misconception is that randomness is less important in large samples than in small samples. Any time readers are tempted to think this they should remember that the *Literary Digest* presidential poll was based on a sample of over two million drawn from lists of automobile and telephone owners. Yet this huge sample, not representative of all economic groups, failed to predict a Roosevelt victory by 19 percentage points. By contrast, the best presidential polls today are based on scientific sampling methods designed to include the proper proportion of voters from each economic stratum, each age stratum, and so forth (see stratified random sampling and cluster sampling in Section 6.10). For example, using fewer than 5,000 people in each sample, the Gallup Poll's final election figures (including campaigns for presidents, senators, and governors) in the 12 national elections from 1950 through 1972 missed the actual vote percentages by approximately 1.6 percentage points on the average.

EXAMPLE 6.12. Find the probability that a given student will appear in the simple random sample drawn in Example 6.11.

We seek the probability that a student number is selected either first or second, ..., or 138th in an ordered sample. The 138 outcomes are mutually exclusive since a number is removed from the population once it has been selected. We have already found that the probability of a particular number being selected first is $\frac{1}{917}$. The probability of a particular student number being selected second is the product of the probability of some other number being first and the probability of the particular number being second; i.e., the required probability is

$$\left(\frac{916}{917}\right)\left(\frac{1}{916}\right) = \frac{1}{917}$$

Similarly, the probability of a particular number being the third number selected is

$$\left(\frac{916}{917}\right)\left(\frac{915}{916}\right)\left(\frac{1}{915}\right) = \frac{1}{917}$$

Continuing in this way, the probability of a particular number being the 138th number selected is

$$\left(\frac{916}{917}\right)\left(\frac{915}{916}\right) \cdots \left(\frac{780}{781}\right)\left(\frac{1}{780}\right) = \frac{1}{917}$$

So the probability of a particular number occurring is the sum of the probabilities of the 138 mutually exclusive events; i.e., the required probability is $\frac{138}{917}$.

We have introduced random numbers to make it easier to select the objects to be used in an experiment. However, random numbers can often be used in place of the actual experiment. For example, we may use random numbers in place of the simple experiment of observing "head" or "tail" when a coin is tossed. If we let the digits 0, 1, 2, 3, 4 represent heads and the digits 5, 6, 7, 8, 9 represent tails, then a sequence of random digits like 94036 85978 stands for *tail, head, head, head, tail, tail, tail, tail, tail, tail*. So, we have used random numbers to simulate an experiment.

The growth of high-speed computers has encouraged the development of increasingly sophisticated techniques for the simulation of a great variety of experiments. Such experiments may be found in any areas where decisions are based on chance fluctuations. These simulation methods, also termed *Monte Carlo Methods*, are based on random numbers (which can be rapidly generated on a computer) and have been used in problems of traffic flow, spread and control of epidemics, engagement of two naval forces, and advertising and marketing of a product.

Already we have noted how random numbers may be used to simulate a simple experiment. Now we give two illustrations to show how the same technique may be applied to the more complicated problem of obtaining experimental sampling distributions.

EXAMPLE 6.13. Three well-balanced dice are tossed together. Find, by simulation, an experimental sampling distribution of the total number of dots T on the three faces.

Since the outcome of the toss of a die is 1, 2, 3, 4, 5, or 6, we use a single digit to represent the outcome of each die. Any time 7, 8, 9, or 0 appears in a random number sequence it is ignored. Starting at an arbitrary point in Table III, we find that the first 20 digits from line 2 are

22368   46573   25595   85393

Then the outcomes of the first five tosses of three dice are taken as

223   646   532   555   533

and the corresponding totals are 7, 16, 10, 15, 11. By continuing in this manner we could simulate many more tosses of three dice. Since T can be any of 16 different numbers, we should probably simulate at least 800 tosses. Then the results should be arranged in a distribution table. (With a high-speed computer this experiment could be done in a few seconds, including the printing of totals and relative frequencies.)

EXAMPLE 6.14. By simulation find an experimental sampling distribution of the mean of random samples of size five drawn from a normal population with mean 50 and standard deviation 10.

Since the random variable of the normal population is continuous, we construct the finite population of size 1,000 shown in Table 6.10 as the approximating population from which we draw random samples. The three-digit random numbers assigned to each integral value of y are indicated in the third column. For example, the random numbers 009, 010, and 011 represent the value 27. Table 6.11 is a condensed frequency table which indicates the goodness of the approximation to the normal distribution and suggests how Table 6.10 was constructed.

Table 6.10. Frequency Distribution of 1,000 Observations Used to Approximate the Normal Distribution of Example 6.14

Observation y	Frequency f	Random Number	Observation y	Frequency f	Random Number
19	1	000	51	40	520–559
20	0	–	52	39	560–598
21	1	001	53	38	599–636
22	1	002	54	37	637–673
23	1	003	55	35	674–708
24	1	004	56	33	709–741
25	2	005–006	57	31	742–772
26	2	007–008	58	29	773–801
27	3	009–011	59	27	802–828
28	4	012–015	60	24	829–852
29	5	016–020	61	22	853–874
30	5	021–025	62	20	875–894
31	6	026–031	63	17	895–911
32	8	032–039	64	15	912–926
33	9	040–048	65	13	927–939
34	11	049–059	66	11	940–950
35	13	060–072	67	9	951–959
36	15	073–087	68	8	960–967
37	17	088–104	69	6	968–973
38	20	105–124	70	5	974–978
39	22	125–146	71	5	979–983
40	24	147–170	72	4	984–987
41	27	171–197	73	3	988–990
42	29	198–226	74	2	991–992
43	31	227–257	75	2	993–994
44	33	258–290	76	1	995
45	35	291–325	77	1	996
46	37	326–362	78	1	997
47	38	363–400	79	1	998
48	39	401–439	80	0	–
49	40	440–479	81	1	999
50	40	480–519			

In the simulated experiment, any three-digit number represents a value of y. For example, the fifteen random digits

    47869   87001   31591

represent the random sample values

    49   55   55   39   52

since they correspond (see Table 6.10) to the random numbers 478, 698, 700, 131, and 591, respectively. Continuing in this way we could find any number of

Table 6.11. Comparison of the Population of Table 6.10 with the Normal Population Frequencies (A Condensed Table)

Approximating Population			Normal Population	
Observation	Frequency	Cumulative Frequency	z	Cumulative Probability Times 1,000
Below 25	5	5	−2.55	5.4
25, 26, 27	7	12	−2.25	12.2
28, 29, 30	14	26	−1.95	25.6
31, 32, 33	23	49	−1.65	49.5
34, 35, 36	39	88	−1.35	88.5
37, 38, 39	59	147	−1.05	146.9
40, 41, 42	80	227	−0.75	226.6
43, 44, 45	99	326	−0.45	326.4
46, 47, 48	114	440	−0.15	440.4
49, 50, 51	120	560	0.15	559.6
52, 53, 54	114	674	0.45	673.6
55, 56, 57	99	773	0.75	773.4
58, 59, 60	80	853	1.05	853.1
61, 62, 63	59	912	1.35	911.5
64, 65, 66	39	951	1.65	950.5
67, 68, 69	23	974	1.95	974.4
70, 71, 72	14	988	2.25	987.8
73, 74, 75	7	995	2.55	994.6
Above 75	5	1000		

samples and from them construct an experimental sampling distribution of the mean. For illustrative purposes, we take a very small number of samples of random normal numbers. (The interested reader may refer to tables of random normal numbers already published and ready for easy use.)

In Table 6.12 we show fifty samples obtained by Monte Carlo Methods. The corresponding experimental sampling distribution of the mean is shown in Table 6.13. By the usual methods, we find that the distribution of Table 6.13 has a mean of

$$\hat{\mu}_{\bar{y}} = 49.85$$

and a standard deviation of

$$\hat{\sigma}_{\bar{y}} = 4.86$$

Considering the small number of samples, these estimates compare favorably with values of the corresponding theoretical parameters which follow:

$$\mu_{\bar{y}} = \mu = 50$$

and

$$\sigma_{\bar{y}} = \frac{\sigma}{\sqrt{n}} = \frac{10}{\sqrt{5}} \doteq 4.47$$

Table 6.12. Fifty Random Samples from the Population of Table 6.10

Sample Number	Sample					Sample Number	Sample				
1	49	55	55	39	52	26	37	37	30	48	50
2	69	47	46	65	49	27	54	43	67	52	55
3	59	42	63	47	47	28	40	40	66	46	35
4	48	60	54	47	50	29	63	61	66	34	50
5	68	58	46	46	37	30	51	46	52	60	67
6	44	40	29	59	44	31	37	54	40	61	44
7	45	54	42	57	60	32	58	53	50	22	55
8	41	39	58	47	59	33	44	30	53	45	54
9	46	46	29	57	32	34	40	52	42	47	49
10	40	38	66	43	33	35	61	44	50	58	46
11	52	61	43	42	51	36	61	67	54	66	78
12	38	50	40	50	39	37	41	55	52	53	41
13	40	48	37	32	61	38	44	38	50	48	45
14	54	30	16	73	60	39	63	47	48	46	53
15	34	58	65	59	51	40	51	57	59	54	47
16	71	63	46	46	41	41	43	60	36	48	62
17	37	50	56	58	55	42	55	55	55	28	67
18	58	53	56	67	65	43	44	28	41	55	55
19	48	51	53	55	61	44	60	73	33	58	52
20	40	56	34	48	44	45	47	52	43	40	48
21	64	67	47	39	45	46	60	40	49	49	49
22	53	45	54	62	54	47	41	58	38	44	59
23	30	57	46	57	44	48	52	60	65	40	75
24	44	33	55	57	51	49	61	58	42	39	59
25	37	56	48	62	38	50	40	42	43	71	42

Table 6.13. Means of 50 Random Samples from the Population of Table 6.10

40.4	42.0	43.2	43.4	43.6	44.0	44.4	44.6	45.0	45.2
45.4	46.0	46.0	46.6	46.8	47.2	47.6	47.6	48.0	48.0
48.2	48.4	48.8	49.4	49.8	49.8	50.0	51.0	51.2	51.4
51.6	51.6	51.8	51.8	51.8	52.0	52.4	53.4	53.4	53.6
53.6	53.6	54.2	55.2	55.2	55.2	55.8	58.4	59.6	65.2

6.10. SAMPLE PLANS

It is not always possible or even desirable to take a simple random sample. Thus, many other methods of sampling have been developed. We briefly describe a few of the more useful sampling plans. *A sampling plan completely specifies how a sample will be obtained from a given population.*

Stratified Random Sampling. First, the population is divided, according to some relevant characteristic, into non-overlapping subpopulations, called **strata**. Then a simple random sample is taken from each stratum. If the same portion of objects is taken from each stratum we have *proportional stratified random sampling;* otherwise, we have *non-proportional stratified random sampling.* The selection of strata is of utmost importance and depends on the population characteristic being investigated. An advantage of stratified random sampling over simple random sampling is illustrated in the following example.

EXAMPLE 6.15. A sample is to be used to estimate the mean number of semester courses completed by full-time students in a certain school by the end of the first semester of the 1975–1976 session. Compare simple random and stratified random sampling methods.

We illustrate the principles with absurdly small population and samples. Suppose three freshmen have completed 5, 5, and 7 semester courses and two seniors have completed 36 and 38 semester courses. The mean number of courses completed is 18.2. The ten simple random samples of size two, each with probability $\frac{1}{10}$, are

$$5,5 \quad 5,7 \quad 5,36 \quad 5,38 \quad 5,7 \quad 5,36 \quad 5,38 \quad 7,36 \quad 7,38 \quad 36,38$$

and their means are

$$5 \quad 6 \quad 20.5 \quad 21.5 \quad 6 \quad 20.5 \quad 21.5 \quad 21.5 \quad 22.5 \quad 37$$

respectively. If we take a random sample of size one from the freshmen and a random sample of size one from the seniors, then the six stratified random samples, each with probability $\frac{1}{6}$, are

$$5,36 \quad 5,38 \quad 5,36 \quad 5,38 \quad 7,36 \quad 7,38$$

and their means are

$$20.5 \quad 21.5 \quad 20.5 \quad 21.5 \quad 21.5 \quad 22.5$$

respectively. Hence, we observe that there is much less variation among the sample means obtained from stratified random sampling. (All stratified sample means are within 4.3 semester courses of the population mean 18.2. However, 40 percent of the random sample means are further away.) Clearly, in general in this kind of problem, *by proper stratification we improve our chances of getting a sample mean close to the population mean.*

Since we sampled $\frac{1}{3}$ of the freshmen and $\frac{1}{2}$ of the seniors in Example 6.15 we used non-proportional stratified sampling. Also, we used only one variable, namely class rank, to stratify. In such cases it might be desirable to use class rank in combination with other variables (such as major, sex, or age) when selecting the strata.

The primary object in stratified random sampling is to select strata which are homogeneous relative to the characteristic of the population being considered. Note that in Example 6.15 the stratum of freshmen had values which were about equal, and the stratum of seniors had another set of values which were about equal.

Cluster Sampling. The population of objects is divided into relatively small subdivisions, called **clusters**, whose objects are typically physically close to each other. Then a specified portion of these clusters is randomly selected. The sample is made up of all (or a random sample of) objects in the clusters selected.

EXAMPLE 6.16. Suppose the Agricultural Extension Division in Virginia wants to study the fertilizer programs of farmers in 15 counties in the western part of the state. Since there are no up-to-date lists of farmers, it would be practically impossible to obtain a simple random sample. Even if it were possible to select a random sample, the time and expense of getting information on widely scattered farmers (sometimes with two or three call-backs) would be formidable.

One way to obtain a satisfactory sample would be to start with detailed maps of the 15 counties. Divide each county into small land areas with clearly marked boundaries (like roads, streams, and ridges) which would be expected to contain five to eight farm families, say. Then randomly select 15 percent of the small land areas and interview each farmer whose residence is found in a selected area. (Such maps with areas already marked can be obtained from the Department of Agriculture in Washington, D.C.) The sampling in this example is also called **area sampling** since the clusters are actually geographic subdivisions.

Cluster sampling may be used for such things as to estimate the yield of a crop, to study the expenditure for food of city families, or to estimate the wild animal count in forest regions. Estimates based on cluster sampling are usually less reliable (i.e., more variable) than those based on random samples. However, they are often more reliable *per unit cost*. Referring to Example 6.16, it should be clear that for the same cost many more farmers from a cluster sample can be interviewed than from a random sample.

Systematic Sampling. If objects are in order (as in a directory, a file, an assembly line), a systematic sample is obtained by randomly selecting one of the first k objects, then every kth object thereafter. For example, to select 10 percent of a student body we could take a corrected student directory, randomly select one of the first 10 names listed, and then every 10th student name which follows.

In a systematic sample every object has the same probability of being included, as in simple random sampling, but the probabilities are not independent. Even though systematic samples are not random samples, in some problems where systematic sampling is used it is reasonable to apply the properties derived from simple random sampling. The primary danger in this practice lies in the fact that some ordered lists have patterns or periodicity. For example, in inspecting every 20th item on an assembly line it may be that every 40th item is faulty. However, it could happen that systematic sampling actually is an improvement over simple random sampling in that the sample is "spread out evenly" over the population.

We have briefly discussed the above three sampling plans so that readers might know that simple random sampling is not the only sampling plan or be led to believe that random sampling is always best. However, for the development of the basic principles of statistical inference it seems desirable to present only arguments based on simple random samples. The primary reasons for this are (1) all probability sampling plans are based on a knowledge of simple random sampling, (2) the concepts of statistical inference can be grasped more easily by way of simple random sampling, and (3) random sampling does apply to a very large number of practical problems.

6.11. EXERCISES

6.35. Use Table III to select a simple random sample of size 13 from the population of the first 82 positive integers.

6.36. Use Table III to obtain a simple random sample of 20 percent of the population of the first 197 positive integers. Note that the sample can be obtained more rapidly if each of the random numbers 001, 201, 401, 601, 801 represents the positive integer 1, each of the random numbers 002, 202, 402, 602, 802 represents the positive integer 2, and so on.

6.37. Use random numbers to select three garages from among those listed in the Yellow Pages of a convenient city telephone directory.

6.38. The 96 students enrolled in a course are to be randomly divided into three equal-size sections—call them I, II, III.
(a) Use random numbers to make the divisions and explain the procedure.
(b) If 33 students are females, randomly divide the students so that 11 females are in each section.

6.39. Suppose 45 entering freshmen girls at a certain university requested single rooms. If there are 4 single rooms in dormitory A, 5 in dormitory B, 2 in dormitory C, and 8 in dormitory D, randomly assign 19 girls to the single rooms. Explain your procedure.

6.40. (a) Without stopping, write a sequence of 300 digits which are supposed to be random. Make a frequency distribution of the digits $0, 1, \cdots, 9$.
(b) Obtain 300 digits from Table III and make a frequency distribution of the digits $0, 1, \cdots, 9$.
(c) Compare the sequences in parts (a) and (b). Look for patterns, cycles, and other regular features.

6.41. Use Table 6.3 to determine which samples are "representative" (in your opinion) of the population, if the statistic to be compared against the corresponding parameter is the

(a) mean
(b) variance
(c) range
(d) harmonic mean

6.42. Find a class in which 50 or more students are enrolled.
(a) Take a random sample of size 10 and estimate the proportion of girls in the class.
(b) Use the first 20 students to enter the classroom to estimate the proportion of girls in the class.
(c) Which of the estimates in parts (a) and (b) is closer to the true proportion of girls in the class? Do you think your results are typical? Explain.

6.43. (a) Simulate 150 tosses of a balanced die by random numbers.
(b) Let a one or six be a success. Calculate the proportion of successes after each sequence of ten tosses and plot the accumulated proportions as in Figure 4.16. What is your estimate of the probability of success?

6.44. Use random numbers to simulate an experiment in which 3 well-balanced coins are tossed 96 times and compare the observed frequencies of 0, 1, 2, and 3 heads with the corresponding expected frequencies of 12, 36, 36, and 12.

6.45. Suppose the probabilities of the responses "yes," "no," and "don't know" to a certain question are 0.40, 0.45, and 0.15, respectively. Use two-digit random numbers with 00 through 39 representing "yes," 40 through 84 representing "no," and 85 through 99 representing "don't know" to simulate 50 responses to the question.

c6.46. The probabilities for a binomial distribution with $n = 10$ and $\pi = 0.4$ are given in Table 5.4. Use four-digit random numbers to simulate "number of successes" 100 times. Compare your frequencies with the expected frequencies.

c6.47. By simulation, find the experimental sampling distribution of the mean of random samples of size three drawn from the distribution with density function

$$f(y) = \frac{8}{9}y \quad \text{for} \quad 0 \leq y \leq \frac{3}{2}$$

Hint: Use a probability histogram as an approximating distribution where the interval from 0 to 1.5 is divided into six intervals of equal length. Then follow the method of Example 6.14 to find 100 samples.

6.48. Suppose the heights in inches of six persons are 63, 65, 67, 69, 71, and 75, the first three being heights of women and the last three heights of men.
(a) Find the sampling distribution of the mean of the 15 random samples of size two.
(b) Find the sampling distribution of the mean of the 9 stratified random samples of size two where one woman and one man are selected.

(c) Suppose the three women represent one cluster and the three men a second cluster. The probability of selecting the first cluster is $\frac{2}{3}$ and a simple random sample of size two is taken from the selected cluster. Find the sampling distribution of the mean of the 6 possible samples.

(d) Find the means of the sampling distributions in parts (a), (b), and (c) and compare them with the population mean.

6.49. The final grades in a statistics course were 80, 89, 66, 88, 83, 60, 86, 63, 77, 79, 84, 77, 77, 78, 71, 93, 74, 52, 78, and 75.

(a) List the five possible systematic samples which can be taken from the list by starting with one of the first five grades.

(b) Find the sampling distribution of the mean of the samples obtained in part (a). Compare the mean of the parent population with the mean of the sampling distribution.

*(c) List the seven possible systematic samples which can be taken from the list by starting with one of the first seven grades. Then do part (b).

^{E}**6.50.** (a) Referring to Exercise 5.26, randomly select without replacement 14 triads from all possible triads which can be formed from the letters a, b, c, d, e, f, g, and make a frequency count of the 42 pair-wise comparisons of letters.

*(b) Select, if possible, 14 triads so that each pair of letters is represented exactly two times.

^{C}**6.51.** A population consists of 7, 10, 13, 16, 22, 25, 28 where the numbers designate incomes (in thousands of dollars) of seven families.

(a) Find the sampling distribution of the mean of random samples of size three drawn without replacement.

(b) Let "low" and "middle" income strata be 7, 10, 13, 16, and 22, 25, 28. Find the sampling distribution of the mean of the stratified random samples of size three where two observations are drawn without replacement from the first stratum and one observation is drawn from the second stratum.

(c) Suppose incomes grouped by three clusters (of houses) to be 7, 13; 10, 16, 25; and 22, 28. The probability of selecting a cluster is $\frac{1}{3}$ and a simple random sample of size two drawn without replacement is taken from the selected cluster. Find the sampling distribution of the mean.

7

ESTIMATION AND TESTS OF HYPOTHESES

7.1. INTRODUCTION

We have studied sampling distributions of simple random samples from a *single population*, and we have examined some of the important properties of these distributions. We learned that a sampling distribution may be used to make probability statements about a single random sample.

In this chapter we describe methods by which one simple random sample may be used to make a statement about a population from which the sample might have been drawn; i.e., *a sample is used to make an inference about a population*.

(7.1)   In general, **statistical inference** is any procedure by which an inference or generalization is made about a population (or populations) based on information obtained from a sample from the population (or populations).

We illustrate a few of these procedures by describing how inferences may be made about unknown population parameters (like the mean, proportion, or standard deviation) by computing statistics from random samples and then applying the appropriate sampling distribution. We do not attempt to study the whole field of statistical inference, but only a few simple aspects which may be considered central to our understanding of the subject.

Statistical inference may be divided into two primary areas, **estimation** and **tests of hypotheses.** To illustrate the difference we consider two simple examples. A pollster takes a stratified random sample of 2,000 potential voters and asks, "Which presidential candidate would you vote for if the election were today?" The proportion of persons in the sample favoring each candidate is considered an estimate of the corresponding true proportion of voters favoring each candidate (see Example 7.2 for further discussion). In the second example, a member of a market research team wants to know whether a new Jello flavor is preferred to the old Jello flavor. The researcher has the *hypothesis* (assumption subject to test) that the new flavor is preferred. After collecting information at random and making the necessary analysis of data, a decision is made as to which flavor is preferred. Note that in the second example, the researcher is not particularly interested in estimating a parameter (or parameters). Instead, he simply wants to know which Jello flavor is preferred. We will show how sampling distributions are used in both types of inference.

7.2. ESTIMATION

Each of us makes estimates every day. Often, perhaps almost always, an estimate is not the result of a conscious act. Certainly we are not usually concerned with properties or methods of estimation. We "just estimate"; i.e., we make a valuation based on opinion or on incomplete data. Some examples, casual or otherwise, are as follows:

1. Before attempting to cross a street each of us estimates our chances of getting across safely.

2. At the end of a school term students may estimate their final grade. Sometimes the teacher makes a different estimate.

3. A builder usually must give "material and cost estimates" before receiving a contract. Many subjective factors may influence his estimates—some could be his guess as to the amount of the low bid, the labor market during the next year, the price and availability of materials, and the risk of losing money.

4. An admissions committee estimates the academic abilities of a candidate before deciding to accept or reject the candidate. Again, many subjective factors may influence their decisions.

5. From a random sample a research group estimates the proportion of people who will vote for a specific candidate.

6. A study on smoking habits of students at Beta College reveals that from 37 to 45 percent of the girls smoke regularly.

7. An agriculture research worker estimates the yield of a variety of corn.

In each illustration the estimate may turn out to be either good or bad. The method of estimation may be desirable or undesirable, objective or not objective, based on sound statistical principles or not.

In Section 7.3 we describe those *statistical properties and methods* which are recognized as being good for a large class of problems. But first we take a closer look at three of the illustrations of estimation given above.

EXAMPLE 7.1. In the first illustration, all of us are likely to wait until our chances of getting across the street safely are very great. We call on our past experience and estimate, usually with a safe margin, whether we have time to cross the street before a car gets us. We do not take time to make specific measurements or to compute the exact number of seconds it will take an approaching car to arrive at the critical spot. Our experience has gradually taught us how to rapidly make the necessary estimates and to act accordingly. (We read about people who make bad estimates.) In this type problem, the estimate is not based on statistical procedure.

EXAMPLE 7.2. An estimate of the proportion of people who will vote for a specific presidential candidate, say candidate J, is often determined by an objective statistical procedure. Key steps include the preparation of questions (or a question), selection of potential voters, making personal interviews, analyzing and testing the data, and making a concluding statement. All of this requires training and skill.

There are some important things to watch for. The questions should be preceded by an introduction and a brief statement of purpose. Each sentence should be clear and to the point. Persons making the interviews should be trained so that they act and react in about the same way. Responses should be recorded in a clear and systematic manner. All interviews should be made over a short period of time. Once the data are collected, the analysis should be completed in a short period of time, and the *conclusion* should be stated in clear simple terms and related directly to the questions asked. Some problems of analysis are discussed briefly in the next four paragraphs.

Suppose 900 of 2,000 persons interviewed said they would vote for candidate J. One might be tempted to report that $\frac{900}{2,000} = 0.45$ of the people would vote for candidate J and, thus, conclude that J will lose the election. However, any conclusion should be based on *all* the responses. To illustrate the reason for this we consider two possibilities.

In the first case, suppose 1,000 persons said they would vote for candidate G and the remaining 100 said they were undecided. Disregarding the undecided responses, one might be led to believe that J has a good chance of losing the election since $\frac{900}{1,900} = .474$ favored J and $\frac{1,000}{1,900} = .526$ favored G. In the second case, suppose 700 persons said they would vote for candidate G and the remaining 400 said they were undecided. Disregarding the undecided responses, one might be led to believe that J has a good chance of winning the election since $\frac{900}{1,600} = .562$ favored J and $\frac{700}{1,600} = .438$ favored G. The undecided responses may be disregarded if it is reasonable to assume that at election time they will split in the same proportions as the decided cases.

Estimates like those of Example 7.2 are called **point estimates**, since each one is a single number; i.e., a single point on the real number scale. Even though an estimate of a parameter is commonly expressed as a point estimate (i.e., as a single value of a statistic), there is a serious disadvantage in this practice. For such an estimate does not indicate how close the value may be to the true unknown parameter value. Normally, when a point estimate is used it should be accompanied by a statement of the precision (variance) of the method of estimation.

It is better in practice to compute, if possible, a set of values, called a **confidence interval** (see Section 7.6), which have high probability of containing the true parameters being estimated. This might lead, in case one, to a statement like "we are 99 percent certain that from .434 to .514 of the votes will be for candidate J and from .486 to .566 will be for candidate G." From such a statement it becomes clear that J could win, but that G is more likely to win.

EXAMPLE 7.3. An agriculture research worker might base his estimates on several variables which affect the yield of corn. The yield in bushels per acre, y, might depend on such variables as amount of fertilizer, type of soil, amount of rain (or water), location of field, depth of planting, time of cultivation, and so forth. To illustrate, we take a very simple and restricted hypothetical situation.

Let x denote pounds of fertilizer per acre. Then the yield for a specific variety of corn might be estimated by the equation

$$y = 100 + 0.5x$$

where x may be any value from 0 to 200 pounds. Using this formula, a farmer who normally applies 100 pounds of fertilizer per acre would obtain the following point estimate of his yield:

$$y = 100 + 0.5(100) = 150 \text{ bushels per acre}$$

In the same way he could estimate the yield for any amount of fertilizer between 0 and 200 pounds per acre; the formula should not be used for more than 200 pounds per acre. As in Example 7.2, it is desirable that a measure of variation in yield be included. (We discuss methods of finding such predictive equations in Chapter 11.)

7.3. PROPERTIES OF POINT ESTIMATION

In Section 6.5 we introduced the expression "t is an unbiased estimator of θ." We observed that $\bar{y}$ is an unbiased estimator of μ, that s^2 is an unbiased estimator of σ^2, and that s is a biased estimator of σ. That is, $\bar{y}$, s^2, and s are estimators of the corresponding parameters. Also, we said that a value of an estimator is an estimate. Thus, $\bar{y} = 20$, $s^2 = 100$, and $s = 10$ are estimates of μ, σ^2, and σ, respectively. In Example 7.2, case one, the sample proportion 0.474 is an estimate of the true proportion μ of persons who will vote for candidate J. In Example 7.3, $y = 150$ is an estimate of the mean yield per acre when 100 pounds of fertilizer per acre are applied.

A *point estimate* is a single value of an estimator. Thus, each of the five estimates of the last paragraph is also a point estimate.

We might have several estimators of a single parameter. For example, the sample mean, sample median, and sample mid-range are all estimators of the population mean μ; a single value of each of these estimators is a point estimate of μ. Are any of these *good estimators*? Which, if any, is the best point estimator of μ? What are the meanings of terms like "good estimator" or "best estimator"?

Such questions are answered in terms of how all values of an estimator are concentrated near the true parameter. (Note that answers cannot be given in terms of a single sample, since the nearness of a particular value depends on the sample drawn.) If we can find a good estimator of a parameter, then we can find a good point estimate. Thus, we must have a clear understanding of what we mean by "good" when referring to an estimator. If possible, it would be desirable to have a *principle of estimation* which gives good estimators (see Exercise 7.15).

We discuss the goodness of an estimator in terms of unbiasedness and efficiency, two of four useful properties of estimation. These basic properties are discussed in Example 7.4 where the concept of best estimator is introduced.

EXAMPLE 7.4. Compare the mean, median, and mid-range of random samples of size three drawn without replacement from the finite population with observations 2, 5, 8, 11, 14.

Each sample of size three and its corresponding mean ($\bar{y}$), median (m), and mid-range (mr) are shown in Table 7.1. (See Exercises 3.10 and 3.12 for a definition and an illustration, respectively, of the mid-range.) The resulting sampling distributions (and columns needed to compute population variances) are shown in Tables 7.2, 7.3, and 7.4.

Table 7.1. Samples of Size Three Drawn Without Replacement from the Population with Observations 2, 5, 8, 11, 14

Sample	Mena ($\bar{y}$)	Median (m)	Mid-range (mr)
2 5 8	5	5	5
2 5 11	6	5	6.5
2 5 14	7	5	8
2 8 11	7	8	6.5
2 8 14	8	8	8
2 11 14	9	11	8
5 8 11	8	8	8
5 8 14	9	8	9.5
5 11 14	10	11	9.5
8 11 14	11	11	11

The population mean is $\mu = 8$. From Table 7.1 the reader should observe that, for the third sample the mid-range is actually equal to the population mean, and for the fourth sample the median is the same as the population mean.

Table 7.2. Sampling Distribution of the Mean of Samples of Size Three Drawn Without Replacement from the Population of Example 7.4 and Some Calculations

$\bar{y}$	f	$u = \bar{y} - 8$	fu	fu^2
5	1	−3	−3	9
6	1	−2	−2	4
7	2	−1	−2	2
8	2	0	0	0
9	2	1	2	2
10	1	2	2	4
11	1	3	3	9
Totals	10		0	30

Table 7.3. Sampling Distribution of the Median of Samples of Size Three Drawn Without Replacement from the Population of Example 7.4 and Some Calculations

m	f	$v = (m - 8)/3$	fv	fv^2
5	3	−1	−3	3
8	4	0	0	0
11	3	1	3	3
Totals	10		0	6

Table 7.4. Sampling Distribution of the Mid-range of Samples of Size Three Drawn Without Replacement from the Population of Example 7.4 and Some Calculations

mr	f	$w = (mr - 8)/1.5$	fw	fw^2
5	1	−2	−2	4
6.5	2	−1	−2	2
8	4	0	0	0
9.5	2	1	2	2
11	1	2	2	4
Totals	10		0	12

For all samples the only cases where the mean is equal to the population mean is when it has the same value as the median and mid-range.

Using the methods of Chapter 6 we find from Table 7.2 that

$$E(\bar{y}) = 8$$

and
$$V(\bar{y}) = V(u) = 30/10 = 3$$

Further, from Tables 7.3 and 7.4 we find that

$$E(m) = 8$$
$$V(m) = 3^2 V(v) = 9(6/10) = 5.4$$
$$E(mr) = 8$$
$$V(mr) = (3/2)^2 V(w) = (9/4)(12/10) = 2.7$$

Since

$$E(\bar{y}) = E(m) = E(mr) = 8 = \mu$$

each of the three estimators under consideration is unbiased in this case (where samples of size three are drawn *without replacement* from a *symmetric and finite* population). The means of the sampling distributions are the same, but the variances are different. Actually, arranged in increasing order of magnitude we have

$$V(mr) < V(\bar{y}) < V(m)$$

Since the mid-range varies less from sample to sample than either the mean or the median we say the mid-range is *more efficient* than either mean or median. Of the three estimators, the mid-range is *most efficient*. (Also, $\bar{y}$ is more efficient than m.) Further, if each estimator in a set is unbiased, we say that estimator with minimum variance is the *best estimator*. It can be shown that similar results hold when samples of size three are drawn *with replacement* from the same population. In particular, when $n = 3$ the reader should be able to show that

$$E(\bar{y}) = E(m) = E(mr) = 8 = \mu$$

and

$$V(\bar{y}) = 6 \qquad V(m) = 11.952 \qquad V(mr) = 5.256$$

We now make more formal (and more general) statements of the above principles. Let t_n and t'_n denote functions of a random sample of size n. As a basis for selecting an estimator of a parameter θ the two following definitions are very useful:

I. The efficiency of an unbiased estimator t_n of θ relative to any other unbiased estimator t'_n of θ is defined by the ratio

$$\frac{V(t'_n)}{V(t_n)}$$

II. If t_n is an unbiased estimator of θ and t'_n is any other unbiased estimator of θ, and if

$$V(t_n) \leq V(t'_n) \text{ for every } t'_n$$

then t_n is called a **best unbiased estimator** of the parameter θ.

Note that Definitions I and II are restricted to unbiased estimators. This is an unnecessary restriction, but for our purposes the definitions are adequate.

In Example 7.4, we indicated which estimators were *more efficient* than others. With Definition I we actually have a measure of relative efficiency of one estimator to another. In Example 7.4, the efficiency of $\bar{y}$ relative to m is

$$\frac{V(m)}{V(\bar{y})} = \frac{5.4}{3} = 1.80$$

the efficiency of mr relative to $\bar{y}$ is

$$\frac{V(\bar{y})}{V(mr)} = \frac{3}{2.7} \doteq 1.11$$

and the efficiency of m relative to mr is

$$\frac{V(mr)}{V(m)} = \frac{2.7}{5.4} = 0.50$$

The first ratio indicates that the sample mean is 1.80 times as efficient as the sample median; the third ratio indicates that the sample median is one-half as efficient as the mid-range.

EXAMPLE 7.5. Compare the mean $\bar{y}$ and median m of random samples of size n drawn from a normal population with mean μ and variance σ^2.

We know that $\bar{y}$ and m are unbiased estimators of the population mean μ and that

$$V(\bar{y}) = \frac{\sigma^2}{n}$$

Further, for large n, it can be shown that

(7.2) $$V(m) \doteq \frac{\pi}{2} \times \frac{\sigma^2}{n}$$

Thus,

$$\frac{V(m)}{V(\bar{y})} \doteq \frac{\pi}{2} \doteq 1.57$$

so the efficiency of $\bar{y}$ relative to m is approximately 1.57 when sampling from a normal population. That is, *the sample mean is relatively more efficient than the sample median for estimating the mean of a normal distribution.*

Put another way, if $\bar{y}$ is based on a sample of size n and m is based on a sample of size n', then

$$\frac{V(m)}{V(\bar{y})} = 1$$

when

$$\frac{\pi\sigma^2}{2n'} = \frac{\sigma^2}{n}$$

or

(7.3) $$n' = \frac{\pi}{2} n$$

In particular, when $n = 64$, $n' \doteq 100$. Thus, if follows that a median based on 100 observations is about as efficient as a mean based on 64 observations.

In Example 7.4 we found that the sample mean is not necessarily the best estimator for the population mean. But it can be shown that when the parent population is normal and the sample is randomly drawn, the sample mean is a best unbiased estimator of the population mean. Thus, since the normal distribution plays such a central role in statistics, it should be clear why the sample mean $\bar{y}$ is such an important measure of location. Also, it can be shown that while $\bar{y}$ is an unbiased estimator of a population mean, both the median and mid-range are unbiased only in special cases.

7.4. EXERCISES

In terms of each of the places indicated in Exercises 7.1 through 7.6 describe an experiment, the parameter (or characteristic) to be estimated, the estimator selected, and the procedure for finding a value of the estimator. Also, indicate how your estimate might be useful.

7.1. A college library

7.2. A classroom

7.3. A parking lot

7.4. A waiting line at meal time

7.5. Mowing of grass outside a classroom

7.6. Waiting room of a dentist

Exercises

7.7. Compare the mean, median, and mid-range of random samples of size three drawn with replacement from the population with observations 1, 4, 7, 10. Using the results of Exercises 6.4(b), 6.8(a), and 6.9(a), compare the three statistics as estimators of the population mean. *Hint:* Start with $E(\bar{y}) = 5.5$, $V(\bar{y}) = \frac{15}{4}$, $E(m) = 5.5$, $V(m) = \frac{63}{8}$, $E(mr) = 5.5$, and $V(mr) = \frac{207}{64}$.

7.8. Compare the sampling distributions of the mean, median, and mid-range of Example 7.4 by
(a) constructing histograms, and
(b) constructing less-than ogives.

c7.9. Random samples of size two are drawn with replacement from the population with observations 1, 3, 5, 7. Let y_1 denote the first sample observation and y_2 the second sample observation. Let

$$t_1 = \frac{y_1 + y_2}{2} \qquad t_2 = \frac{y_1 + 2y_2}{3}$$

$$t_3 = \frac{y_1 + 3y_2}{4} \qquad t_4 = y_2$$

(a) Which, if any, of the estimators t_1, t_2, t_3, and t_4 are unbiased estimators of the population mean?
(b) Which of the unbiased estimators is the best estimator?
(c) Find the efficiency of t_1 relative to each of the other unbiased estimators; find the efficiency of t_2 relative to each of the other unbiased estimators.

c7.10. Random samples of size two are drawn with replacement from the population with values 1, 3, 5, 7 (see Exercise 7.9). Let u_1 denote the smaller sample value and let u_2 the larger sample value. Let

$$t'_1 = \frac{u_1 + u_2}{2} \qquad t'_2 = \frac{u_1 + 2u_2}{3} \qquad t'_3 = u_2$$

(a) Which, if any, of the estimators t'_1, t'_2, t'_3 are unbiased estimators of the population mean μ?
(b) The *second moment of the variable* t'_1 *about the population mean* μ is defined by

$$M_1^2 = \frac{\sum_{j}^{N}(t'_{1j} - \mu)^2}{N}$$

The second moments M_2^2 and M_3^2 of t'_2 and t'_3, respectively, are defined in a similar way. Find M_1^2, M_2^2, and M_3^2 and rank them.
(c) On page 218 the definition of relative efficiency required that both estimators of θ be unbiased. If one or both estimators are biased, relative efficiency may be similarly defined as the ratio of second moments. Use the second moments to discuss the relative efficiency of the estimators t'_1, t'_2, and t'_3.
(d) For each estimator verify that $M^2 = \sigma^2 + [E(t') - \mu]^2$.

E7.11. Random samples of size three are drawn without replacement from the population with observations 2, 5, 8, 14, 23. The sample mean, median, and mid-range are to be compared.

(a) Which, if any, of the three statistics is an unbiased estimator of the population mean? The population median? The population mid-range?

(b) Compare the efficiency of the three statistics as estimators of the population mean. *Hint:* Use second moments of Exercise 7.10.

(c) In the definition of an unbiased statistic we have preferred the mean (with good reasons) over all other measures of location. But other definitions of an unbiased statistic might be acceptable. Suppose we had said a statistic t is an *unbiased prime estimator* of a parameter θ of the parent population provided the *median of the sampling distribution* of t is equal to the parameter θ. Which, if any, of the three statistics is an unbiased prime estimator of the population mean? The population median? The population mid-range?

(d) If the mid-range of a sampling distribution of a statistic t is equal to a parameter θ of the parent population, we say t is an unbiased double prime estimator of θ. Which, if any, of the three statistics is an unbiased double prime estimator of the population mean? The population median? The population mid-range?

(e) The *second moment of a statistic* t *about the parent population median* M is defined by

$$\frac{\sum_{i=1}^{N}(t_i - M)^2}{N}$$

where N denotes the number of observations in the sampling distribution of t. Use this definition of variation to compare the efficiency of the three statistics as estimators of the population median.

(f) Replace "median" by "mid-range" in part (e) and make the comparisons.

E7.12. Random samples of size three are drawn with replacement from the population with observations 1, 4, 10, 19. The sample mean, median, and mid-range are to be compared. Do parts (a), (b), (c), (d), (e), and (f) of Exercise 7.11 given the following sampling distributions:

$\bar{y}$	1	2	3	4	5	6	7	8	9	10	11	13	14	16	19
f	1	3	3	4	6	3	6	9	3	7	6	6	3	3	1

m	1	4	10	19
f	10	22	22	10

mr	1	2.5	4	5.5	7	10	11.5	14.5	19
f	1	6	1	12	6	19	12	6	1

*7.13. (a) Show that the variance of the population with the three values a, b, c is

$$\sigma^2 = \frac{(a-b)^2 + (b-c)^2 + (c-a)^2}{9}$$

(b) Let

$$t = wy_1 + (1-w)y_2$$

where w is any real number in the interval $0 \leq w \leq 1$ and y_1 and y_2 denote the first and second sample observations, respectively, of a random sample of size two drawn with replacement from the population a, b, c. Show that

$$E(t) = \mu = \frac{a+b+c}{3}$$

and

$$V(t) = [w^2 + (1-w)^2]\sigma^2$$

(c) Let

$$t' = wu_1 + (1-w)u_2$$

where w is any real number in the interval $0 \leq w \leq 1$ and u_1 and u_2 are defined as in Exercise 7.10. For $a \leq b \leq c$ show that

$$E(t') = \frac{(1+4w)a + 3b + (5-4w)c}{9}$$

and

$$V(t') = [4(2 - 2w + 5w^2)a^2 + 18(1 - 2w + 2w^2)b^2$$
$$+ 4(5 - 8w + 5w^2)c^2 + 6(-1 + 2w - 6w^2)ab$$
$$+ 6(-5 + 10w - 6w^2)bc + 2(-5 + 2w - 2w^2)ac]/81$$

(d) Let $w = \frac{1}{2}$ in (b) and (c). Show that t' is an unbiased estimator of μ. Also, show that

$$V(t) = V(t') = \frac{\sigma^2}{2}$$

(e) Let $w = 0$ in (b) and (c). Discuss the estimators t and t' in terms of the means and variances of their distributions.

E7.14. The goodness of an estimator is often defined in terms of the following four criteria:

    unbiasedness   efficiency
    consistency   sufficiency

We introduced the first two criteria in Section 7.3. Now, we briefly consider *consistency* and *sufficiency*.

An estimator t of a population parameter θ is said to be a *consistent esti-mator* of θ if the probability that t has a value which differs from θ by more than any arbitrary positive constant approaches zero when n approaches infinity.

Roughly speaking, t is a consistent estimator of θ provided almost all values of t are concentrated arbitrarily close to θ when n becomes large. Clearly, $\bar{y}$ is a consistent estimator of μ. [For as n approaches infinity the standard error $\sigma/\sqrt{n}$ approaches zero and almost all values are concentrated in an arbitrarily small interval about μ, since $E(\bar{y}) = \mu$.] In fact, when the parent population is normal, $\bar{y}$ is a consistent estimator of the population median M, since $M = \mu$.

An estimator t is said to be a *sufficient estimator of a population parameter* θ provided it "utilizes all the information a sample contains about θ."

The precise meaning of the expression "utilizes all the information a sample contains about θ" is usually expressed in rather complicated mathematical terms and must be omitted here. However, a practical way to understand the concept of sufficiency is to note that in order to utilize all available sample information a statistic must include in its formula all the sample values. So the range, median, and mid-range cannot be sufficient because they utilize only the largest and smallest or the middle sample values. On the other hand, the mean $\bar{y}$ and variance s^2 may be sufficient because they include in their formulas all the sample values.

The criterion of consistency is not very useful in practice since it is defined in terms of samples of very large size. Also, for example, if t_n is a consistent estimator of θ, then so is

$$\left(\frac{n+a}{n+b}\right)t_n$$

where $a \neq -n$ and $b \neq -n$ are any fixed real numbers a and b.

(a) Show that $\left(\dfrac{n+100}{n}\right)\bar{y}$ is a consistent estimator of μ.

(b) Compute the statistic in part (a) for $n = 2, 3, \cdots, 10$.

E7.15. A question often asked is, "How does one find an estimator of a given population parameter with some or all of the desirable properties listed in Exercise 7.14?" An answer could lead to the comparison of many different methods of finding estimators. We briefly consider two primary methods of estimation, the **method of least squares** (see Chapter 11) and the **method of maximum likelihood.**

Generally, the maximum likelihood method yields estimators which are consistent, efficient, or sufficient when the parameter being estimated has an estimator with one or more of these properties. For example, if the parameter being estimated has a sufficient estimator, the maximum likelihood estimator will be a sufficient estimator. However, the method of maximum likelihood does not usually yield unbiased estimators.

We illustrate the method with an extremely simple problem. Suppose coins 1, 2, and 3 look exactly alike, but have probabilities of 0.3, 0.4, and 0.5, respectively, of a head showing on a single toss. If one head and one tail are obtained in two tosses of one coin, which coin was tossed? That is, what is the maximum likelihood estimate of the parameter π, the probability of getting a head?

The *method of maximum likelihood selects that value of π for which the probability of obtaining a given sample is a maximum.* Clearly, we will need the probabilities of simple events of the sample space of each coin. Let HH denote head on toss one and head on toss two, HT denote head on toss one and tail on toss two, and so on. Since the two trials are independent, the probabilities for the four simple events of coin 1 are

$$P(HH) = P(H)P(H) = (0.3)(0.3) = 0.09$$
$$P(HT) = P(H)P(T) = (0.3)(0.7) = 0.21$$
$$P(TH) = 0.21 \qquad P(TT) = 0.49$$

Similarly the probabilities for coin 2 are

$$P(HH) = 0.16 \qquad P(HT) = P(TH) = 0.24 \qquad P(TT) = 0.36$$

and the probabilities for coin 3 are

$$P(HH) = 0.25 \qquad P(HT) = P(TH) = 0.25 \qquad P(TT) = 0.25$$

Thus, the probability of one head and one tail is

$$P(HT) + P(TH) = 0.21 + 0.21 = 0.42 \text{ for coin 1}$$

Similarly, for coins 2 and 3 the corresponding probabilities are $0.24 + 0.24 = 0.48$ and $0.25 + 0.25 = 0.50$, respectively. Since the maximum probability is for coin 3, our estimate of the probability of heads on a single toss is 0.5.

(a) Suppose four similar coins have probabilities 0.2, 0.3, 0.4, and 0.7, respectively, of a head showing on a single toss. If one head and one tail are obtained in two tosses of one coin, what is the maximum likelihood estimate of π?

(b) Use the four coins of part (a). What is the maximum likelihood estimate of π if one coin is tossed three times and one head and two tails are obtained?

7.5. RELATIONSHIPS OF ESTIMATES TO SAMPLE SIZE, ERROR, AND RISK

We have already observed that a point estimate does not indicate how close the estimate is to the true value. Thus, it is highly desirable that supporting information be given so that the merits of the estimate may be assessed. For example, the size of the sample and the standard deviation of the parent population would be useful in judging the precision of an estimate of the population mean. The next five examples should be useful in understanding how supplementary information relates to the problems of estimation. (These illustrations are in terms of the sample mean since we presently know more about the sampling distribution of the mean than we do about other sampling distributions. However, the methods presented may easily be extended to other sampling distributions as they are introduced later.)

EXAMPLE 7.6. Suppose the population of verbal scores of a large group of college students has a mean of 580 and a standard deviation of 60. (a) What is the probability that a random sample of 36 students will have a mean score less than 560? (b) What is the probability that the sample mean will deviate more than 20 units from the population mean?

Since $n = 36$ is fairly large, we may use the Central Limit Theorem and assume the sample mean is approximately normally distributed with mean 580 and standard error $60/\sqrt{36} = 10$. Thus, it follows that

$$P(\bar{y} < 560) = P\left(\frac{\bar{y} - \mu}{\sigma_{\bar{y}}} < \frac{560 - 580}{10}\right)$$
$$= P(z < -2) = 0.0228$$

So for part (a) the **error e in estimation** is

$$e = \bar{y} - \mu = 560 - 580 = -20$$

and the probability is 0.0228 that the sample mean will be more than 20 units below the population mean. We may also say that there are approximately 2 chances in 100 that the sample mean is below 560.

Due to symmetry we know the probability is 0.0228 that the sample mean will be more than 20 units above the population mean. Thus we may combine the two statements and say the probability is

$$0.0228 + 0.0228 = 0.0456$$

that the **error in the estimate** will be greater than 20 units.

In Example 7.6 we illustrated the relationship between the error in the estimate and the **risk** (i.e., chance) of such an error when the population mean is known. In practice, μ is not known, and often σ is not known. However, the sample standard deviation s may often be used as an approximation of σ when the sample size is greater than 30. The following examples are more realistic than Example 7.6.

EXAMPLE 7.7. If a random sample of size 49 is taken from a population with standard deviation 28, what is the chance that the point estimate of the population mean will be in error by more than 6 units?

The sample size is large enough that we may assume the sample mean to be approximately normally distributed. The standard error of the mean is

$$\sigma_{\bar{y}} = \frac{\sigma}{\sqrt{n}} = \frac{28}{\sqrt{49}} = 4$$

Since the absolute value of the minimum error is

$$|e| = |\bar{y} - \mu| = 6$$

the absolute value of the minimum error in standard units is

$$z' = \frac{|\bar{y} - \mu|}{\sigma_{\bar{y}}} = \frac{6}{4} = 1.50$$

Thus, the required probability is

$$P(z < -1.50 \text{ or } z > 1.50) = 2P(z < -1.50)$$
$$= 2(0.0668) = 0.1336$$

So the risk is approximately 0.1336 that the error in estimating the population mean will be 6 or more units.

An experimenter often asks what sample size is recommended. It is possible to find the minimum sample size required if both the size of the error in estimating the population mean and the risk of such an error are specified. Example 7.8 illustrates the method.

EXAMPLE 7.8. Suppose the standard deviation of the diameter of a stand of Douglas fir trees is 1.5 inches. A random sample of trees is to be used to estimate the mean diameter of all trees in the stand. What is the smallest sample one can take if the probability is 0.95 that the error in the estimate of the population mean will not exceed 0.3 inches?

Assuming normality, we know that if

$$P(-z' < z < z') = 0.95$$

then

$$P(z < -z') = 0.025$$

and

$$z' = 1.96$$

Since

$$z' = \frac{\bar{y} - \mu}{\sigma_{\bar{y}}} = \frac{e}{\frac{\sigma}{\sqrt{n}}} = \frac{e\sqrt{n}}{\sigma}$$

it follows that

$$1.96 = \frac{0.3\sqrt{n}}{1.5}$$

Thus

$$\sqrt{n} = \frac{(1.5)(1.96)}{0.3} = 9.80$$

and

$$n = 96.04$$

So the required sample size is 97, rounded up to the next whole number.

Not every statistic has a symmetric sampling distribution. Futher, the maximum specified error of estimation is not always the same in both directions. So we give two illustrations to suggest how the above methods may be extended to both symmetric and skewed distributions.

EXAMPLE 7.9. A random sample of size 64 was drawn from a normal population with mean 47 and standard deviation 12. If the sample mean is used to estimate the population mean, what is the probability that it does not underestimate by more than 2 units or overestimate by more than 3 units?

The standard error of $\bar{y}$ is

$$\sigma_{\bar{y}} = \frac{\sigma}{\sqrt{n}} = \frac{12}{\sqrt{64}} = 1.5$$

Thus, the probability that the error $e = \bar{y} - \mu$ is between -2 and 3 units is

$$P(-2 < \bar{y} - \mu < 3) = P\left(\frac{-2}{1.5} < \frac{\bar{y} - \mu}{\sigma_{\bar{y}}} < \frac{3}{1.5}\right)$$
$$\doteq P(-1.3 < z < 2.0)$$
$$= P(2.00) - P(-1.3)$$
$$= .9772 - .0968$$
$$\doteq .88$$

That is, the probability is approximately .88 that the sample mean will fall between $47 - 2 = 45$ and $47 + 3 = 50$. Further, $1 - .88 = .12$ is the chance (risk) of the error of estimation of the population mean being less than or equal to -2 or greater than or equal to 3 units.

EXAMPLE 7.10. A dichotomous population has 40 percent successes. What is the chance that in a random sample of size 100 the proportion of successes will be less than .50 and more than .30? Such a result will make the point estimate of the proportion of successes in the population in error by less than 0.1.

Relationships of Estimates to Sample Size, Error, and Risk

Suppose the number of successes T is distributed as a binomial distribution (not symmetric unless $\pi = \frac{1}{2}$). Assume the sample size is large enough to use the normal approximation. Then the standardized statistic

$$\frac{T - n\pi}{\sqrt{n\pi(1-\pi)}} = \frac{\frac{T}{n} - \pi}{\sqrt{\frac{\pi(1-\pi)}{n}}}$$

is approximately distributed as a standard normal variable z (see Section 5.11). Let $p = T/n$ denote the *proportion of successes in a sample*. The reader should note that for a fixed n a value of the statistic p has the same probability of occurring as the corresponding value of T. Thus, the proportion p is also distributed as a binomial distribution with mean

(7.4) $$\mu_p = \pi$$

and variance

(7.5) $$\sigma_p^2 = \frac{\pi(1-\pi)}{n}$$

We require

$$P(-0.1 < p - \pi < 0.1)$$

Since

$$\mu_p = \pi = 0.40$$

it follows that

$$\sigma_p = \sqrt{\frac{\pi(1-\pi)}{n}} = \sqrt{\frac{(.40)(.60)}{100}} \doteq 0.0490$$

and

$$P(-0.1 < p - \pi < 0.1) = P\left(\frac{-0.1}{0.0490} < \frac{p - \mu_p}{\sigma_p} < \frac{0.1}{0.0490}\right)$$
$$\doteq P(-2.04 < z < 2.04)$$
$$= 1 - 2P(z < -2.04)$$
$$= 1 - 2(0.0208), \text{ by interpolation}$$
$$= .9584$$

That is, the probability is approximately 0.96 that the sample proportion will fall between .30 and .50. [Further, $1 - .96 = .04$ is the chance (risk) of the error of estimation of the population proportion being greater than or equal to 0.1.]

Now we summarize and generalize the primary procedures of these last four exercises. Let z_1 denote the value of the standard normal random variable

z for which the area to its left is α_1, and let z_2 denote the value of z for which the area to its right is α_2 where z_2 is greater than z_1. Then $1 - \alpha_1 - \alpha_2$ is the area under the normal curve between z_1 and z_2. That is

(7.6) $$P(z_1 < z < z_2) = 1 - \alpha_1 - \alpha_2$$

Due to the symmetry of the normal distribution we usually let $\alpha_1 = \alpha_2 = \alpha/2$. In which case it is customary to let

$$z_2 = z_{\alpha/2}$$

Hence

$$z_1 = -z_2 = -z_{\alpha/2}$$

So Equation (7.6) becomes

(7.7) $$P(-z_{\alpha/2} < z < z_{\alpha/2}) = 1 - \alpha$$

Since

(7.8) $$z = \frac{\bar{y} - \mu}{\sigma/\sqrt{n}} = \frac{e}{\sigma/\sqrt{n}}$$

we may write Equation (7.7) as

(7.9) $$P\left(-z_{\alpha/2} < \frac{e}{\sigma/\sqrt{n}} < z_{\alpha/2}\right) = 1 - \alpha$$

or as

(7.10) $$P\left(-z_{\alpha/2}\frac{\sigma}{\sqrt{n}} < e < z_{\alpha/2}\frac{\sigma}{\sqrt{n}}\right) = 1 - \alpha$$

Equation (7.10) indicates that

(7.11) *The probability is $1 - \alpha$ that the error in the estimate of the mean is less than $z_{\alpha/2}(\sigma)/\sqrt{n}$; i.e., the probability is $1 - \alpha$ that the estimate $\bar{y}$ differs from the population mean μ by less than $z_{\alpha/2}(\sigma)/\sqrt{n}$ units.*

In particular, if $\alpha = 0.05$ we know that

$$z_{\alpha/2} = z_{0.025} = 1.96$$

and therefore we can say with probability $1 - 0.05 = 0.95$ that the error of estimating μ is less than $1.96(\sigma)/\sqrt{n}$.

Referring to Example 7.8, we might want to find the smallest size of a random sample which will allow us with probability $1 - \alpha$ to estimate the population mean with error less than e. This can easily be done by substituting $z_{\alpha/2}$ for z in Equation (7.8) and solving to get

$$z_{\alpha/2} = \frac{e\sqrt{n}}{\sigma}$$

or

$$\sqrt{n} = \frac{z_{\alpha/2}\sigma}{e}$$

or

(7.12) $$n = \left(\frac{z_{\alpha/2}\sigma}{e}\right)^2$$

The value of n given by Equation (7.12) will seldom be an integer. When the right-hand side of Equation (7.12) is not an integer, then n is taken as the next highest integer (see solution to Example 7.8).

In many practical problems σ is not known. However, when n is greater than 30 the sample standard deviation s may often be used as a satisfactory approximation of σ in Equations (7.8), (7.9), (7.10), and (7.12). Otherwise, the reader should use the methods of Chapter 10.

7.6. CONFIDENCE INTERVAL

So far we have considered estimation of a parameter θ with a single value. Now we introduce estimation of a parameter θ by an interval of the form

(7.13) $$\hat{\theta}_1 < \theta < \hat{\theta}_2$$

where $\hat{\theta}_1$ and $\hat{\theta}_2$ are values obtained from sample observations. Such estimation is called **interval estimation** and is quite common in practice.

When a probability statement is associated with an interval estimate as in Expression (7.13), we call it a **confidence interval.** The values $\hat{\theta}_1$ and $\hat{\theta}_2$ are called **confidence limits**, and the probability, say $1 - \alpha$, associated with Expression (7.13) is called the **degree of confidence** or **confidence coefficient.** The **length of the interval** is $\hat{\theta}_2 - \hat{\theta}_1$.

At the end of Example 7.2 we find a typical statement of confidence interval estimation. It is "we are 99 percent certain that from 0.434 to 0.514 of the votes will be for candidate J." The degree of confidence is 0.99, and the confidence limits are 0.434 and 0.514. If π denotes the true proportion who will vote for candidate J, then

$$0.434 < \pi < 0.514$$

is the 99 percent confidence interval of π. Our particular confidence interval was based on a particular random sample. Another random sample would most likely give a different confidence interval. Since an interval varies with a random sample we might call an interval a **random interval.** Some intervals will contain the parameter π; others will not. Since the degree of confidence is 0.99, the limits of the interval are selected so that in the long run 99 out of every 100 intervals contain the true π. Of course, a particular interval either does or does not contain the true parameter π, but we know we have a 0.99 chance of selecting a sample with a 99 percent confidence interval that contains π. Thus, it is natural to say we are 99 percent confident that the parameter falls in a particular interval.

Now we use the sample mean and the normal distribution to illustrate the procedure for getting a $(1 - \alpha)100$ percent confidence interval. From Equation (7.10) we can say with probability $1 - \alpha$ that the difference $e = \bar{y} - \mu$ falls in the interval

(7.14) $$-z_{\alpha/2} \frac{\sigma}{\sqrt{n}} < \bar{y} - \mu < z_{\alpha/2} \frac{\sigma}{\sqrt{n}}$$

By applying simple properties of inequalities to Inequality (7.14) we obtain

(7.15) $$\bar{y} - z_{\alpha/2} \frac{\sigma}{\sqrt{n}} < \mu < \bar{y} + z_{\alpha/2} \frac{\sigma}{\sqrt{n}}$$

and we can say with probability $1 - \alpha$ that Inequality (7.15) is satisfied for a given random sample. Since the confidence limits of μ in Inequality (7.15) are symmetric about $\bar{y}$ we *call Inequality (7.15) a symmetric $(1 - \alpha)100$ percent confidence interval of μ.*

EXAMPLE 7.11. A random sample of 49 male college students has a mean height of 70.4 inches and a standard deviation of 2.2 inches. Find a 99 percent confidence interval of the mean of the population from which the sample was drawn.

Since the sample size is greater than 30, we assume the sample mean $\bar{y}$ is approximately normally distributed. Further, we feel that the population standard deviation σ may be replaced by the sample standard deviation s. Thus, the standard error of the mean is approximately

$$\sigma_{\bar{y}} \doteq \frac{s}{\sqrt{n}} = \frac{2.2}{\sqrt{49}} \doteq 0.314$$

Since $1 - \alpha = 0.99$, $\alpha/2 = 0.005$, $z_{0.005} = 2.576$, and

$$z_{0.005} \sigma_{\bar{y}} = (2.58)(0.314) \doteq 0.810$$

The 0.99 confidence limits are

$$\bar{y} \pm z_{0.005} \sigma_{\bar{y}} = 70.4 \pm 0.81 = 69.59 \text{ and } 71.21$$

So an approximation to the required symmetric 99 percent confidence interval is

$$69.6 \text{ inches} < \mu < 71.2 \text{ inches}$$

The length of a symmetric $(1 - \alpha)100$ percent confidence interval of the mean as obtained from Inequality (7.15) is

(7.16) $$\left(\bar{y} + z_{\alpha/2}\frac{\sigma}{\sqrt{n}}\right) - \left(\bar{y} - z_{\alpha/2}\frac{\sigma}{\sqrt{n}}\right) = 2z_{\alpha/2}\frac{\sigma}{\sqrt{n}}$$

For Example 7.11, the length of the symmetric 0.99 confidence interval is $2(2.58)(0.314) = 2(0.810) = 1.62$, and the length of the symmetric 0.95 confidence interval is

$$2z_{0.025}\frac{\sigma}{\sqrt{n}} = 2(1.96)(0.314) = 1.23$$

This illustrates a general fact, *that as the degree of confidence increases, the confidence interval becomes longer and thus gives less information on the specific value of the parameter being estimated.*

Most of the sampling distributions studied in the remaining chapters are nonsymmetric, and this may lead to the application of nonsymmetric confidence intervals. To illustrate any difference in procedure we now take a look at nonsymmetric confidence intervals for means of normal populations.

Starting with Equation (7.6), we may obtain

(7.17) $$P\left(z_1\frac{\sigma}{\sqrt{n}} < \bar{y} - \mu < z_2\frac{\sigma}{\sqrt{n}}\right) = 1 - \alpha_1 - \alpha_2$$

in much the same way that Equation (7.10) was derived. Then by applying properties of inequalities we find (see Exercise 7.31)

(7.18) $$\bar{y} - z_2\frac{\sigma}{\sqrt{n}} < \mu < \bar{y} - z_1\frac{\sigma}{\sqrt{n}}$$

and we can say with probability $1 - \alpha_1 - \alpha_2$ that Inequality (7.18) is satisfied for a given random sample. We call Inequality (7.18) a *nonsymmetric* $(1 - \alpha_1 - \alpha_2)100$ *percent confidence interval of the population mean* μ whenever $\alpha_1 \neq \alpha_2$. [If $\alpha_1 = \alpha_2 = \alpha/2$, Inequality (7.15) is a special case of Inequality (7.18).]

Figure 7.1 shows the confidence interval (heavy line) and illustrates how the distributions corresponding to the two *confidence limits* relate to the sample mean. The sample mean $\bar{y}$ was used to determine the limiting population means, and the mean of every normal distribution between the two curves drawn is a candidate for the true mean. Of course, those in the vicinity of $\bar{y}$ are more likely candidates than those near the ends of the interval. Note that we have constructed a figure with α_1 smaller than α_2 and that there is a z-scale axis

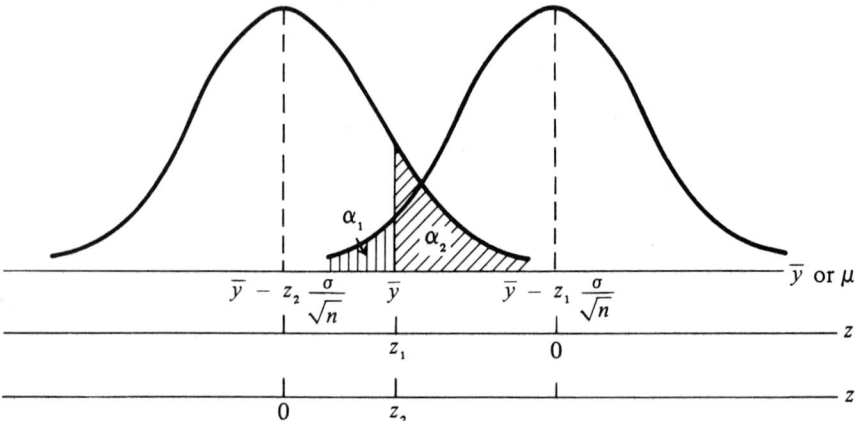

Figure 7.1. Limit Curves for a Confidence Interval of Means

corresponding to each normal curve. Also, note why the standardized values of z, namely z_1 and z_2, have switched sides in going from Statement (7.17) to Statement (7.18).

EXAMPLE 7.12. Let $\alpha_1 = 0.01$ and $\alpha_2 = 0.04$, and use the data of Example 7.11 to find a nonsymmetric 95 percent confidence interval of the population mean. Also, compare the lengths of the symmetric and nonsymmetric 95 percent confidence intervals.

From Example 7.11 we know that $\sigma_{\bar{y}} \doteq 0.314$. From Table II we find that

$$z_1 = -z_{0.01} = -2.33 \quad \text{and} \quad z_2 = z_{0.04} = 1.75$$

Thus

$$z_1 \sigma_{\bar{y}} \doteq -0.732 \qquad z_2 \sigma_{\bar{y}} \doteq 0.550$$
$$\bar{y} - z_1 \sigma_{\bar{y}} = 70.4 - (-0.73) = 71.13 \qquad \bar{y} - z_2 \sigma_{\bar{y}} = 70.4 - 0.55 = 69.85$$

So the required 95 percent confidence interval of μ is

$$69.85 < \mu < 71.13$$

The length of the nonsymmetric 95 percent confidence interval is

$$z_2 \sigma_{\bar{y}} - z_1 \sigma_{\bar{y}} = 0.55 - (-0.73) = 1.28$$

Thus, the nonsymmetric 0.95 confidence interval is longer than the symmetric 0.95 confidence interval (which has length 1.23). This illustrates a general property. For it can be shown that *for a fixed confidence coefficient and a fixed* sample size, the symmetric confidence interval of the mean of a normal distribution is shorter than a nonsymmetric confidence interval. This explains why we use symmetric intervals for means of normal distributions.

7.7 EXERCISES

7.16. A random sample of size 16 was drawn from a normal population with standard deviation 5. If the sample mean is 32.4, find the population mean for which $P(\bar{y} < 32.4) = 0.10$.

7.17. A random sample of size 40 was drawn from a population with variance 10. If the sample mean is 52.8 and the probability of getting a larger sample mean is 0.05, what is the population mean?

7.18. If a random sample of size 49 was taken from a population with standard deviation 28, what is the chance that the point estimate of the mean will be in error by less than 8 units?

7.19. The standard deviation of a standardized mathematics test is 100 and the mean is unknown. What is the chance that a random sample of 50 students will have a mean in error by more than 20 points?

7.20. A random sample of size 8 was drawn from a normal population with variance 72. The sample mean is 75. What is the chance that the sample was drawn from a population with mean no larger than 82 or no smaller than 70?

7.21. A manufacturer of a certain small foreign automobile claims that their cars get, on the average, between 29 and 30 miles per gallon of gasoline with standard deviation of 1.8 miles per gallon. Suppose a random sample of 36 of these cars was tested under the same conditions and it was found that the mean number of miles per gallon was 29.2. What is the chance that the sample was drawn from a population with mean no larger than 30 or no smaller than 29 miles per gallon?

7.22. A population has a standard deviation of 12.4. A random sample is used to estimate the population mean. What is the smallest sample one can take if the probability is 0.90 that the error in the estimate of the population mean will not exceed 4?

7.23. Suppose the standard deviation of the income per family in suburbia is $1,500. Assuming a given city to be fairly typical, what is the smallest sample one can take if the probability is 0.99 that the error in the estimate of the population mean will not exceed $300?

7.24. There were 15 failures in a random sample of size 75 drawn from a dichotomous population with $\pi = .3$. What is the chance that the point estimate of the proportion of successes in the population will be in error by less than 0.05 units?

7.25. In a random sample of 100 machine parts 5 were defective. What is the chance that the point estimate of the proportion of defectives in the population with $\pi = .02$ will be in error by more than 0.03 units?

*7.26. As part of a research project a student wishes to estimate the proportion of male students who smoke. The student wishes to keep the error within 3 percent, with a risk not greater than 0.05. How large a sample must be taken? *Hint:* First show that

$$\pi(1 - \pi) = \frac{1}{4} - \left(\frac{1}{2} - \pi\right)^2$$

and then notice that $\pi(1 - \pi)$ has the maximum value of $\frac{1}{4}$ when $\pi = \frac{1}{2}$.

7.27. The variance of a normal population is 100. If a random sample of 25 observations from this population has a mean of 53, find the symmetric 0.95 confidence interval of the population mean.

7.28. A random sample of 81 cars passing a checkpoint had a mean speed of 53.5 miles per hour with standard deviation of 6.3 miles per hour. Find two 95 percent confidence intervals of the mean speed of cars passing the checkpoint. What is the length of the shortest .95 confidence interval?

7.29. In a certain city 400 babies were born in March and 160 were girls. Assuming these 400 babies represent a random sample from the population, estimate by a symmetric 95 percent confidence interval the proportion of male births in the population. *Hint:* Use $\sqrt{p(1-p)}$ as an estimate of $\sqrt{\pi(1-\pi)}$.

7.30. Suppose two manufacturers, A and B, make the same gauge of copper wire and that their population standard deviations are 30 and 40 pounds, respectively. A random sample of size 60 from manufacturer A had a mean tensile strength of 5,092 pounds, and a random sample of size 40 from manufacturer B had a mean tensile strength of 5,106 pounds. Find a 99 percent confidence interval for the difference in mean tensile strengths of A and B. *Hint:* Use Equation (6.15).

E7.31. Derive Inequality (7.18).

7.8. TESTS OF HYPOTHESES

Now we investigate **tests of hypotheses,** the second primary area of statistical inference. As used in science, an hypothesis is an assertion which is subject to testing by experimental verification.

A statistical hypothesis is a statement about one or more **populations.**
A test of a statistical hypothesis is a procedure, based on sample information, for deciding whether to reject or fail to reject the hypothesis.

There are many different types of statistical hypotheses, and for each hypothesis there are many possible tests. Thus, we restrict our attention to a few of the basic hypotheses which seem suitable for an introduction to such a large area of inference.

In this section, we assume the form of the probability distribution to be known and that the statistical hypothesis specifies the value (or values) of one or more parameters of the population. That is, we consider parametric statistical hypotheses or, briefly, **parametric hypotheses** which specify one member (or a subset of members) of a family of distributions. Three typical illustrations of such hypotheses follow:

1. The coin is well-balanced.
2. The mean income per family in city A is equal to or less than $7,500.
3. The new machine makes fewer defective parts than the old machine.

Statement 1 is equivalent to the statement that "for a dichotomous distribution the probability of head on a single trial is $\frac{1}{2}$"; that is, $\pi = 0.5$. Statement 2 may be restated as $\mu \leq \$7,500$. For Statement 3 the hypothesis may be rewritten as $\pi_1 < \pi_2$. Clearly, all these involve statements about a parameter (or parameters) of a family of distributions.

Before examining the details of hypothesis testing we briefly discuss some important terms. Next, the principles are gradually introduced by six examples. Finally, in Section 7.10 a general six-step procedure for testing hypotheses is presented and some important aspects of each step are discussed.

To test a parametric hypothesis we first select a suitable statistic for (or estimator of) the parameter under test. A primary part of the test procedure is a rule which divides the set of all possible values of the test statistic into two sets, one being the region of rejection, called the **critical region**, and the other the **noncritical region**. If, as a result of an experiment (which includes the selection of a random sample), the value of the test statistic is a value of the critical region, we *reject the statistical hypothesis*; otherwise, we *fail to reject the statistical hypothesis*. If the statistical hypothesis subject to test, also termed **null hypothesis**, is rejected we *accept some alternative value* (or values) *of the parameter* associated with the choice of the critical region. The parameter values accepted when we reject the null hypothesis belong to an hypothesis called the **alternative hypothesis** of the test. The null and alternative hypotheses are denoted by H_o and H_a, respectively. An example should clarify the meaning of these terms.

EXAMPLE 7.13. Test the hypothesis of Statement 1 on this page; i.e., test $H_o: \pi = 0.5$ where π denotes the probability of head on a single trial.

Suppose we suspect the coin is biased in favor of tails (or against heads). Then the alternative hypothesis is $H_a: \pi < 0.5$ and is called a **one-sided hypothesis**. If we toss the coin n times and count the number of times, T, heads occur, then $p = T/n$ may be used as an estimator of π. (Either p or T could be used as the test statistic.) Clearly, any sample value of p near 0.5 could be explained as a random fluctuation from $\pi = 0.5$ and as a result we would not want to reject H_o. However, for a sample value of p sufficiently far away from 0.5 in either direction we would be inclined to believe the coin is not well-balanced and therefore conclude that $\pi \neq 0.5$ which is a **two-sided hypothesis**. Only values of p sufficiently smaller than 0.5 would lead us to the acceptance of the one-sided alternative hypothesis H_a: $\pi < 0.5$ of this example. To be specific, suppose that in 50 tosses of the coin we decide that fewer than 20 heads would be grounds to reject H_o and thus accept H_a. Hence, when $n = 50$ the critical region, denoted CR, is that set of values of p for which $p < \frac{20}{50} = 0.4$; that is, the

$$CR = \{p \mid p < 0.4\}$$

Then the noncritical region is $\{p \mid p \geq 0.4\}$.

Suppose the specific coin under consideration is actually vigorously tossed 50 times. If only 18 heads occurred, then p would be 0.36 and we would reject H_o, leading to the acceptance of the alternative hypothesis H_a: $\pi < 0.5$. If 22 heads occurred, then p would be 0.44 and we would fail to reject H_o: $\pi = 0.5$. But the test procedure would not lead to the acceptance of the null hypothesis H_o, as explained later. From practical considerations the real point is that even though we may not be able to conclude firmly that $\pi = 0.5$, we can feel reasonably secure in the belief that whatever amount π differs from 0.5 is not large enough to worry about. Note that in case 35 heads occurred, a very unlikely case under the circumstances, we would still fail to reject H_o, but this certainly would not lead us to accept $\pi = 0.5$; however, it might lead us to wish the null hypothesis had been stated as $\pi \geq 0.5$.

The statistic T has a binomial distribution. Under the temporary assumption that H_o is correct, we find that $n\pi = n(1 - \pi) = 25$. Thus, since $25 > 5$, a normal distribution may be used to approximate the binomial distribution with mean

$$\mu = n\pi_o = 25$$

and standard deviation

$$\sigma = \sqrt{n\pi_o(1 - \pi_o)} = \sqrt{50\left(\frac{1}{2}\right)\left(\frac{1}{2}\right)} \doteq 3.536$$

where π_o is the value specified by the null hypothesis. In Example 7.13 the probability that a sample value of p falls in the critical region when $\pi = 0.5$ is

$$P(p < 0.4) = P(T < 20) \doteq P(y < 19.5)$$
$$= P\left[z < \frac{19.5 - 25}{3.536}\right] = P(z < -1.56)$$
$$\doteq 0.0596 = 0.06$$

Thus, even when H_o is correct there is approximately a 0.06 chance of accepting the wrong conclusion that H_a: $\pi < 0.5$.

In Example 7.13 we have shown, among other things, how the critical region depends on the hypotheses and how to compute the probability of falsely accepting the alternative hypothesis. These two concepts are stated in more general terms in the following definitions:

> The **critical region** is based on a specific distribution obtained by temporarily assuming the null hypothesis is true.
>
> The probability that a value of the test statistic falls in the critical region when the null hypothesis is true is called the **significance level of the test** and is denoted by α.

In a test of H_o the significance level is usually selected first and then the critical region is determined. When a sample value of the test statistic falls in the critical region the investigator is faced with the two choices

1. The null hypothesis is correct, but a rare event has occurred.
2. The null hypothesis is not correct.

It is customary to select the second of these two alternatives and declare that the sample result is *significant*—knowing that there is a small known maximum chance of error in making such a decision. Otherwise, the sample result is *not significant*—meaning the experimenter does not have enough evidence to say the sample value is significant. Clearly, the emphasis of the test procedure and the probability control of the experimental results are based on the rejection of the null hypothesis.

The choice of the significance level depends upon the nature of the experiment itself and especially the consequences which would arise if one erroneously rejected the null hypothesis. However, since it is commonly accepted convention to let $\alpha = 0.05$, that value should be used unless there is good reason for another choice. When $\alpha = 0.05$ and *the null hypothesis is correct*, chance fluctuations alone will make us erroneously reject the null hypothesis one time out of 20 experiments on the average. If we are not willing to make the wrong conclusion so frequently, we should select a smaller α. In any case, when selecting α the following two points should be taken into account:

1. There is always a chance of making an error no matter how small α is.
2. In actual practice, conclusions of many experimental results are based on a significance level of $\alpha = 0.05$.

We have already noted that the choice of critical region is associated with the alternative hypothesis and that the alternative hypothesis depends on our understanding (or our expectation) of the experimental situation. In Example 7.13, if we had expected the coin to be biased in favor of heads, then the critical region would have been to the right side (or tail) of the distribution. On the other hand, if we had no information about the coin, we would let the alternative hypothesis be $\pi \neq 0.5$ and use both sides (or tails) of the distribution to define the critical region. In testing the null hypothesis against each of these three alternative hypotheses the critical region would be located in the tails of the binomial distribution (or in the tails of the approximating normal distribution). Since this is usually the case, we state the following:

> When the critical region of a test procedure consists of only one side of a distribution, it is referred to as a **one-tailed test**; when the critical region consists of both sides of a distribution it is referred to as a **two-tailed test**.

The experimenter should give careful consideration to the choice of alternative hypothesis. Unless there is reason to select an alternative hypothesis which leads to a one-tailed test the two-tailed test should normally be applied when working with a single parameter from a binomial or a normal distribution.

EXAMPLE 7.14. Compare the one-tailed and two-tailed 0.05 level tests of a sample mean using the critical regions of the standard normal random variable z.

From Table II we find

$$P(z < -1.960) = P(z > 1.960) = 0.025$$

and

$$P(z < -1.645) = P(z > 1.645) = 0.05$$

Thus, the critical region to the left for a one-tailed test is

$$CR_l = \{z | z > -1.645\}$$

the critical region to the right for a one-tailed test is

$$CR_r = \{z | z < 1.645\}$$

and the critical region for the two-tailed test is

$$CR = \{z | z < -1.960 \quad \text{or} \quad z > 1.960\}$$

These three critical regions are shown in Figure 7.2 and are indicated by wide lines. Clearly, these tests can lead to different conclusions. For example, if a sample value of z falls between -1.960 and -1.645 or to the right of 1.960,

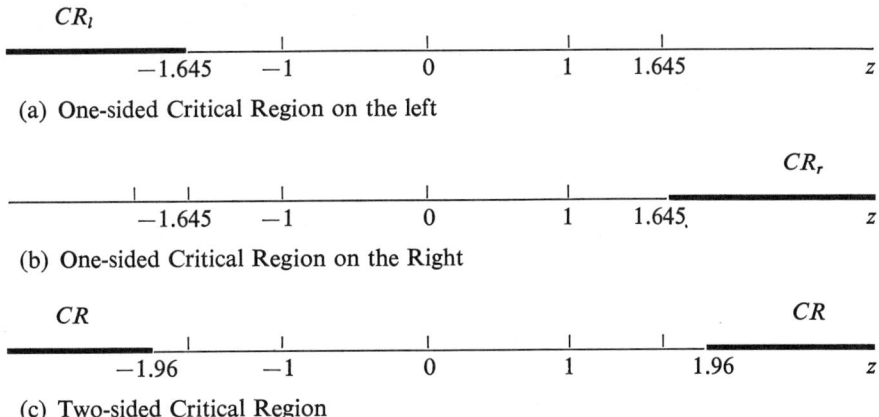

Figure 7.2. Critical Regions of Three 0.05 Level Tests

the first and third tests have different conclusions [see Figure 7.2, parts (a) and (c)].

EXAMPLE 7.15. Use a 5 percent level test of the hypothesis of Statement 2 on page 237.
The one-sided null hypothesis is

$$H_o: \mu \leq \$7,500$$

and the one-sided alternative hypothesis is

$$H_a: \mu > \$7,500$$

Thus, the *mean income* of a random sample of families in city A will be used in the test procedure. Since the distribution of family incomes is likely to be skewed to the right, we must choose a sample size large enough so that the sample mean is approximately normally distributed. We assume that $n = 25$ is large enough. Further, since the economy of the city has been fairly stable over the last 10 years, we assume the standard deviation of the population of incomes to be essentially the same as it was in the census report of five years ago. That is, $\sigma = \$1,800$. (Since there has been a steady increase in wages and salaries, the mean income is greater than in the census report.) In case the sample size is small and σ is unknown use the methods of Section 10.2.

The critical region is determined from a distribution based on the null hypothesis. When the null hypothesis specifies a single parameter value, as in Example 7.13, we use that value to locate the critical region. When there is an interval of values we use that parameter value which is nearest the parameter values of the alternative hypothesis. For this and other reasons the equality

242   Estimation and Tests of Hypotheses

relation is normally included in the statement of the null hypothesis. Thus, to locate the critical region in our example, we use the normal population with mean $\mu = \$7{,}500$ (from the null hypothesis) and standard error of the mean

$$\sigma_{\bar{y}} = \frac{1{,}800}{\sqrt{25}} = \$360$$

We know that $P(z > 1.645) = 0.05$ and that

$$z = \frac{\bar{y} - 7{,}500}{360}$$

Hence

$$\bar{y} = 360z + 7{,}500 = 360(1.645) + 7{,}500 = \$8{,}092.20$$

is the *critical point*, and the critical region is

$$CR = \{\bar{y} \mid \bar{y} > \$8{,}092.20\}$$

In general, a **critical point** is a boundary point between the critical and non-critical regions.

In practice, we usually express the critical region in standard units (i.e., in units of a distribution found in standard tables) and convert the sample statistic to standard units for comparison with critical values. Thus, as indicated in Figure 7.3, the critical region might also be expressed as

$$CR = \{z \mid z > 1.645\}$$

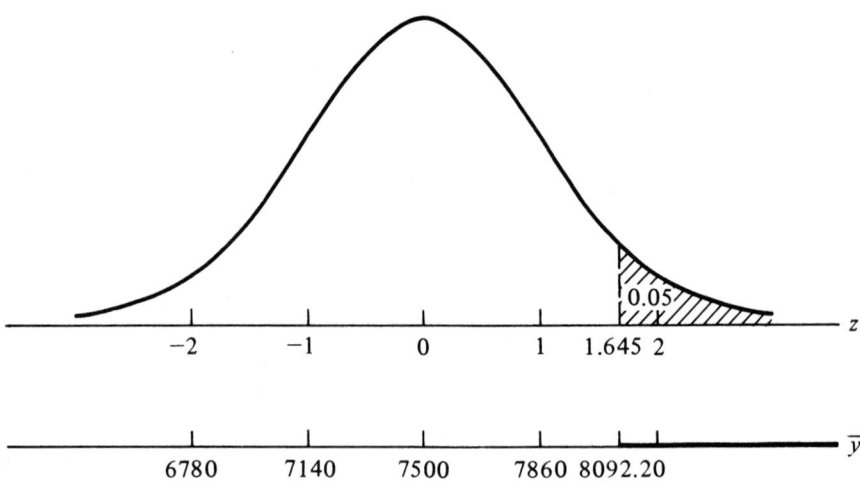

Figure 7.3.   Critical Region and Significance Level for Example 7.15

Note that we have done all the above work on Example 7.15 without knowing the sample values. Readers should follow the practice of making decisions on the sample size, significance level, statistic to be used in the test, and critical region before obtaining observations for the random sample. Otherwise, they might be tempted to "adjust" some of these things to obtain the desired conclusion. In so doing they would likely lose a measure of experimental objectivity as well as their reasons for using statistics in the first place.

Suppose the sample of size 25 is drawn and the sample mean is found to be $8,103.50. Since $8,103.50 falls in the critical region, we reject the null hypothesis and conclude that the mean family income in city A is greater than $7,500. In making this decision we know there is less than 0.05 chance of error. (Usually in practice we find the standardized value of $\bar{y}$, namely

$$z = \frac{\bar{y} - 7,500}{360} = \frac{8,103.50 - 7,500}{360} \doteq 1.68$$

Of course, we would make the same conclusion. *Note:* In this case, the test statistic z is not an estimator of μ. However, z is defined in terms of $\bar{y}$ which is an estimator of μ.)

Note that we could have stated "the sample mean $8,103.50 *is significantly larger* than $7,500.00" instead of "the population mean *is larger* than $7,500.00." Readers should realize the difference and never make a statement like "the population mean is significantly larger than $7,500.00." The word "significantly" is used with the sample mean, but not with population mean.

One final observation about tests of hypotheses. When we *accept a statement*, like the population mean is larger than $7,500, it does not mean that we have proved the truth of the statement. It simply means that the evidence is so strong with an α chance of error that we accept the conclusion so that a course of action may be taken.

7.9. TESTS OF HYPOTHESES: TWO KINDS OF ERRORS

As a result of the test procedure described in Section 7.8 we either *reject the null hypothesis* or *fail to reject the null hypothesis*. Unfortunately, with either decision we might make an error.

> If the null hypothesis is true and we conclude that it is false, we make an error known as the **type I error**.

The probability of making this error is the same as the significance level of the test and is sometimes referred to as the *size of the type I error*.

> If the null hypothesis is false and we fail to reject it, we make an error known as the **type II error**.

The probability of making the type II error is unknown (unless the alternative hypothesis specifies a single value—see Exercises 7.44 and 7.45) since it depends on an unknown parameter.

We say we reject H_o and thus *accept H_a since we can attach a specific chance of making an error* in such a statement. On the other hand, when we fail to reject H_o, we do not say "accept the null hypothesis" *because we cannot usually attach a specific chance of making an error* in such a statement. Clearly, we cannot accept the null hypothesis in the same sense in which we accept the alternative hypothesis. However, there are times when experimenters feel they must act as though they accept the null hypothesis—even though they do not know their chances of making an error. Thus, experimenters might actually "accept the null hypothesis," being aware that there might be a very large chance that they are committing a type II error, whereas the statistician "fails to reject the null hypothesis."

The probability of making the type II error, denoted by $\beta(\theta)$, is the same as the probability of a sample value of the test statistic t falling in the noncritical region when the true value of the parameter θ is different from any specified by the null hypothesis. In symbols,

(7.19) $\quad\quad \beta(\theta) = P(t \text{ falls in non-}CR \mid \theta \text{ different from those specified by } H_o)$

Whenever a value of θ not specified by the null hypothesis is necessarily specified by the alternative hypothesis we may write

(7.20) $\quad\quad\quad\quad \beta(\theta) = P(t \text{ falls in non-}CR \mid \theta \text{ specified by } H_a)$

EXAMPLE 7.16. Use the data of Example 7.15 to illustrate the computations and applications of probabilities of the type II error.

The noncritical region of Example 7.15 is

$$\text{non-}CR = \{\bar{y} \mid \bar{y} \leq \$8{,}092.20\}$$

and the values of μ outside H_o are those of H_a: $\mu > \$7{,}500$. We replace t and θ in Equation (7.20) by $\bar{y}$ and μ, respectively, to get

(7.21) $\quad\quad\quad\quad \beta(\mu) = P(\bar{y} \leq \$8{,}092.20 \mid \mu > \$7{,}500)$

where it is understood that $\bar{y}$ is distributed normally with standard error $\sigma_{\bar{y}} = \$360$. We now show how to compute $\beta(\mu)$ for two different values of μ.

When $\mu = \$8{,}000$, we find

$$\beta(8{,}000) = P(\bar{y} \leq 8{,}092.20 \mid \mu = 8{,}000)$$

$$= P\left(z \leq \frac{8{,}092.20 - 8{,}000}{360}\right) = P(z \leq 0.26) \doteq 0.6025$$

When $\mu = \$8,500$, we find

$$\beta(8,500) = P\left(z \le \frac{8,092.20 - 8,500}{360}\right)$$
$$= P(z \le -1.13) \doteq 0.1295$$

Figure 7.4 shows the relationship between parameters and probabilities of the type II error. The area shaded by dots is the probability of the type II error when the true population mean is $8,500, and the area shaded by slanted lines is the probability of the type II error when the true population mean is $8,000.

Using the methods illustrated in the last paragraph, we find probabilities of the type II error for values of μ at intervals of $250. These and other probabilities (found by similar methods) are shown in Table 7.5. Observe that as μ moves further away from the values of the null hypothesis, the probability of the type II error gets smaller, as we expected. That is, we are less likely to fail to reject H_o. Put another way, as μ increases the probability of accepting the alternative hypothesis becomes greater. Furthermore, by observing the magnitude of the probabilities and the rate of change of the probabilities in the vicinity of a particular μ we can decide whether the risks connected with a particular decision are acceptable or not.

Since the true value of μ is not known, it is useful to have a curve that gives at a glance the probability of failure to reject the null hypothesis. Such a

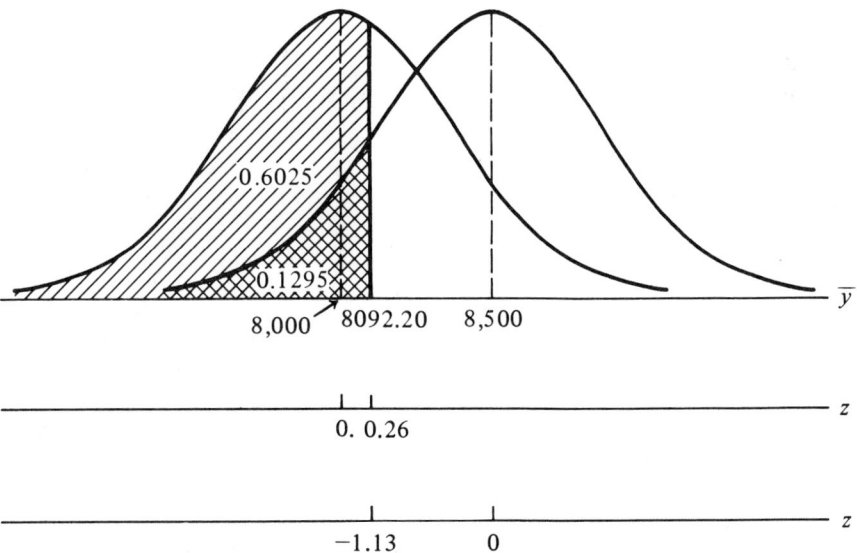

Figure 7.4. Probabilities of Type II Error

Table 7.5. Probabilities Associated with Example 7.16

Parameter μ	Type II Error	Probability of Failure to Reject H_o	Rejecting H_o
7250		0.990	0.010
7500		0.950	0.050
7750	0.829	0.829	0.171
8000	0.602	0.602	0.398
8250	0.330	0.330	0.670
8500	0.130	0.130	0.870
8750	0.034	0.034	0.966
9000	0.006	0.006	0.994

curve is called an **operating characteristic curve** or simply an **OC-curve**. The graph of the *OC*-curve for Example 7.16 is shown in Figure 7.5. Note that the *OC*-curve is drawn for all values of the parameter and that it includes the probability of the type II error for values of the alternative hypothesis.

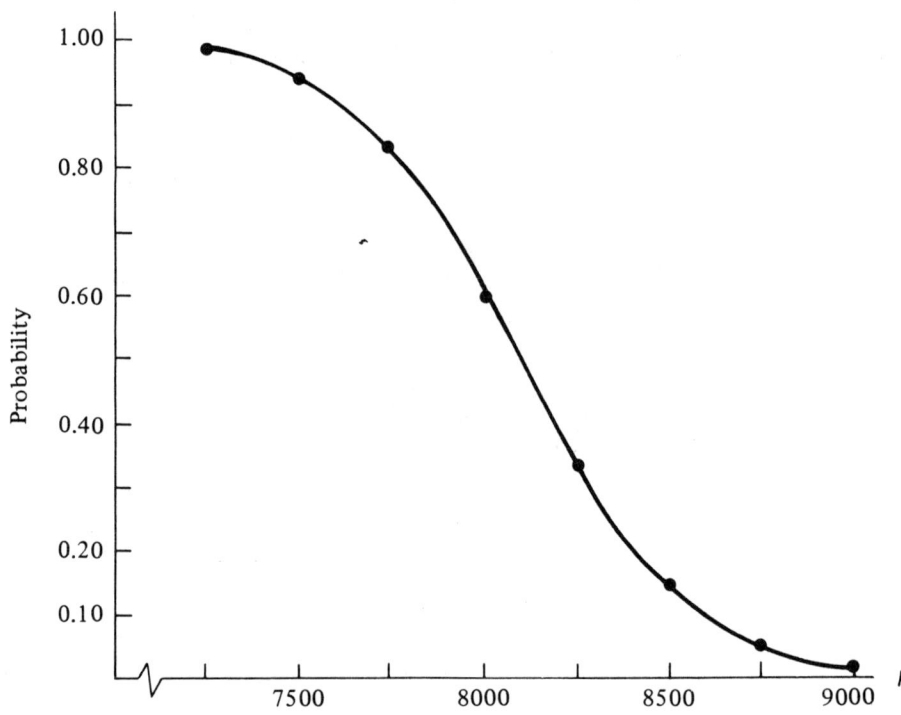

Figure 7.5. Operating Characteristic Curve for Example 7.16

If we had plotted the probabilities of rejecting H_o, the resulting graph would have been the **curve of the power function of the test.** Since the **power of a test** at a parameter point is *one minus the operating characteristic value at that point,* only one of these curves need be considered. It has become fairly standard to use the OC-curve in practical applications, particularly in industrial problems, and to refer to the power function curve in theoretical work. With the OC-curve, emphasis is placed on the probability of a sample value falling in the noncritical region; with the power function curve, emphasis is on the probability of a sample value falling in the critical region.

By now the reader should have observed that α and $\beta(\mu)$ do not change independently of each other. For a fixed sample size n, if we let α become smaller, then $\beta(\mu)$ at a particular μ becomes larger. That is, if we lower the chance of making the type I error we increase the chance of making the type II error. Thus, in a given problem it is up to the experimenter to decide on the relative importance of α and β. An example should make the relationship clear.

EXAMPLE 7.17. Use the data of Example 7.15 to compare α and $\beta(\mu)$. For the test procedure of Example 7.15 we had

$$H_o: \mu \leq \$7{,}500 \qquad H_a: \mu > \$7{,}500$$
$$n = 25 \qquad \alpha = 0.05 \qquad \sigma_{\bar{y}} = \$360$$
$$\bar{y} \text{ normally distributed}$$

Now we keep everything the same except we allow α to change. We wish to find how $\beta(\$8{,}000)$ changes as α takes values 0.10, 0.05, and 0.01, respectively. The choice of \$8,000 for the mean μ is more or less arbitrary since we would find similar results at any other value of μ.

Since

$$P(z > 1.28) \doteq 0.10$$

the critical point is

$$\bar{y}_c = z_{0.10}\, \sigma_{\bar{y}} + \mu = (1.28)(360) + 7{,}500 = \$7{,}960.80$$

and the critical region for $\alpha = 0.10$ is

$$CR = \{\bar{y} \mid \bar{y} > 7{,}960.80\}$$

Thus, the probability of the type II error at $\mu = \$8{,}000$ is

$$\beta(8{,}000) = P(\bar{y} \leq 7{,}960.80 \mid \mu = 8{,}000)$$
$$= P\left(z \leq \frac{7{,}960.80 - 8{,}000}{360}\right)$$
$$= P(z \leq -0.11) = 0.4562$$

In a similar way we can show that

$$\beta(8{,}000) = 0.8261 \quad \text{when} \quad \alpha = 0.01$$

The relationship between α and $\beta(\mu)$ is summarized in Table 7.6 (the entry for $\alpha = 0.05$ was computed in Example 7.16), i.e., for a fixed sample size and a fixed μ, as α increases $\beta(\mu)$ decreases or conversely.

Table 7.6. Values of α and $\beta(\mu)$ when $n = 25$
(Computed for Data of Example 7.15)

α	0.10	0.05	0.01
$\beta(8000)$	0.46	0.60	0.83

When α is fixed, it can be shown that $\beta(\mu)$ decreases as n increases. It will be left as an exercise for the student to show the results of Table 7.7. Again the choice of $8,000 for the mean μ is arbitrary since similar results hold for other values of μ.

Table 7.7. Values of n and $\beta(\mu)$ when $\alpha = 0.05$
(Computed for Data of Example 7.15)

n	25	49	64	100	144
$\beta(8000)$	0.60	0.38	0.28	0.13	0.046

From the results of the last two very brief tables it should be clear that n, α, and $\beta(\mu)$ are so related that if any two are given the third one can be computed. Usually n and α are specified and $\beta(\mu)$ may be determined for the given test. We illustrate how the sample size may be computed when α and $\beta(\mu)$ are specified.

EXAMPLE 7.18. Use the family income data of Example 7.15 to find n if $\alpha = 0.05$ and $\beta(\mu) \leq 0.10$ when $\mu \geq \$8{,}000$.

This is a fairly typical problem. Often the experimenter wishes the type II error to be small whenever the true parameter deviates more than a specified amount from the null hypothesis value (values). In this problem, if the true mean is $500 or more above the maximum value of the null hypothesis, we want the probability of the type II error to be no greater than 0.10. That is, if the true mean exceeds $8,000, we want to accept the alternative hypothesis with high probability of being correct, namely 0.90 or more.

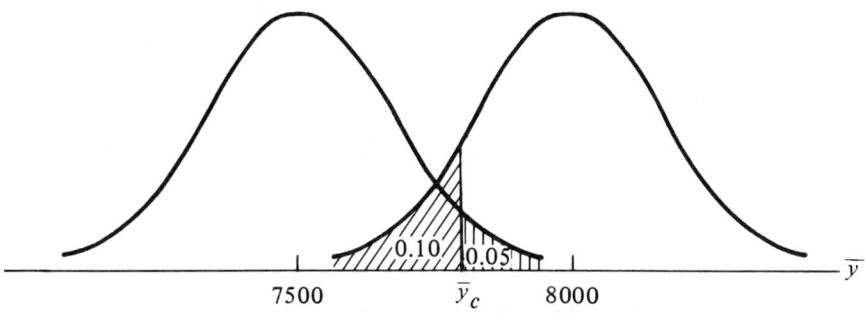

Figure 7.6. $\alpha = 0.05$ and $\beta(8000) = 0.10$ for Example 7.18

The standard error of the mean is $\sigma_{\bar{y}} = 1800/\sqrt{n}$. From Figure 7.6 we can observe that the critical point $\bar{y}_c$ must be between $7,500 and $8,000. In fact $\bar{y}_c$ must have a value such that

$$P(\bar{y} > \bar{y}_c) = 0.05 \quad \text{when} \quad \mu = \$7,500$$

and

$$P(\bar{y} \leq \bar{y}_c) = 0.10 \quad \text{when} \quad \mu = \$8,000$$

Since

$$P(z > 1.645) = 0.05, \quad P(z < -1.282) = 0.10$$

and

$$z = \frac{\bar{y}_c - \mu}{\sigma_{\bar{y}}}$$

it follows that

$$1.645 = \frac{\bar{y}_c - 7,500}{\frac{1,800}{\sqrt{n}}} \quad \text{and} \quad -1.282 = \frac{\bar{y}_c - 8,000}{\frac{1,800}{\sqrt{n}}}$$

Simplifying the last two equations gives

$$1.645 \left(\frac{1,800}{\sqrt{n}}\right) = \bar{y}_c - 7,500 \quad \text{and} \quad -1.282\left(\frac{1,800}{\sqrt{n}}\right) = \bar{y}_c - 8,000$$

or

$$1.645\left(\frac{1,800}{\sqrt{n}}\right) + 7,500 = \bar{y}_c \quad \text{and} \quad -1.282\left(\frac{1,800}{\sqrt{n}}\right) + 8,000 = \bar{y}_c$$

Thus

$$1.645\left(\frac{1,800}{\sqrt{n}}\right) + 7,500 = -1.282\left(\frac{1,800}{\sqrt{n}}\right) + 8,000$$

or
$$(1.645 + 1.282)\left(\frac{1,800}{\sqrt{n}}\right) = 500$$

or
$$\sqrt{n} = \frac{(2.927)(1,800)}{500} = 10.5372 \doteq 10.54$$

So
$$n = 110.09$$

Thus, the required sample size is 110. (Note that this value is in agreement with Table 7.7.)

7.10. TESTS OF HYPOTHESES: SUMMARY

We have presented, in terms of means and proportions, the method of testing hypotheses. But what we have done is illustrative of tests of hypotheses involving any parameter (or parameters). For easy reference, we now *outline the general procedure for testing hypotheses*. In other examples in this and later chapters we follow this procedure.

1. Formulate the null and the alternative hypotheses.
2. State the assumptions and specify the significance level and the sample size. (Ideally the probability of making a type II error should be taken into account when specifying the sample size.)
3. Specify the statistic to be used.
4. Construct the critical region.
5. Compute the statistic value from the sample.
6. State the conclusion in terms of the hypotheses.

In Step 1 of the general test procedure, the "equals" relation should be in the null hypothesis, and the statement the experimenter really "wants to accept" should be in the alternative hypothesis, if possible. The alternative hypothesis should be stated so that the rejection of the null hypothesis is equivalent to the acceptance of the alternative hypothesis. If in doubt about the statement of the alternative hypothesis, it should include all parameter values not specified in the null hypothesis.

Step 2 is often difficult for the experimenter, but it is very important. The results of the test depend on this step. The more accurate and specific the assumptions, the better the test. The choice of the size of α should depend on the consequences of committing a type I error—the more serious the consequences, the smaller α should be. However, the smaller α is made, the larger the size of the type II error. Fortunately, the probabilities of the type I and II errors can usually be kept in control and in balance by choosing the sample size large enough (see Example 7.18). The importance of the control of the probabilities of errors of type I and II can be seen by looking at an analogous (but not similar) situation in the affairs of humans. Suppose a man is on trial for murder. He is assumed innocent until proved guilty. So the null hypothesis is that he is innocent; the alternative hypothesis is that he is guilty (to some degree). The sample size is like the evidence that is brought to bear in the case. A type I error is made if the jury pronounces him guilty when he is actually innocent. A type II error is made if he is pronounced innocent when he is actually guilty. Clearly, in the long run errors of both types will be made, but it is hoped that both can be kept small by bringing *all the evidence* before the court in as objective a manner as possible. Similar examples where statistics may be used are in testing a new drug that might be fatal or in testing a new machine that might be seriously defective.

It seems appropriate at this time to summarize the interlocking relations among the sample size, significance level, and probability of committing a type II error which are illustrated in Section 7.9. These relationships are usually exhibited by the tests of parameters presented in this book.

(a) For *a fixed sample size*, decreasing the significance level will result in an increase in the probability of committing a type II error.

(b) For a *fixed significance level*, increasing the sample size will result in decreasing the probability of committing a type II error.

(c) The probabilities of committing both kinds of errors may be reduced by increasing the sample size. Of course, there are limits (due to time, economics, facilities, etc.) on how much the sample size can be increased.

The three statements above are based on the assumption that the standard deviation of the population is known. In most cases the standard deviation must be estimated, and the control of the estimate can affect the probabilities of both kinds of errors.

Step 3 is usually straightforward if a good job has been done in Step 2. [In practice, the standardized statistic is typically used since computed values can be compared with those found in prepared tables. For example, in tests about a population mean, the statistic $(\bar{y} - \mu)/\sigma_{\bar{y}}$ may be used in place of the statistic $\bar{y}$.]

In Step 4 common sense should usually guide an experimenter to the proper critical region—provided the experimenter does not lose sight of the alternative hypothesis. This is especially true of the relatively simple test procedures presented in this book. In more advanced treatments of hypotheses testing, principles and methods are presented for finding "a best critical region" of a test.

The computations of Step 5 are made according to the specifications in Step 3. For simple problems a desk calculator may be desirable; for more complicated analyses computers should be used when available. Sometimes tables, similar to Table II in the Appendix, show cumulative probabilities. When such tables exist it is often desirable that we find the probability of getting a value equal to or more extreme than the value computed from the data. With such a probability the readers of the report of an experiment can better make their own assessments of the experimental results.

If in Step 6 we reject the null hypothesis, then the alternative hypothesis is accepted with knowledge that the probability of committing a type I error is no larger than α. Otherwise, we fail to reject the null hypothesis. From the statistical viewpoint we do not say "accept the null hypothesis" because we do not know the chance of a type II error, except in that rare case where H_a specifies a single value of the parameter. However, there are situations (see Example 7.19) in which experimenters feel they must accept (from a practical point of view) the null hypothesis whenever they cannot reject it. In these cases, they accept the null hypothesis with the strong hope that they are not exposing themselves to high probabilities of making type II errors, especially serious type II errors. The point is that we (as statisticians) use the term "accept" only in case the probability of making an error is known.

7.11. TESTS OF HYPOTHESES: DIFFERENCES BETWEEN TWO MEANS

One of the most common hypotheses tested involves two means. Some typical illustrations are that we may want to know how a new teaching method compares with a traditional method, how men and women compare in their abilities to perform a certain task, whether a new drug is more effective than an older drug, what effect a new diet has on the weight of beef cattle, whether rats can perform as well under a restricted condition as with the usual control condition, and so on.

In this section we consider only one of several conditions under which two means may be compared. We apply Equation (6.15) which assumes the two populations are normal with known standard deviations. If the sample sizes are both over 30, these two assumptions are not necessary since (1) due to the Central Limit Theorem the difference in the two sample means will be approximately normally distributed, and (2) the sample variances are not likely to vary much from the population variances. See Section 10.4 in case the sample sizes are small and the population standard deviations are unknown.

The null hypothesis is that the difference $\mu_1 - \mu_2$ in two population means μ_1 and μ_2 is equal to some constant k. Usually the alternative hypothesis is either one-sided to the left (that is, $\mu_1 - \mu_2 < k$), one-sided to the right ($\mu_1 - \mu_2 > k$), or two-sided ($\mu_1 - \mu_2 \neq k$). Often $k = 0$. In this restricted and important case, the null hypothesis states that the population means are equal (that is, $\mu_1 = \mu_2$) and the alternative hypothesis takes a corresponding simple form.

EXAMPLE 7.19. The manufacturers of two types of foreign compact cars, makes 1 and 2, in competition for the American market make conflicting claims about the gasoline consumption of their cars. A large distributor of foreign cars wishes to determine which, if either, make of car actually does give better gasoline mileage. He tests 40 cars of each make under the same conditions. Make 1 has a mean of $\bar{y}_1 = 27.54$ miles per gallon with standard deviation of $s_1 = 1.8$ miles per gallon. Make 2 has a mean of $\bar{y}_2 = 28.38$ miles per gallon with standard deviation of $s_2 = 2.3$ miles per gallon. Using a 5 percent level two-sided test, what can the distributor conclude about gasoline mileage of the two makes of car?

Let μ_1 and μ_2 be the true population means of makes 1 and 2, respectively. Applying the general test procedure of Section 7.10, we have the following six steps:

1. $H_o: \mu_1 - \mu_2 = 0$ or $\mu_1 = \mu_2$
   $H_a: \mu_1 - \mu_2 \neq 0$ or $\mu_1 \neq \mu_2$

2. Random samples of sizes $n_1 = 40$ and $n_2 = 40$ are independently drawn from two populations. The significance level of the test is $\alpha = 0.05$.

3. The sample sizes are large enough to assume the statistic $\bar{y}_1 - \bar{y}_2$ is approximately normally distributed with mean $\mu_1 - \mu_2$ and standard deviation $\sqrt{\sigma_1^2/n_1 + \sigma_2^2/n_2}$. Further, the large sample sizes allow us to believe that no serious error is committed by letting the sample variances replace the population variances. Thus, under the temporary assumption that the null hypothesis is true we use

$$\frac{(\bar{y}_1 - \bar{y}_2) - 0}{\sqrt{\frac{s_1^2}{n_1} + \frac{s_2^2}{n_2}}}$$

as the statistic which is approximately distributed as a standard normal random variable z.

4. Since

$$P(z < -1.96) = P(z > 1.96) = 0.025$$

the critical region is

$$CR = \{z \mid z < -1.96 \text{ or } z > 1.96\}$$

5. Since

$$n_1 = n_2 = 40 \qquad \bar{y}_1 = 27.54 \qquad \bar{y}_2 = 28.38$$
$$s_1 = 1.8 \qquad s_2 = 2.3$$

the value of the statistic is

$$\frac{(\bar{y}_1 - \bar{y}_2) - 0}{\sqrt{\frac{s_1^2}{n_1} + \frac{s_2^2}{n_2}}} = \frac{27.54 - 28.38}{\sqrt{\frac{(1.8)^2}{40} + \frac{(2.3)^2}{40}}} = \frac{-0.84}{0.462} = -1.82$$

6. The sample statistic falls in the noncritical region since -1.82 is between -1.96 and 1.96. Thus, we fail to reject the null hypothesis. That is, on the basis of the experiment we cannot say one make of car gives better gas mileage than the other. (Clearly, we would not want to conclude that the two makes of car give *equal* gas mileage. The point is that, based on the experiment, whatever difference exists is not large enough to fuss over one way or the other.)

Note: Had we use the alternative hypothesis that $\mu_1 - \mu_2 < 0$ we would have rejected the null hypothesis since the critical region would have been

$$CR = \{z | z < -1.645\}$$

Thus, we would have concluded that make 2 gives better gasoline mileage than 1. This points up one reason for deciding on the test procedure in advance of the analysis.

It may be interesting to check our thinking on the results of the experiment in Example 7.19 if we use exactly the same numbers but change the names. Let makes 1 and 2 refer to two machines that are used to cut off yarn which is rolled on bobbins. Suppose 40 bobbins are randomly selected from the production of each machine. Let each measurement denote the number of excess yards on a bobbin. Then $\bar{y}_1 = 27.54$ represents the mean number of excess yards on the 40 bobbins at machine 1, and so on. The conclusion of the test would then translate to the statement "on the basis of the experiment we cannot say one machine leaves less excess yards than the other machine." With this experiment would we feel more like accepting the null hypothesis? Would we be more likely to state a one-sided alternative hypothesis? Suppose you knew that a new machine would have been bought to replace machine 2 had the results of the experiment been significant—would this make you want to accept the null hypothesis?

7.12. EXERCISES

7.32. Suppose the probability of success π on each of 36 independent trials is the same. Find a critical region, in terms of p, for a 0.05 level test of $H_o: \pi = 0.3$ against $H_a: \pi > 0.3$.

7.33. A random sample of size 25 was drawn from a normal population with standard deviation 10. Use the critical region $\{\bar{y} | \bar{y} > 55\}$ in a test of $H_o: \mu = 50$ against $H_a: \mu > 50$. What is the significance level of the test?

Exercises

7.34. A random sample of size 35 was drawn from a normal population with variance 70. Find, in terms of $\bar{y}$, the critical region for a 0.05 level test of $H_o: \mu = 40$ against $H_a: \mu \neq 40$.

7.35. The probability of success π is the same for each of 24 independent trials. To test $H_o: \pi = 0.4$ against $H_a: \pi \neq 0.4$ use the critical region $\{p \mid p < 0.2 \text{ or } p > 0.6\}$. What is the significance level of the test?

7.36. Given $\bar{y} = 68.2$, $\sigma = 10$, $n = 100$, and $\alpha = 0.05$, test the hypothesis that $\mu = 70$.

7.37. Given $\bar{y} = 68.2$, $\sigma = 10$, $n = 100$, and $\alpha = 0.05$, test the hypothesis that $\mu \geq 70$.

7.38. Given $\bar{y} = 68.2$, $\sigma = 10$, $n = 144$, and $\alpha = 0.05$, test the hypothesis that $\mu = 70$.

7.39. Given $T = 60$, $n = 200$, and $\alpha = 0.05$, test the hypothesis $\pi = 0.25$ against the hypothesis $\pi \neq 0.25$.

7.40. In 120 tosses of a coin heads occurred 50 times. Is this enough evidence to indicate the coin is biased in favor of tails?

7.41. After five years experience at a university it was found that a mathematics placement examination has a mean score of 73 and a standard deviation of 6. It is claimed that the incoming freshman class will have a higher mean score than 73. Use a random sample of 30 freshmen with a mean score of 75 to test $H_o: \mu = 73$ against the appropriate alternative hypothesis.

7.42. A random sample of size 4 is drawn from a normal population with variance 25 and unknown mean. The critical region for the null hypothesis $\mu = 10$ is $\{\bar{y} \mid \bar{y} > 14\}$. What is the probability of making the type II error when the true population mean is 12?

7.43. In exercise 7.42 suppose the critical region is $\{\bar{y} \mid \bar{y} < 6 \text{ or } \bar{y} > 14\}$. What is the probability of making the type II error when the true mean is 10.5?

7.44. A particular coin is tossed three times. Let π denote the probability of obtaining a head on a single toss and let T denote the number of heads obtained in three tosses. Use T (which has a binomial distribution) to test $H_o: \pi = 0.5$ against the alternative $H_a: \pi = 0.3$. If the critical region is $\{T \mid T = 0\}$, find the sizes of the two types of error.

7.45. In Exercise 7.44 suppose the critical region had been $\{T \mid T = 3\}$. Calculate the sizes of the two types of error for this critical region. Compare your results with those of Exercise 7.44, indicating which critical region you prefer.

C7.46. The bacteria content of a canned food product must be less than 30.0 to be acceptable. Assume that long experience indicates that the bacteria content is approximately normally distributed with standard deviation 0.6. A random sample of 16 cans is used to test, at the 0.05 level, the null hypothesis $\mu \geq 30.0$ against the alternative hypothesis $\mu < 30.0$.

(a) Compute at least five values and draw the operating characteristic curve to determine the chance of nonrejection of a lot which has mean bacteria content of 29.6.
(b) Draw on the same graph with part (a) the operating characteristic curve for the above test when samples of size 64 are used in place of samples of size 16.
(c) Suppose we require that the probability of a type II error not be greater than 0.10. Find the largest bacteria count for which $\alpha = 0.05$, $\beta = 0.10$, and $n = 16$; for which $\alpha = 0.05$, $\beta = 0.10$, and $n = 64$. Use this information to make a statement about the effect of sample size on the test.
(d) Construct curves for power of the tests in parts (a) and (b).

7.47. Suppose family income in a city is approximately normally distributed with mean μ and standard deviation $2,000$. A random sample of size n is used to test $H_o: \mu \geq \$8,000$ against $H_a: \mu < \$8,000$. Find the smallest value of n for which $\alpha = 0.05$, $\beta(\mu) \leq 0.01$, and $\mu \leq \$7,000$.

7.48. The first random sample of size 4 is drawn from a normal population with standard deviation 8 and the second random sample of size 5 is drawn from a normal population with variance 45. Test $H_o: \mu_1 = \mu_2$ against $H_a: \mu_1 > \mu_2$ when the sample means have values $\bar{y}_1 = 87$ and $\bar{y}_2 = 78$, respectively.

7.49. Two groups of school children were taught to read by two different methods, called method 1 and method 2. After six weeks' instruction the same reading test was given to both groups. The results were as follows: $n_1 = 40$, $\bar{y}_1 = 71.2$, $s_1 = 8$, $n_2 = 50$, $\bar{y}_2 = 74.7$, $s_2 = 11$. Test the hypothesis $H_o: \mu_1 = \mu_2$.

7.50. For Exercise 7.30 use a 0.10 level test of $H_o: \mu_A = \mu_B$.

7.51. Suppose random samples are independently drawn from two normal populations with variances $\sigma_1^2 = 88$ and $\sigma_2^2 = 144$. Test $H_o: \mu_1 = \mu_2 + 6$ against $H_a: \mu_1 < \mu_2 + 6$ at the .05 level when the sample results are $n_1 = 8$, $\bar{y}_1 = 75$, $n_2 = 10$, and $\bar{y}_2 = 77$.

8

CHI-SQUARE DISTRIBUTIONS AND THEIR APPLICATIONS

8.1. CHI-SQUARE DISTRIBUTIONS

In Chapter 7 we discussed problems of inference in terms of means and proportions. Now we introduce a new family of distributions, called **chi-square distributions**, to show how inferences can be made about the variance and standard deviation. Also, we show how these chi-square distributions may be used in investigations involving frequency counts (instead of measured observations).

Suppose random samples of size n are drawn from a normal population with mean μ and variance σ^2. We know that the sampling distribution of the mean is normally distributed with mean μ and variance σ^2/n. However, the sampling distribution of the variance s^2 is definitely not normally distributed.

In problems with the sample mean $\bar{y}$ we soon learned to apply the statistic $(\bar{y} - \mu)/\sigma_{\bar{y}}$, the standardized version of $\bar{y}$. Two primary reasons for using $(\bar{y} - \mu)/\sigma_{\bar{y}}$ were that (1) probability tables are prepared for the standardized statistic, and (2) the unit of measure is in pure (or abstract) units. (By using the standardized statistic we take out the effect of the local unit of measure. For example, we can compare mean height of Texans in inches with mean height of Frenchmen in meters.)

It is obvious that we should "standardize" s^2 in some way. (For example, we might find it difficult to think in terms of square feet or square gallons or square temperature.) For a sample of size n this is typically done by using

(8.1) $$\frac{(n-1)s^2}{\sigma^2} \quad \text{or} \quad \frac{SS}{\sigma^2}$$

where $SS = (n-1)s^2$ and is called "sum of squares of sample deviates about the sample mean" or "corrected sum of squares."

The nature of the sampling distribution of $[(n-1)s^2]/\sigma^2$ depends on the parent population.

(8.2) For random samples of size n drawn from a normal population with mean μ and variance σ^2 it can be shown that $[(n-1)s^2]/\sigma^2$ has the chi-square distribution with $n - 1$ degrees of freedom. The continuous random variable $[(n - 1)s^2]/\sigma^2$ has mean $n - 1$ and variance $2(n - 1)$.

The Greek letter ν, nu, is often used in place of $n - 1$ and refers to *degrees of freedom* as well as to population mean. The random variable of the chi-square distribution with ν degrees of freedom is denoted by χ_ν^2 or, when there is no danger of confusion, by χ^2. Thus, for the chi-square distribution of Statement (8.2) we have

(8.3) $$E(\chi^2) = \nu = n - 1$$

and

(8.4) $$V(\chi^2) = 2\nu = 2(n - 1)$$

The curves of three typical chi-square distributions are shown in Figure 8.1. Clearly, the shape of a distribution changes with the *degrees of freedom*

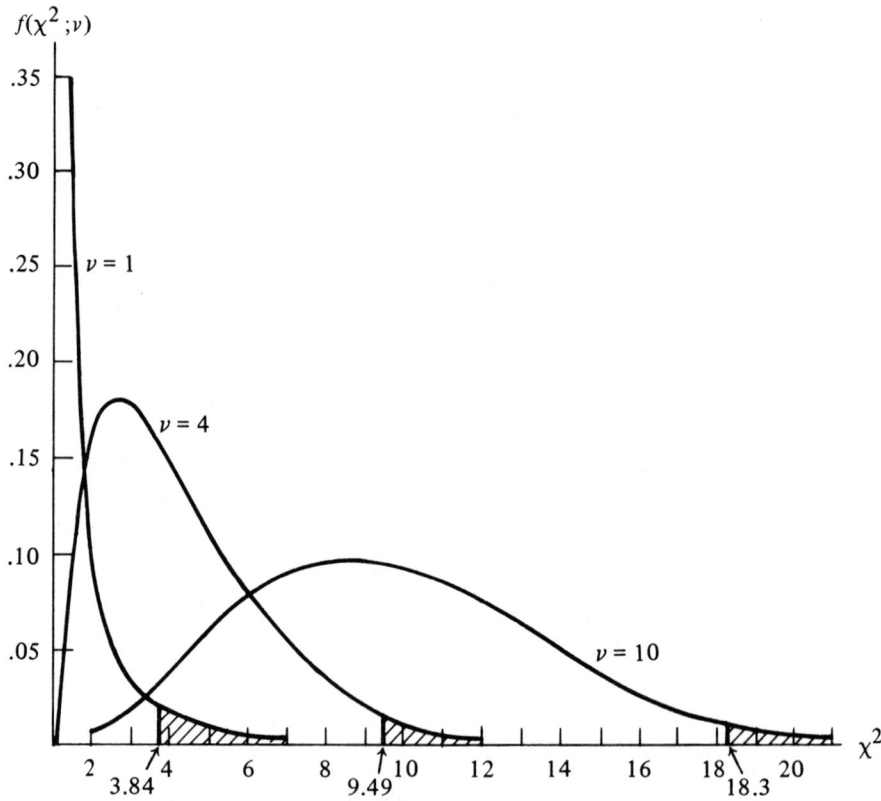

Figure 8.1. Chi-square Distributions for $\nu = 1, 4,$ and 10. The 5 Percent Points, $\chi_{.05}^2$, are Indicated

(i.e., with the mean). Thus, in working with s^2, the standardization process does not lead to a single simple distribution as with the normal distribution.

Let χ_α^2 be that value of χ^2 (the random variable of a chi-square distribution with ν degrees of freedom) for which the area under the curve and to its right is α. That is, for any α in the interval $0 < \alpha < 1$, χ_α^2 is that value of χ^2 for which

$$P(\chi^2 > \chi_\alpha^2) = \alpha$$

Figure 8.1 shows three different values of $\chi_{.05}^2$. Values of χ_α^2 for selected values of α and ν are shown in Table IV in the Appendix. Since a chi-square distribution is nonsymmetric, it is necessary that α level points, denoted χ_α^2 and $\chi_{1-\alpha}^2$, in both the upper and lower tails be given. (The area to the right of a χ^2 point is used instead of that to the left since this is typically the way a chi-square distribution is applied.) For an illustration with $\alpha = 0.05$ and $\nu = 4$ see Figure 8.2.

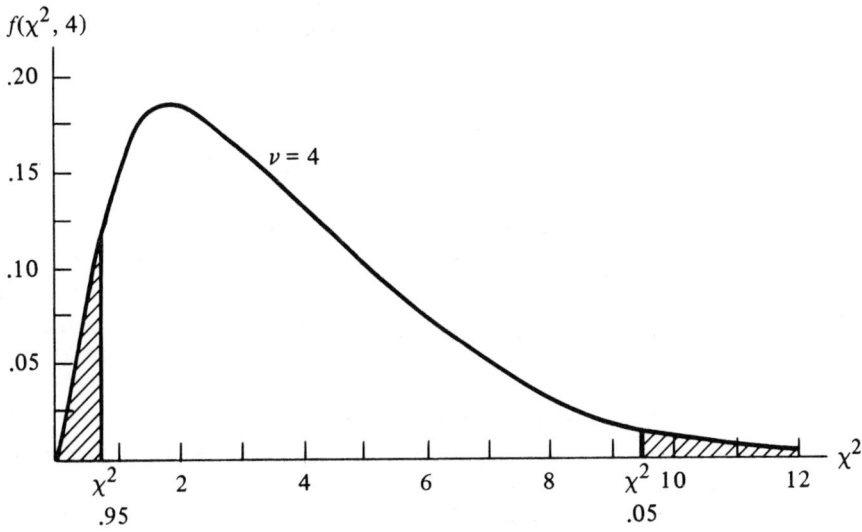

Figure 8.2. Greater Than Percentiles for the Chi-square Distribution with 4 Degrees of Freedom

8.2. INFERENCES ABOUT VARIANCES AND STANDARD DEVIATIONS

The general principles of point and interval estimation and tests of hypotheses have already been presented in Chapter 7. Now we give some of the details associated with inferences about variances and standard deviations.

From Section 6.6 we learned that the sample variance s^2 of a random sample is an unbiased estimator of the population variance σ^2, but that the sample standard deviation s underestimates σ. However, for a large sample size the bias in the standard deviation is relatively insignificant.

(8.5) Actually, for sufficiently large samples it can be shown that the sampling distribution of s can be approximated by a normal distribution with mean σ and standard deviation $\sigma/\sqrt{2n}$.

Further, the sample variance and standard deviation are preferred to other measures of variation because they are generally more reliable; i.e., their sampling distributions generally have smaller standard deviations.

The value of a sample standard deviation is taken as the point estimate of the population standard deviation σ. Thus, for example, if we found that the length of life of a random sample of 60 similar bulbs had a standard deviation of 23.2 hours we would estimate the standard deviation of the population of this type bulb to be 23.2 hours. In this case we could assert with probability 0.95 that the true standard deviation does not deviate more than approximately

$$1.96 \frac{23.2}{\sqrt{2(60)}} = 4.15 \text{ hours}$$

from 23.2 hours.

If random samples of size n are drawn from a normal population with mean μ and standard deviation σ, we learned in Section 8.1 that the statistic $[(n-1)s^2]/\sigma^2$ is distributed as chi-square with $n-1$ degrees of freedom. We will illustrate how a chi-square distribution may be used to test the null hypothesis that a population standard deviation equals a specified constant and to establish a $(1 - \alpha_1 - \alpha_2)100$ percent confidence interval for σ. (The reader should note that the statistic to be used does not require any knowledge of the population mean μ—recall that in working with the mean we required that the standard deviation be known or that the sample size be large enough so that the sample standard deviation s could replace σ.)

There are many situations in which we want to be reasonably certain that a material or a process is uniform. As examples, we may want to know that (1) portions taken from a large vat are homogeneous, (2) the fertility of each test row of an agricultural plot is about the same, (3) members of a group of trainees have about the same potential to master a skill, and so on.

EXAMPLE 8.1. A new machine is to be compared with a standard for the precision with which it cuts off pieces. The variance of the standard is 0.060. We wish to determine at the 0.05 level of significance whether the new machine has a larger variance. Suppose a sample of 20 pieces cut off by the new machine has a variance of 0.085.

Use the general procedure outlined in Section 7.10 as follows:

1. $H_o: \sigma^2 = 0.060$ and $H_a: \sigma^2 > 0.060$

If the statement of the problem had been in terms of the standard deviation, then the hypotheses would have been too. (In other problems a one-sided to the left or two-sided alternative hypothesis might be required.)

2. Assume the 20 sample values were randomly taken from a normal population with unknown variance σ^2. Let the significance level be $\alpha = 0.05$.

3. The statistic to be used is

$$\chi^2 = \frac{(n-1)s^2}{\sigma^2} = \frac{19s^2}{0.060}$$

and it has the chi-square distribution with 19 degrees of freedom. The statistic χ^2 is generally easier to apply than s^2 since the critical value (or values) can be read directly from Table IV in the Appendix. Note that had the hypotheses been stated in terms of the standard deviation we still would have used the same statistic since a test on a variance σ^2 is equivalent to a test on a standard deviation σ.

4. Since only very large values of s^2 would make us want to reject H_o and accept H_a, we use the upper (right) tail of the chi-square distribution with 19 degrees of freedom. From Table IV we find that the upper 0.05 level value of χ^2 is $\chi^2_{.05} = 30.1$. Thus, the critical region is

$$CR = \{\chi^2 | \chi^2 > 30.1\}$$

Note that we could have substituted 30.1 for χ^2 in the equation $\chi^2 = 19s^2/(0.060)$, and then solved for s^2 to get

$$s^2 = (0.060)(30.1)/19 = 0.095$$

Thus, in terms of s^2 the critical region is

$$CR = \{s^2 | s^2 > 0.095\}$$

5. Typically, the sample values would be listed in some order and the variance would be computed. Since we already know the value of s^2, we find directly that the sample statistic is

$$\chi^2 = \frac{19(0.085)}{0.060} = 26.9$$

6. Since the sample value of 26.9 does not fall in the critical region, we fail to reject the null hypothesis. That is, we do not have enough evidence to say the population variance is greater than 0.060 or that the population standard deviation is greater than $\sqrt{0.060} \doteq 0.245$. Put another way, the sample variance is not significantly greater than 0.060. In terms of the new machine, we would say that there is no conclusive evidence that it is less precise in cutting off pieces than the standard.

A confidence interval for a normal population variance may be derived by use of the χ^2 statistic. Let χ^2_1 denote the value of the random variable χ^2 for which the area to its left and under the chi-square curve with ν degrees of freedom is α_1. Let χ^2_2 denote the value of the random variable χ^2 for which the area to its right and under the chi-square curve with v degrees of freedom is α_2 where χ_2^2 is greater than χ_1^2. Then $1 - \alpha_1 - \alpha_2$ is the area under the chi-square curve with v degrees of freedom between χ_1^2 and χ_2^2. That is, the probability is $1 - \alpha_1 - \alpha_2$ that

(8.6) $$\chi_1^2 < \frac{SS}{\sigma^2} < \chi_2^2$$

See Figure 8.3 for a diagram.

By applying simple algebra to Inequality (8.6) we obtain

(8.7) $$\frac{SS}{\chi_2^2} < \sigma^2 < \frac{SS}{\chi_1^2}$$

which is a $1 - \alpha_1 - \alpha_2$ confidence interval of σ^2. In case $\alpha_1 = \alpha_2 = \alpha/2$, the corresponding $1 - \alpha$ confidence interval for σ^2 may be written as

(8.8) $$\frac{SS}{\chi_{\alpha/2}^2} < \sigma^2 < \frac{SS}{\chi_{1-\alpha/2}^2}$$

Since the standard deviation is the positive square root of the variance, a $1 - \alpha_1 - \alpha_2$ confidence interval of σ is

(8.9) $$\sqrt{\frac{SS}{\chi_2^2}} < \sigma < \sqrt{\frac{SS}{\chi_1^2}}$$

The length of the $1 - \alpha_1 - \alpha_2$ confidence interval of σ^2 given by Inequality (8.7) is

$$SS\left[\frac{1}{\chi_1^2} - \frac{1}{\chi_2^2}\right]$$

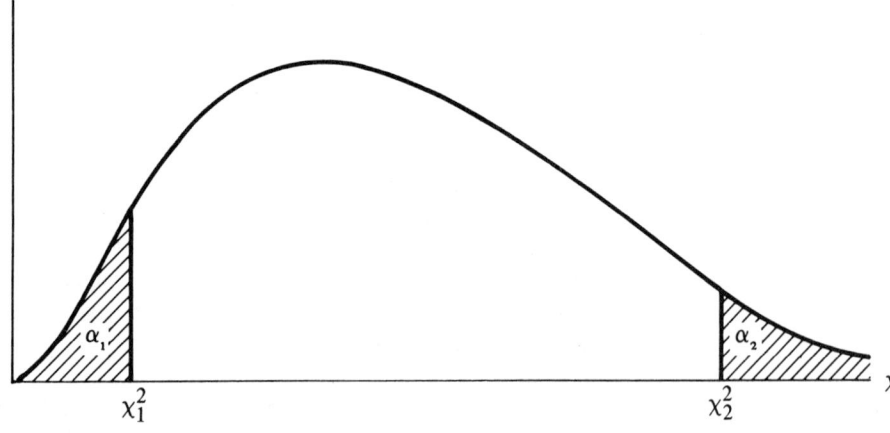

Figure 8.3. Percentiles of a Chi-square Distribution

and is shortest when

(8.10) $$\frac{1}{\chi_1^2} - \frac{1}{\chi_2^2}$$

is a minimum, SS being fixed for a particular sample. If appropriate tables were available, the values of χ_1^2 and χ_2^2 which make Expression (8.9) a minimum could be found by inspection, but in the absence of such tables a $1 - \alpha_1 - \alpha_2$ confidence interval for σ^2 is usually found by Inequality (8.8). Such a choice gives an interval with length very near the minimum unless ν is small. Sometimes in practice zero is taken as the lower limit of the interval for σ^2 and the upper limit is computed for $\chi_1^2 = \chi_{1-\alpha}^2$. Some further considerations of length of intervals for σ^2 and σ are given in the next illustration.

EXAMPLE 8.2. Use the data of Example 8.1 to find 95 percent confidence intervals for σ^2 and σ. Compare lengths of intervals for σ^2; for σ.

Using Table IV with 19 degrees of freedom we find

$$\chi_1^2 = \chi_{0.975}^2 = 8.91$$

and

$$\chi_2^2 = \chi_{0.025}^2 = 32.9$$

Since $n = 20$ and $s^2 = 0.085$, $SS = (20 - 1)(0.085) = 1.615$. Thus, by substituting in Inequality (8.7), we find that

$$\frac{1.615}{32.9} < \sigma^2 < \frac{1.615}{8.91}$$

or

$$0.0491 < \sigma^2 < 0.181$$

Further, since $\chi_1^2 = \chi_{0.95}^2 = 10.1$ and $SS/\chi_1^2 = 0.160$, another 95 percent confidence interval of σ^2 is

$$0 < \sigma^2 < 0.160$$

The second interval of σ^2 is longer than the first, but neither is the shortest 0.95 confidence interval. The first interval is preferred to the second interval and, generally, to any other .95 confidence interval of σ^2.

By taking the positive square roots of the limits of the above two intervals, we find the following two 95 percent confidence intervals of σ, respectively

$$0.22 < \sigma < 0.43$$

and

$$0 < \sigma < 0.40$$

Clearly, the first 0.95 confidence interval of σ is shorter. The reader should be cautioned against expecting the same results with every similar problem.

EXAMPLE 8.3. The length of life of a random sample of 60 similar bulbs had a standard deviation of 23.2. For degrees of freedom not found in Table IV, Statement (8.5) may be used to find an approximate 95 percent confidence interval of σ.

We assume the sample size of 60 is large enough so that we may apply Statement (8.5). Thus, the statistic

$$\frac{s - \sigma}{\frac{\sigma}{\sqrt{2n}}}$$

is distributed approximately as the random variable z of the standard normal distribution. Thus, the probability is approximately $1 - \alpha$ that

(8.11) $$s - z_{\alpha/2}\frac{\sigma}{\sqrt{2n}} < \sigma < s + z_{\alpha/2}\frac{\sigma}{\sqrt{2n}}$$

By using simple algebra on Inequality (8.11) we obtain

(8.12) $$\frac{s}{1 + \frac{z_{\alpha/2}}{\sqrt{2n}}} < \sigma < \frac{s}{1 - \frac{z_{\alpha/2}}{\sqrt{2n}}}$$

which is *an approximate $1 - \alpha$ confidence interval of σ when the sample size is large*. Since $z_{\alpha/2} = z_{0.025} = 1.96$, we find by substituting in Inequality (8.12) that

$$\frac{23.2}{1 + \frac{1.96}{\sqrt{2(60)}}} < \sigma < \frac{23.2}{1 - \frac{1.96}{\sqrt{2(60)}}}$$

or

$$19.7 < \sigma < 28.3$$

Thus, an approximate 95 percent confidence interval for the true standard deviation σ of the length of life of bulbs of a certain type is from 19.7 hours to 28.3 hours.

8.3. EXERCISES

8.1. The values 24, 7, 19, 10, 25 were randomly drawn from a normal population with standard deviation 10. Find the value of χ^2.

8.2. (a) A random sample of size 17 is drawn from a normal population with variance 30. If $\sum y_i = 163.7$ and $\sum y_i^2 = 2345.67$, find the value of χ^2.
(b) What is the expected value of χ^2 for random samples of size 17 drawn from the population in part (a)?

8.3. A random sample of size 14 is drawn from a normal population with mean μ and standard deviation 5.

(a) What is the expected value of s^2?

(b) Find the variance of s^2. *Hint:* Recall that $V(ky) = k^2 V(y)$ where k is any constant and y is a random variable.

8.4. (a) What value of s^2 corresponds to each of the six values of χ^2 shown in Figure 8.2?

(b) For four degrees of freedom, find the 5 percentile value of s^2. That is, find the value of s^2 corresponding to $\chi^2_{.95}$.

8.5. A random sample of size 11 is drawn from a normal population. The sample variance is found to be 0.5678. Find a .95 confidence interval for the population variance.

8.6. A sample of size 72 is drawn from a normal population with standard deviation 48. Sketch a curve which approximates the curve for the distribution of s.

8.7. A random sample of 50 objects has a standard deviation of 34.5. If we assert with probability 0.90 that the true standard deviation does not deviate more than k units from 34.5, what is the value of k?

8.8. A random sample of size 31 drawn from a normal population has variance 12.34. Find a .90 confidence interval of the population standard deviation σ by (a) Formula (8.9); (b) Formula (8.12).

8.9. For a fixed confidence coefficient and a fixed sample variance the confidence interval of σ^2 given by Expression (8.8) gets shorter as the sample size increases. Suppose $s^2 = 4.56$ and the confidence coefficient is 0.95. Find the confidence interval of σ^2 when $\alpha_1 = \alpha_2 = 0.025$ and (a) $n = 5$; (b) $n = 10$; (c) $n = 30$.

8.10. A sample of size 13 has variance 191.7. Use a two-tailed .05 level test to determine whether the true variance differs from $\sigma^2 = 100$. Write a complete report.

8.11. A type of paper packing bag commonly used by a grocery store has tensile strength which is normally distributed with mean 24.7 pounds and standard deviation 1.5 pounds. A new salesman claims that his company uses an improved process to make paper bags which are stronger and have less variability in tensile strength. If a random sample of 25 paper bags made by the "improved process" has a standard deviation of 1.2 pounds, would the manager of the grocery store be justified in accepting the salesman's claim that his company does make bags with standard deviation of tensil strength less than 1.5 pounds? Use a .05 level test and write a complete report.

8.12. Suppose a canning company packs and distributes number $2\frac{1}{2}$ cans of whole unpeeled apricots in light syrup. The net weight per can is 29 ounces with the apricots accounting for about half the weight. In 12 randomly selected cans the weight of the syrup in ounces was as follows:

13.8   14.1   13.9   14.0   14.1   14.2
13.9   13.6   13.7   13.5   13.7   13.6

(a) Find a .95 confidence interval of the standard deviation of the weight of syrup.
(b) Test the hypothesis $\sigma = .18$ against the alternative hypothesis $\sigma > .18$. Let $\alpha = .05$ and write a complete report of the test procedure.

8.4. DEGREES OF FREEDOM

A chi-square distribution with ν degrees of freedom is a member of a family of distributions, there being one member for each positive integer value of ν. The random variable χ^2 of a chi-square distribution with ν degrees of freedom is said to have ν degrees of freedom. We know the statistic SS/σ^2 has a chi-square distribution with $n - 1$ degrees of freedom provided SS is computed from a random sample of size n taken from a normal population. Thus, since $SS = \sigma^2 \chi^2$, we say SS has $n - 1$ degrees of freedom. Also, since $s^2 = SS/(n - 1)$, we say s^2 has $n - 1$ degrees of freedom. In fact, SS and s^2 are said to have $n - 1$ degrees of freedom regardless of the nature of the parent population.

Sometimes independent random samples of sizes n_1 and n_2 are taken from two populations with a common variance σ^2. Then the sample variances s_1^2 and s_2^2 computed from samples of sizes n_1 and n_2 have $n_1 - 1$ and $n_2 - 1$ degrees of freedom, respectively. Further, it can be shown that the statistic

(8.13) $$s_p^2 = \frac{(n_1 - 1)s_1^2 + (n_2 - 1)s_2^2}{(n_1 - 1) + (n_2 - 1)} = \frac{SS_1 + SS_2}{n_1 + n_2 - 2}$$

called the "pooled" variance estimator of two samples, is an unbiased estimator of the common variance σ^2 and has $n_1 + n_2 - 2$ degrees of freedom. In general, when $SS_1 = (n_1 - 1)s_1^2, \cdots, SS_k = (n_k - 1)s_k^2$ are computed from independent random samples from k populations with a common variance σ^2, it can be shown that

(8.14) $$s_p^2 = \frac{SS_1 + \cdots + SS_k}{(n_1 - 1) + \cdots + (n_k - 1)}$$

called the "pooled" variance estimator of k samples, is an unbiased estimator of the common variance σ^2 and has $n_1 + \cdots + n_k - k$ degrees of freedom. If the k populations are normally distributed, then

(8.15) $$\frac{SS_1 + \cdots + SS_k}{\sigma^2} = \frac{(n_1 + \cdots + n_k - k)s_p^2}{\sigma^2}$$

is distributed as chi-square with $n_1 + \cdots + n_k - k$ degrees of freedom. Note that for each variance estimator the number of degrees of freedom is the divisor required to make the sample variance an unbiased estimator of the population variance σ^2.

We have indicated how the number of degrees of freedom ν of a variance estimator is associated with a chi-square distribution with ν degrees of freedom, but this does not necessarily give the reader any insight into a real understanding of degrees of freedom. Perhaps it would be better to look directly at a variance estimator.

In the simplest case, write

(8.16) $$SS = \sum_{u=1}^{n} (y_u - \bar{y})^2 = \sum_{u=1}^{n} e_u^2$$

where $e_u = y_u - \bar{y}$ is the deviate about the sample mean of the uth observation. Since the sum of the deviates about the sample mean is known to be zero (i.e., $\sum_{u=1}^{n} e_u = 0$), there are only $n - 1$ arbitrary choices (degrees of freedom) possible on the deviates e_u. That is, if $n - 1$ of the deviates are known the nth one must also be known. For example, if $n = 3$ and $e_1 = 2$ and $e_2 = -1$, then $e_3 = -e_1 - e_2 = -2 - (-1) = -1$.

The restriction on the deviates is really due to the fact that they are expressed in terms of the *sample mean*, which in turn is a linear relation among the sample values. So, referring to the last paragraph, the sum of squares SS and corresponding variance estimator s^2 have $n - 1$ degrees of freedom due to the presence of one linear relation among the sample values. Thus, in general, we can say

(8.17) A variance estimator loses one degree of freedom for each independent linear relation used in the definition.

For example, in Equation (8.13) there are two independent linear relations in the expression $SS_1 + SS_2$. So the number of degrees of freedom of $SS_1 + SS_2$ (or of s_p^2) is $n_1 + n_2 - 2$. We will learn more about degrees of freedom in later sections.

8.5. DISTRIBUTION OF z^2

In Section 5.11 we explained how the normal distribution can be used to approximate the binomial distribution of the total number of successes T. Examples 7.10 and 7.13 illustrated how the normal approximates the binomial distribution of the proportion of successes $p = T/n$. Actually, since a proportion p may be viewed as a simple type of mean $\bar{y}$, we may apply the normal approximation [see Formula (6.15) of Section 6.7 and Example 8.9] to inference problems involving two proportions. However, we often wish to generalize to problems with more than two proportions, and for this the chi-square distribution is often very useful.

In this section we relate the chi-square distribution with one degree of freedom to the binomial distribution and illustrate how it applies to a single proportion. In Sections 8.6, 8.8, and 8.9 we show how a chi-square distribution may be applied in many other practical problems where the experimenter has results in terms of frequency counts of mutually exclusive categories.

(8.18) If z is the random variable of a standard normal distribution, it can be shown that z^2 has the chi-square distribution with one degree of freedom.

Since the normal distribution approximates the binomial distribution under certain restrictions, it follows from Statement (8.18) that the chi-square distribution with one degree of freedom also approximates the binomial distribution. Example 8.4 illustrates the relationship between the distribution of z and the distribution of z^2, and Example 8.5 shows how to apply the chi-square approximation.

EXAMPLE 8.4. Use the standard normal distribution to graph the chi-square distribution with one degree of freedom.

First, obtain a histogram of the z^2 distribution from the normal curve probabilities (see Table 8.1). Then connect the mid-points of the tops of the rectangles by a smooth curve (see Figure 8.4). The curve of Figure 8.4 approximates the chi-square distribution with one degree of freedom. (Table 8.1 should be useful in making the transformation from the probabilities of the normal distribution to the probabilities of the chi-square distribution with one degree of freedom. Note that two intervals symmetric about the mean of the normal values correspond to one interval of the chi-square values. The rectangles of the histogram in Figure 8.4 are constructed so that the areas are proportional to the probabilities in the last column of Table 8.1. Since the interval from 6.25 to

Table 8.1. Relation Between the Probabilities of z and the Probabilities of z^2

Interval for z	Interval for z^2	Probability for Interval of z	Interval of z^2 (Combined z Intervals)	Probability for Interval of z^2
$-\infty$ to -2.5	∞ to 6.25	0.0062		
-2.5 to -2.0	6.25 to 4.00	0.0166		
-2.0 to -1.5	4.00 to 2.25	0.0440		
-1.5 to -1.0	2.25 to 1.00	0.0919		
-1.0 to -0.5	1.00 to 0.25	0.1498		
-0.5 to 0.0	0.25 to 0.00	0.1915		
0.0 to 0.5	0.00 to 0.25	0.1915	0.00 to 0.25	0.3830
0.5 to 0.0	0.25 to 1.00	0.1498	0.25 to 1.00	0.2996
1.0 to 1.5	1.00 to 2.25	0.0919	1.00 to 2.25	0.1838
1.5 to 2.0	2.25 to 4.00	0.0440	2.25 to 4.00	0.0880
2.0 to 2.5	4.00 to 6.25	0.0166	4.00 to 6.25	0.0332
2.5 to ∞	6.25 to ∞	0.0062	6.25 to ∞	0.0124

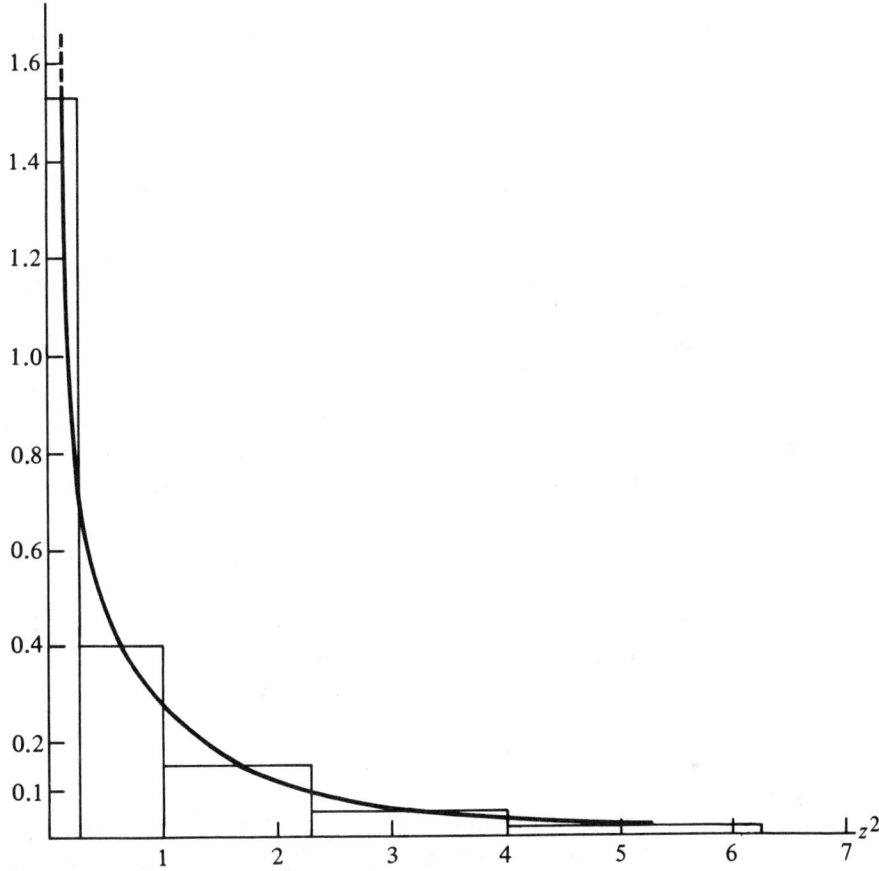

Figure 8.4. Histogram and Curve for Chi-square Distribution with 1 Degree of Freedom

infinity does not have a mid-value, we select the point above $z^2 = 7.625$ as the one through which we draw the curve—the z^2 values of the other mid-values being 0.125, 0.625, 1.625, 3.125, and 5.125.)

EXAMPLE 8.5. In a random sample of 300 students at a university, 135 said they smoked cigarettes regularly. Test the null hypothesis that 50 percent of the students in the university smoke regularly against the alternative hypothesis that the percent is different from 50.

Following the general procedure outlined in Section 7.10 we have:

1. $H_o: \pi = 0.50$ and $H_a: \pi \neq 0.50$

2. Assume the sample of size 300 is randomly taken from the population, and that the two categories are mutually exclusive. Let $\alpha = 0.05$.
3. The statistic is

$$\left(\frac{p - \pi}{\sqrt{\frac{\pi(1-\pi)}{n}}}\right)^2 = \frac{(p - 0.50)^2}{\frac{(0.50)(0.50)}{300}} = 1{,}200\,(p - 0.5)^2$$

and it has the chi-square distribution with 1 degree of freedom. [The sample size is large enough that we need not use the correction factor except as indicated in step 6 below. However, if n were less than 250 we should replace $p - \pi$ in the numerator by $|p - \pi| - 1/(2n)$.]

4. Clearly, the closer the agreement between the hypothesized value of $\pi = 0.50$ and the observed proportion p, the smaller will be the value of χ^2. Thus, only large values of χ^2 would make us want to reject H_o and accept H_a. From Table IV we find that the upper 0.05 level value of χ^2 is 3.84. Thus, the critical region is

$$CR = \{\chi^2 \mid \chi^2 > 3.84\}$$

Note that we test a null hypothesis with a two-sided alternative hypothesis by using one tail of the chi-square distribution.

5. Since $p = \frac{135}{300} = 0.45$, the sample statistic is

$$\chi^2 = 1{,}200(0.45 - 0.50)^2 = 3.00$$

6. Since the sample value of $\chi^2 = 3.00$ does not fall in the critical region, we fail to reject the null hypothesis. Thus, we say the proportion of smokers in the sample is not significantly different from 50 percent. [Any time the test statistic falls just inside the critical region the correction factor $1/(2n)$ should be used since it makes the value of the statistic smaller.]

From Example 8.5 it should be clear that for tests on proportions *only a null hypothesis with a two-sided alternative hypothesis can be tested with the chi-square statistic*. If a one-sided alternative hypothesis is required, then the normal approximation must be used. Further, *the chi-square distribution is not useful in finding a confidence interval for π.*

The above test can be extended to any number of mutually exclusive categories, but the *computing form* of the statistic χ^2 is different from Step 3 of Example 8.5. To develop the new computing form for a k-category problem, first suppose the n random observations are classified into two mutually exclusive categories C_1 and C_2 with corresponding true proportions π_1 and $\pi_2 = 1 - \pi_1$ of observations. Let f_1 and $f_2 = n - f_1$ denote the frequencies of sample values; i.e., observed frequencies, which fall in categories C_1 and C_2, respectively. Let

$h_1 = n\pi_1$ and $h_2 = n\pi_2 = n - h_1$ denote the hypothetical or expected frequencies in categories C_1 and C_2, respectively.

Since $f_1 = T$ is a random variable with a binomial distribution, we know that for large n the statistic

(8.19) $$\frac{f_1 - n\pi_1}{\sqrt{n\pi_1\pi_2}}$$

is approximately normally distributed with mean zero and variance one, provided $n\pi_1 \geq 5$ and $n\pi_2 \geq 5$. [If n is less than 250, say, we should replace f_1 in Expression (8.19) by $f_1 - \frac{1}{2}$ whenever $f_1 > n\pi_1$; otherwise we replace f_1 by $f_1 + \frac{1}{2}$.] Thus, for large n the statistic

(8.20) $$\frac{(f_1 - n\pi_1)^2}{n\pi_1\pi_2}$$

is approximately distributed as chi-square with one degree of freedom, provided $h_1 = n\pi_1 \geq 5$ and $h_2 = n\pi_2 \geq 5$. Further, since it can be shown (see Exercise 8.20) that

(8.21) $$\frac{(f_1 - h_1)^2}{h_1} + \frac{(f_2 - h_2)^2}{h_2} = \frac{(f_1 - n\pi_1)^2}{n\pi_1\pi_2}$$

it follows that for large n the statistic

(8.22) $$\sum_{i=1}^{2} \frac{(f_i - h_i)^2}{h_i}$$

denoted by χ'^2 and read "chi prime square", is approximately distributed as chi-square with one degree of freedom, provided both h_1 and h_2 are equal to or greater than 5. Expression (8.22) is denoted by χ^2 in most places, but we use χ'^2 to indicate that it is at best only approximately distributed as chi-square with one degree of freedom. In case $n < 250$, replace $f_i - h_i$ by $|f_i - h_i| - \frac{1}{2}$ in Formula (8.22).

The reader should recognize that Expression (8.22) has certain computational advantages over Expression (8.19) and that it is preferred to Expression (8.20) because of its symmetry in terms of the frequencies. But the primary reason for introducing Expression (8.22) is that it can be easily generalized to any number of categories (see Section 8.6).

Referring to Example 8.5, we illustrate the computation of χ'^2. Since $h_1 = n\pi_1 = 300(0.50) = 150$ and $h_2 = n - h_1 = 300 - 150 = 150$, it follows that

$$\chi'^2 = \frac{(135-150)^2}{150} + \frac{(165-150)^2}{150} = \frac{(-15)^2}{150} + \frac{(15)^2}{150} = \frac{2(225)}{150} = 3.00$$

Thus, we have also verified Equation (8.21).

8.6. GOODNESS OF FIT

In Section 8.5 we used the chi-square distribution with one degree of freedom to determine whether the observed frequencies of two categories matched (or fitted) the corresponding hypothetical frequencies sufficiently well. That is, in the simple case with only two categories we tested the goodness of the fit of observed frequencies to hypothetical frequencies. These basic ideas can be extended to test the **goodness of fit** of observed data when there are any number of categories. Applications of the chi-square goodness-of-fit test are almost as extensive as problems in which counting occurs. Some examples may be found in

1. The analysis of data from opinion polls
2. The comparison of the grade distribution this year with that of former years
3. A study of defective parts made by different machines
4. Checking characteristics of experimental animals (or insects) against theoretical values
5. Studies which relate to our daily lives—weather, health, news, traffic, amount of sleep, type of diet, insurance, working conditions, budget, and so on

Before illustrating the test procedure, we describe the notation and statistic to be applied. Suppose each value of a population (quantitative or qualitative, discrete or continuous) falls in exactly one of k mutually exclusive categories $C_1, C_2, \cdots, C_k$. Let π_i denote the true proportion of values (or probability of a value) falling in category C_i ($i = 1, 2, \cdots, k$) where

$$(8.23) \qquad \sum_{i=1}^{k} \pi_i = \pi_1 + \pi_2 + \cdots + \pi_k = 1$$

For a random sample of n observations let f_i and $h_i = n\pi_i$ denote the observed and hypothetical frequencies, respectively, in category C_i where

$$(8.24) \qquad \sum_{i=1}^{k} f_i = n \quad \text{and} \quad \sum_{i=1}^{k} h_i = n$$

Then, for large n, it can be shown that the statistic

$$(8.25) \qquad \chi'^2 = \sum_{i=1}^{k} \frac{(f_i - h_i)^2}{h_i}$$

is approximately distributed as chi-square with $k - 1$ degrees of freedom, provided each h_i is equal to or greater than 5.

The statistic of Statement (8.25) is useful in testing the null hypothesis that the expected frequencies are $h_1, h_2, \cdots, h_k$ against the alternative hypothesis that at least one is not as specified. In case any category C' has an expected frequency less than 5 the experimenter should either (1) combine C' with another category or (2) simply delete C' from the experiment, and then proceed to test the goodness of fit of the data to the resulting distribution containing the remaining $k - 1$ categories. If each of two or more categories has an expected frequency less than 5, the same principle should be applied (see Example 8.7 for an illustration). The correction factor $\frac{1}{2}$ is not applied to Equation (8.25) except when $k = 2$ *and* $n < 250$.

EXAMPLE 8.6. A random sample of 1,000 undergraduate students represents 4 percent of the undergraduate enrollment at a large university. Of the undergraduate enrollment it is known that 35 percent are freshmen, 25 percent are sophomores, 21 percent are juniors, and 19 percent are seniors. Suppose 340 of the students in the sample were freshmen, 245 were sophomores, 195 were juniors, and 220 were seniors. We wish to determine, at the 0.05 level, if the sample proportions of students are significantly different from the actual proportions of students in the four classes. (This may also be considered a test of whether or not a random sample is a representative one.)

Let the subscripts 1, 2, 3, and 4 denote freshmen, sophomores, juniors, and seniors, respectively. Thus, for example, π_1 denotes the true proportion of freshmen. Now, by the general test procedure we have:

1. $H_o: \pi_1 = 0.35, \pi_2 = 0.25, \pi_3 = 0.21$, and $\pi_4 = 0.19$
   H_a: At least one π_i is different from the value specified in H_o.
   Note that the null hypothesis could be stated as $h_1 = n\pi_1 = 350, h_2 = 250, h_3 = 210$, and $h_4 = 190$. (Normally we do not know the values of $\pi_1, \cdots, \pi_k$.)

2. Assume that each of 1,000 random observations can be placed in exactly one of four categories. (The reader should note that no statement of normality is made.) Let $\alpha = 0.05$.

3. Under the temporary assumption that the null hypothesis is true, the statistic to be used is

$$\chi'^2 = \frac{(f_1 - 350)^2}{350} + \frac{(f_2 - 250)^2}{250} + \frac{(f_3 - 210)^2}{210} + \frac{(f_4 - 190)^2}{190}$$

We assume that $n = 1,000$ is large enough for χ'^2 to be distributed approximately as a chi-square distribution with three degrees of freedom.

4. When the observed frequencies are identical to the expected frequencies $\chi'^2 = 0$. The more the observed frequencies deviate from the expected frequencies the larger χ'^2 becomes. Thus, the upper tail of the chi-square distribu-

tion with three degrees of freedom is used to test the null hypothesis. Since $\chi^2 = 7.81$, the critical region is

$$CR = \{\chi'^2 \mid \chi'^2 > 7.81\}$$

5. The observed frequencies are

$$f_1 = 340 \quad f_2 = 245 \quad f_3 = 195 \quad f_4 = 220$$

Thus

$$\chi'^2 = \frac{(-10)^2}{350} + \frac{(-5)^2}{250} + \frac{(-15)^2}{210} + \frac{(30)^2}{190} \doteq 6.19$$

6. Since $\chi'^2 = 6.19$ falls in the noncritical region, we fail to reject the null hypothesis. Thus, we conclude that the sample proportions of students are not significantly different from the actual proportions of students in the four classes.

If the value of χ'^2 had fallen in the critical region, we would have rejected the null hypothesis and concluded that at least one observed proportion was significantly different from the corresponding true proportion (or, in most other situations, hypothetical proportion). In this case, since we know the hypothetical proportions are the true proportions, we would have made a type I error, and we would likely attribute the error to the large number of seniors in the sample.

The use of Equation (8.25) requires that each $h_i = n\pi_i$ be known or be specified by the null hypothesis. Often this is not the case. Instead, the observed frequencies are used to determine estimates $\hat{\pi}_i$ of the parameters π_i which in turn replace the parameters in Equation (8.25).

Then, for large n, it can be shown that the statistic

(8.26) $$\chi''^2 = \sum_{i=1}^{k} \frac{(f_i - n\hat{\pi}_i)^2}{n\hat{\pi}_i}$$

is distributed approximately as chi-square with $k - 1 - c$ degrees of freedom, where c denotes the number of independent parameters of a distribution which must be estimated in order to determine the estimators $\hat{\pi}_i$.

This is illustrated in the following example.

EXAMPLE 8.7. How well does a normal curve fit the distribution of 125 test scores of Table 2.4 (see page 17)?

Since we wish to compare the test score data with a particular normal distribution, we must first estimate two parameters, μ and σ^2, in order to determine which member of the family of normal curves should be used to fit the data. Assuming the 125 scores to be a random sample from a normal population with mean μ and variance σ^2, we find from Example 3.6 that

$\bar{y} = 68.640$ and from Example 3.17 that

$$s^2 = \frac{1361100}{(125)(124)} = 87.8129$$

and

$$s = 9.371$$

Thus, we use $\hat{\mu} = 68.64$ and $\hat{\sigma} = 9.371$ as estimates of μ and σ.

Table 8.2 illustrates how the normal curve with mean 68.64 and standard deviation 9.371 is fitted to the data. The entries in column 4 are differences of successive entries in column 3, it being understood that the extreme values are .0000 and 1.0000. All other values are obtained as indicated.

By adding the small areas in the tails of the normal distribution (shown in Table 8.2) to those of the extreme classes of Table 2.4 we obtain the hypothetical frequencies shown in Table 8.3. Thus, we are able to compare the observed and

Table 8.2. Fitting of Normal Curve to Test Score Histogram

Upper Class Boundary y	$\dfrac{y - 68.64}{9.371}$ z'	Area to Left of z' $P(z < z')$	Area for class Interval $\hat{\pi}$	Hypothetical Frequency $125\hat{\pi}$
39.5	−3.11	.0009	.0009	0.11
44.5	−2.58	.0049	.0040	0.50
49.5	−2.04	.0207	.0158	1.98
54.5	−1.51	.0655	.0448	5.60
59.5	−0.98	.1635	.0980	12.25
64.5	−0.44	.3300	.1665	20.81
69.5	0.09	.5359	.2059	25.74
74.5	0.63	.7357	.1998	24.98
79.5	1.16	.8770	.1413	17.66
84.5	1.69	.9545	.0775	9.69
89.5	2.23	.9871	.0326	4.08
94.5	2.76	.9971	.0100	1.25
	Area to right of 2.76		.0029	0.36

Table 8.3. Comparison of Observed and Hypothetical Test Score Frequencies

Class Boundaries	Observed Frequency	Hypothetical Frequency	Class Boundaries	Observed Frequency	Hypothetical Frequency
39.5–44.5	1	0.61	69.5–74.5	34	24.98
44.5–	3	1.98	74.5–	12	17.66
49.5–	7	5.60	79.5–	13	9.69
54.5–	8	12.25	84.5–	1	4.08
59.5–	16	20.81	89.5–94.5	2	1.61
64.5–69.5	28	25.74			

hypothetical frequencies for the 11 classes of scores. The greatest difference in frequencies is for class 69.5–74.5, but the fit does not generally appear to be satisfactory.

Four of the eleven classes have hypothetical frequencies less than 5. Thus, before testing the goodness of fit we decide to combine the first three classes to make a new first class and the last two classes to make a new last class, the eighth class. Now, since the resulting eight classes are mutually exclusive and all classes have hypothetical frequencies in excess of 5, we may use Equation (8.26) to compute an approximate random value of a chi-square distribution with $8 - 1 - 2 = 5$ degrees of freedom. We find

$$\chi'''^2 = \frac{(11 - 8.2)^2}{8.2} + \frac{(8 - 12.2)^2}{12.2} + \cdots + \frac{(13 - 9.7)^2}{9.7} + \frac{(3 - 5.7)^2}{5.7}$$
$$= 11.2$$

Since the upper 0.05 value of the chi-square distribution with 5 degrees of freedom is 11.1, the critical region is

$$CR = \{\chi^2 | \chi^2 > 11.1\}$$

If χ'''^2 were distributed as chi-square with 5 degrees of freedom, we would reject the null hypothesis and conclude, with a 5 percent chance of error, that the data did not come from a normal population. But, as we shall see shortly, the evidence is doubtful.

Following the general procedure outlined in Section 7.10 we summarize the test as follows:

1. H_o: the data came from a normal population.
   H_a: the data did not come from a normal population.

2. Assume that 125 observations are randomly drawn from a population with 8 mutually exclusive categories. Let $\alpha = 0.05$.

3. The statistic is

$$\chi'''^2 = \sum_{i=1}^{8} \frac{(f_i - 125\hat{\pi}_i)^2}{125\hat{\pi}_i}$$

and it is approximately distributed as the chi-square distribution with $k - 1 - c = 8 - 1 - 2 = 5$ degrees of freedom. Since the sample size is not large, the statistic χ'''^2 may not be a very good approximation to χ^2. (However, in case the computed value of χ'''^2 is just inside the critical region, we shall *not* consider any specific correction as we did with two categories.)

4. The critical region for a χ^2 statistic is given above, and it is assumed the critical region for χ'''^2 would be approximately the same; that is

$$CR \doteq \{\chi'''^2 | \chi'''^2 > 11.1\}$$

(In the absence of a correction factor, the critical point should actually be larger.)

5. See the computations in Tables 8.2 and 8.3 which led to the value $\chi''^2 = 11.2$. As we have observed in examples with two categories, the computed value of the statistic χ''^2 would be smaller if a correction were made for continuity. In this case, a corrected value of χ''^2 would very likely be less than 11.1.

6. For reasons already noted, it is unlikely that we should reject the null hypothesis. Even if we knew the sample statistic value fell in the noncritical region and we failed to reject the null hypothesis that the data were drawn from a normal population with mean $\mu = 68.64$ and standard deviation $\sigma = 9.371$, it would be a gross error to conclude that the data did come from this distribution. For there are numerous distributions with similar shapes that would have approximately the same proportions in each of the eight categories selected. In case the null hypothesis is not rejected, the real point to note is that *the data very likely came from a distribution not greatly different from the particular distribution specified by the null hypothesis.*

We have illustrated how the chi-square goodness-of-fit test may be applied to a single variable of classification. The same procedures may be applied to two or more variables. The following is a simple illustration of a famous problem (Mendelian theory of inheritance) with two variables.

EXAMPLE 8.8. Suppose each of two characteristics is inherited in the ratio 3 to 1, written $3:1$. Then when these two characteristics are crossed the progeny should have, according to the Mendelian theory, frequencies in the ratios $9:3:3:1$. That is, $\frac{9}{16}$ should have both dominant characteristics, $\frac{3}{16}$ a dominant first characteristic and recessive second characteristic, $\frac{3}{16}$ a recessive first characteristic and dominant second characteristic, and $\frac{1}{16}$ both recessive characteristics.

Suppose that in a hypothetical experiment with peas, the results of two crossed characteristics ("round" and "yellow") were that 274 were round and yellow, 93 round and green, 96 wrinkled and yellow, and 37 wrinkled and green. Do these frequencies "fit" the theoretical assumption that "round" is the dominant shape and "yellow" is the dominant color in which the true ratios are $9:3:3:1$?

Let π_1 denote the true proportion of peas which are round and yellow, π_2 denote the true proportion of peas which are round and green, and so forth. Now by the general test procedure we have:

1. $H_o: \pi_1 = \frac{9}{16}, \pi_2 = \frac{3}{16}, \pi_3 = \frac{3}{16}, \pi_4 = \frac{1}{16}$
H_a: At least one π_i is different from the value specified in H_o.

2. Assume that each of 500 random observations can be placed in exactly one of four categories. Let $\alpha = 0.05$.

3. Under the temporary assumption that the null hypothesis is true, $h_1 = 500(\frac{9}{16}) \doteq 281.2$, $h_2 = 500(\frac{3}{16}) \doteq 93.8$, $h_3 = 93.8$, $h_4 = 31.2$, and the statistic is

$$\chi'^2 = \frac{(f_1 - 281.2)^2}{281.2} + \cdots + \frac{(f_4 - 31.2)^2}{31.2}$$

with 3 degrees of freedom.

4. The 0.05 point of the chi-square distribution with 3 degrees of freedom is 7.81. Thus

$$CR \doteq \{\chi'^2 \mid \chi'^2 > 7.81\}$$

5. Substituting the observed frequencies in the formula of step 3 gives

$$\chi'^2 = \frac{(-7.2)^2}{281.2} + \frac{(-.8)^2}{93.8} + \frac{(2.2)^2}{93.8} + \frac{(5.8)^2}{31.2} \doteq 1.32$$

6. Since $\chi'^2 = 1.32$ falls in the noncritical region, we fail to reject the null hypothesis. So we conclude that the sample proportions of peas are not significantly different from those given by Mendelian theory.

8.7. EXERCISES

8.13. Find the estimate s_p^2 of the common variance based on three samples if $n_1 = 10$, $n_2 = 15$, $n_3 = 20$, $s_1 = 12$, $s_2^2 = 289$, and $SS_3 = 4{,}275$.

8.14. For two samples it is found that $s_p = 4.40$ with 23 degrees of freedom. Find s_1 if $n_2 = 15$ and $s_2 = 4.16$.

8.15. Suppose random samples of sizes 12, 15, and 8 are independently drawn from normal populations $N(\mu_1, \sigma)$, $N(\mu_2, \sigma)$, and $N(\mu_3, \sigma)$, what can you say about the distribution of $2(SS_1 + SS_2 + SS_3)$ when $\sigma^2 = \frac{1}{2}$? If $s_p^2 = 1.2$, what is the value of $2(SS_1 + SS_2 + SS_3)$? Note: $N(\mu_1, \sigma)$ denotes "normal with mean μ and standard deviation σ."

8.16. Suppose χ^2 has a chi-square distribution with one degree of freedom. Use the standard normal tables to find
(a) $P\{\chi^2 < 0.36\}$
(b) $P\{\chi^2 > 1.69\}$
(c) $P\{\chi^2 < 3.84\}$
(d) χ_0^2 such that $P\{\chi^2 > \chi_0^2\} = .20$
(e) χ_0^2 such that $P\{\chi^2 < \chi_0^2\} = .925$

8.17. A six appeared 60 times in 300 tosses of a die.
(a) Use the method of Example 8.5 to decide whether the probability of getting a six on a single toss is $\frac{1}{6}$.
(b) Use both Formulas (8.20) and (8.22) to verify the conclusion of part (a).

8.18. (a) During the month of April, 270 out of 600 babies born in a certain city were girls. Use the chi-square distributions to test the hypothesis that of babies born in the city 0.5 are male.

(b) Do part (a) if 135 out of 300 babies born are girls. Compare conclusions of parts (a) and (b) in terms of sample sizes and proportions of girls born.

8.19. Past experience at College A has shown that 30 percent of the students who take a mathematics placement examination are encouraged to enroll in Calculus I. If 30 out of 72 students in the most recent freshman class were encouraged to enroll in Calculus I, would we be justified in concluding that this new group of students is not different in mathematics ability (as determined by the placement examination) from past groups?

E8.20. Prove Equation (8.21).

8.21. A die was tossed 120 times. The numbers (faces with dots) 1, 2, $\cdots$, 6 occurred with frequencies 24, 17, 26, 16, 19, 18, respectively. Could the die be considered well-balanced?

8.22. In a poll, 200 television viewers were asked to indicate which of four regular programs was generally preferred. The results were as follows:

Program	A	B	C	D
Number Preferring Program	62	43	56	39

On the basis of these data, can one deny the claim that the programs are about equally popular with all viewers?

8.23. Among white Europeans the percentages of the various blood groups among the general population are approximately as follows:

Blood Group	A	B	AB	O
Percentage	45	10	5	40

Suppose the observed frequencies of blood groups of a sample of white persons in Country M are 300, 55, 45, 300, respectively. Test to determine whether Country M possesses the same distribution of blood types as the white Europeans.

8.24. According to a genetic model, characteristic 1 is inherited in the ratio 3 to 1, written 3:1, and characteristic 2 is inherited in the ratio 2:1. Then when these two characteristics are crossed the progeny should have, according to the Mendelian theory, frequencies in the ratios 6:2:3:1. In an experiment in which these two characteristics were crossed the corresponding observed frequencies were 96, 36, 38, 10, respectively.

(a) Are these data consistent with the model?

(b) Are these data consistent with the model for characteristic 2 that the ratio is 2 to 1?

E8.25. For a group of well-known experiments it is accepted that progeny have colors a, b, c in the ratios $9:4:3$. For a similar new experiment a research worker believes that offspring will occur with frequencies in the ratios $8:4:4$. If the research worker actually obtained frequencies of $n/2$, $n/4$, $n/4$, how many offspring must there be in the sample to reject the usual hypothesis of $9:4:3$ ratios? Use a 0.05 level test.

8.26. Each student in a class of 80 performed an experiment involving three independent trials in which the probability of success on each trial was 0.6. The following results were obtained:

Number of successes	0	1	2	3
Frequency	6	24	35	15

Test to determine whether the observed frequencies deviate significantly from expected frequencies.

8.27. Draw a random sample of 400 digits from Table III in the Appendix and list the observed frequencies for the following classes:

$$\{0, 1, 2, 3\} \quad \{4, 5, 6\} \quad \{7, 8\} \quad \{9\}$$

Use a 0.05 level test to determine whether the sample frequencies are compatible with the theoretic frequencies.

8.28. Is the experimental sampling distribution of Table 6.1 consistent with the theoretical sampling distribution of Table 6.2?

8.29. Is the distribution of the 50 sample means of Table 6.13 compatible with the normal distribution which has mean 49.85 and standard deviation 4.86. *Hint:* Let classes have common length 3 with the first class being 39.45–42.45.

8.8. CONTINGENCY TABLES

The χ'^2 statistic can be extended to many other tests dealing with frequency counts. In this section we consider some ways in which χ'^2 is useful in reaching a decision as to whether two variables of classification are related or not.

8.8.1. Difference between Two Means from Binomial Populations

A manufacturer claims that his newest insecticide (which does not contain DDT) is as effective as the old DDT insecticide. To check the claim of the manufacturer, an investigator conducted an experiment and found the numbers of insects "killed" or "not killed" as indicated in Table 8.4. Such a table of frequencies is called a **2 × 2 contingency table**.

Table 8.4. Two Samples From Binomial Populations

Effect on Insect	Type of Insecticide		Total
	Old	New	
Killed	130	320	450
Not Killed	70	80	150
Sample Size	200	400	600

Let the true proportion of insects killed by the old and new insecticides be denoted by π_1 and π_2, respectively. The null hypothesis is $\pi_1 = \pi_2$ (or $\pi_1 - \pi_2 = 0$), and we shall test it against the alternative hypothesis $\pi_1 \neq \pi_2$. An estimator of $\pi_1 - \pi_2$ is $p_1 - p_2$ where p_1 and p_2 denote the sample proportions of insects killed by the old and new insecticides, respectively.

We know that when a sample of size n_i is randomly drawn from a dichotomous population with true proportion of success equal to π_i, then the sample proportion of success p_i has a binomial distribution with mean π_i and variance $[\pi_i(1 - \pi_i)]/n_i$ where $i = 1$ or 2. Further, whenever both $n_i\pi_i \geq 5$ and $n_i(1 - \pi_i) \geq 5$ are satisfied, p_i is approximately normally distributed with mean π_i and variance $[\pi_i(1 - \pi_i)]/n_i$. If in addition to the above the samples are independently drawn, then we know that $p_1 - p_2$ is approximately normally distributed with mean $\pi_1 - \pi_2$ and variance

$$\frac{\pi_1(1 - \pi_1)}{n_1} + \frac{\pi_2(1 - \pi_2)}{n_2}$$

Thus, by substituting in Equation (6.15), it follows that

(8.27)
$$\frac{(p_1 - p_2) - (\pi_1 - \pi_2)}{\sqrt{\frac{\pi_1(1 - \pi_1)}{n_1} + \frac{\pi_2(1 - \pi_2)}{n_2}}}$$

is distributed approximately as the standard normal variable z.

The statistic in Expression (8.27), with a slight change in the standard error of $p_1 - p_2$, is the one used in testing the equality of π_1 and π_2. Letting $\pi_1 = \pi_2 = \pi$, the standard error of $p_1 - p_2$ reduces to

$$\sqrt{\pi(1 - \pi)\left[\frac{1}{n_1} + \frac{1}{n_2}\right]}$$

Then, if we replace π by p, the proportion of successes in the combined samples, it can be shown that the statistic

(8.28) $$\frac{p_1 - p_2 - 0}{\sqrt{p(1-p)\left[\frac{1}{n_1} + \frac{1}{n_2}\right]}}$$

is distributed approximately as the standard normal variable z. (The reader should see Exercise 8.32 to find how much $\sqrt{p(1-p)}$ varies about $\sqrt{\pi(1-\pi)}$ as p changes.)

Returning to the test of $H_o: \pi_1 - \pi_2 = 0$ against $H_a: \pi_1 - \pi_2 \ne 0$ we find that

$$CR \doteq \{z \,|\, z < -1.96 \text{ or } z > 1.96\}$$

when $\alpha = 0.05$. Using the data of Table 8.4 and Expression (8.28) we obtain

$$p_1 = \tfrac{130}{200} = 0.65 \qquad p_2 = \tfrac{320}{400} = 0.80$$
$$p_1 - p_2 = -0.15 \qquad p = \tfrac{450}{600} = 0.75$$

$$\sqrt{p(1-p)\left[\frac{1}{n_1} + \frac{1}{n_2}\right]} = \sqrt{(.75)(.25)\left[\frac{1}{200} + \frac{1}{400}\right]} = 0.0375$$

$$\frac{p_1 - p_2}{\sqrt{p(1-p)\left[\frac{1}{n_1} + \frac{1}{n_2}\right]}} = \frac{-0.15}{0.0375} = -4.00$$

Since $-4.00 < -1.96$, the value of the statistic falls in the CR. Thus, we reject H_o and accept H_a. That is, since $p_2 > p_1$ we conclude that the new insecticide is actually better than the old. So, the proportion of insects killed actually *depends* on the type of insecticide used.

Since the statistic in Expression (8.28) is approximately normally distributed, its square

(8.29) $$\frac{(p_1 - p_2)^2}{p(1-p)\left[\frac{1}{n_1} + \frac{1}{n_2}\right]}$$

is distributed approximately as *chi-square with 1 degree of freedom*. This suggests, recalling the discussion following Example 8.5, that we may use

(8.30) $$\chi'^2 = \sum_{i=1}^{4} \frac{(f_i - h_i)^2}{h_i}$$

as an alternative way to compute a value given by Expression (8.29). The four observed and four hypothetical frequencies of a 2×2 contingency table are denoted by the symbols f_i and h_i, respectively, where $i = 1, 2, 3, 4$.

For Table 8.4 the observed frequencies are $f_1 = 130$, $f_2 = 320$, $f_3 = 70$, and $f_4 = 80$. The corresponding hypothetical frequencies are obtained from p, an estimate of the true proportion of success in the combined populations. The hypothetical frequency of *insects killed* by the old insecticide is the estimate

$$h_1 = 200p = 200(.75) = 150$$

and the hypothetical frequency of *insects killed* by the new insecticide is the estimate

$$h_2 = 400p = 400(.75) = 300$$

By subtracting these hypothetical frequencies from their respective sample sizes we obtain

$$h_3 = 200 - 150 = 50 \quad \text{and} \quad h_4 = 400 - 300 = 100$$

as the hypothetical frequencies of *insects not killed* by the old and the new insecticides. Substituting in Equation (8.30) gives

$$\chi'^2 = \frac{(130-150)^2}{150} + \frac{(320-300)^2}{300} + \frac{(70-50)^2}{50} + \frac{(80-100)^2}{100} = 16$$

which is what we get by squaring the -4.00 obtained from Expression (8.28). For short-cut formulas of χ'^2 see Exercise 8.36. An exact test is described in Exercise 8.37.

Since the statistic in Expression (8.29) is distributed approximately as chi-square with one degree of freedom and since Expressions (8.29) and (8.30) give the same values for a given set of data, it is not surprising that χ'^2 of Expression (8.30) is also distributed approximately as chi-square with one degree of freedom. That χ'^2 has one degree of freedom could be observed from the fact that when one hypothetical frequency in a 2×2 contingency table is computed the other three may be found by subtraction from the sample sizes or totals.

The primary purpose for introducing χ'^2 in this section was to make the next generalization easier to follow. Actually, when χ'^2 is used we are limited to the specific type of test described in connection with Table 8.4. However, the statistic in Expression (8.28) may be applied to tests with one-sided alternative hypotheses, and Expression (8.27) may be changed so as to test hypotheses of the form $\pi_1 - \pi_2 = k$, k being a constant, or to establish a confidence interval for $\pi_1 - \pi_2$.

If the means of two dichotomous populations are different (i.e., $\pi_1 \neq \pi_2$), then their variances are certainly different [i.e., $\pi_1(1 - \pi_1) \neq \pi_2(1 - \pi_2)$] unless $\pi_1 + \pi_2 = 1$. Hence, we find the appropriate standardized statistic if in Expression (8.27) we replace $\pi_1(1 - \pi_1)$ by $p_1(1 - p_1)$ and $\pi_2(1 - \pi_2)$ by $p_2(1 - p_2)$. Thus, the resulting statistic

(8.31) $$\frac{p_1 - p_2 - (\pi_1 - \pi_2)}{\sqrt{\frac{p_1(1-p_1)}{n_1} + \frac{p_2(1-p_2)}{n_2}}} \doteq z$$

is used to find a confidence interval for $\pi_1 - \pi_2$ or to test $H_0: \pi_1 - \pi_2 = k$ against either a one-sided or a two-sided alternative hypothesis.

EXAMPLE 8.9. Find an approximate symmetric 0.95 confidence interval for the true difference in the proportion of insects killed by the old and new insecticides.

The appropriate values of the standard normal are $z = \pm 1.96$. Thus, by the methods of Chapter 7 we find the confidence limits of $\pi_1 - \pi_2$ to be

$$p_1 - p_2 \pm 1.96 \sqrt{\frac{p_1(1-p_1)}{n_1} + \frac{p_2(1-p_2)}{n_2}}$$

For the data of Table 8.4, the limits are

$$0.65 - 0.80 \pm 1.96 \sqrt{\frac{(0.65)(0.35)}{200} + \frac{(0.80)(0.20)}{400}}$$

or

$$-0.15 \pm 1.96 \, (0.0392)$$

or

$$-0.15 \pm 0.08$$

Thus, the required 0.95 confidence interval is

$$-0.23 < \pi_1 - \pi_2 < -0.07$$

or

$$0.07 < \pi_2 - \pi_1 < 0.23$$

That is, we have high confidence that the new insecticide can kill from approximately 7 to 23 percent more insects that the old.

8.8.2. Equality of Two or More Means from Binomial Populations

If we wish to test the hypothesis

$$\pi_1 = \pi_2 = \cdots = \pi_c$$

where $c > 2$, the approximate standard normal statistic of Expression (8.28) or (8.31) cannot be used. However, an extension of the χ'^2 statistic of Equation (8.30) may be applied. We illustrate the procedure by

EXAMPLE 8.10. Suppose a random sample of 20 percent of the students at three liberal arts colleges were asked the following question: Should men ever be drafted for military service when war has not been declared by Congress? The responses are shown in Table 8.5, a 2 × 3 contingency table. Counts for responses other then "yes" or "no" have been omitted. Hence, conclusions will be based on only those students who responded with yes or no.

Table 8.5. Number of Responses to Draft Question

Response	Liberal Arts College			Total
	1	2	3	
Yes	180	125	105	410
No	120	125	145	390
Total	300	250	250	800

Let π_1, π_2, and π_3 denote the true proportion of students (based on only yes or no answers) at schools 1, 2, and 3, respectively, who would respond yes, and let p_1, p_2, and p_3 be the corresponding sample proportions of yes responses. Let the proportion of yes responses for the three populations combined and the three samples combined be denoted by π and p, respectively. Designate the six observed frequencies f_i by the same rule as was used in Section 8.8.1. Now, we use the general test procedure to illustrate the method:

1. $H_o: \pi_1 = \pi_2 = \pi_3 = \pi$ or the proportion of yes responses at the three schools is the same. That is, the *response* is *independent* of the *college*.

    H_a: At least one pair of π's is different. That is, the *response* is *not independent* (i.e., dependent) of the *college*.

2. Assume three independent samples are randomly drawn from three dichotomous populations and that each response recorded can be placed in exactly one of two categories. Let $\alpha = 0.05$.

3. Use the statistic

$$\chi'^2 = \sum_{i=1}^{6} \frac{(f_i - h_i)^2}{h_i}$$

which is distributed approximately as the chi-square distribution with two degrees of freedom.

4. Since only large differences $|f_i - h_i|$ would make us want to reject H_o, we use the critical region

$$CR \doteq \{\chi^2 | \chi^2 > 5.99\}$$

5. The proportion of yes responses in the three combined samples is

$$p = \frac{180 + 125 + 105}{300 + 250 + 250} = \frac{410}{800} = 0.5125$$

Thus, by the method of Section 8.8.1, the hypothetical frequencies are

$$h_1 = 300p = 153.75 \doteq 153.8$$
$$h_2 = 250p = 128.125 \doteq 128.1$$
$$h_3 = 250p = 128.125 \doteq 128.1$$
$$h_4 = 300(1 - p) \doteq 146.2$$
$$h_5 = h_6 \doteq 121.9$$

However, since the totals for responses and for colleges are fixed, we could have used h_1 and h_2 and computed the other four h's by subtraction from these totals. This illustrates why χ'^2 has only two degrees of freedom. Finally

$$\chi'^2 = \frac{(180 - 153.8)^2}{153.8} + \frac{(125 - 128.1)^2}{128.1} + \cdots + \frac{(145 - 121.9)^2}{121.9}$$
$$\doteq 17.86$$

Since each of the six hypothetical frequencies is 5 or greater we use χ'^2 as a satisfactory approximation to χ^2.

6. Since 17.86 falls in the critical region we reject H_0 and conclude that the true proportion of yes responses among the three colleges do differ. That is, the proportion of *yes responses is not independent of* (or is related to or depends on or associated with) *the college*.

EXAMPLE 8.11. In Example 8.10, suppose college 1 is a men's college and both colleges 2 and 3 are women's colleges. Is the proportion of yes responses independent of type of college?

One of the attractive things about contingency tables is that we may combine two or more columns and still use the same test procedure, provided the number of degrees of freedom is changed.

To test

$$H_o : \pi_1 = \frac{\pi_2 + \pi_3}{2} \quad (\text{or } 2\pi_1 - \pi_2 - \pi_3 = 0)$$

against

$$H_a : \pi_1 \neq \frac{\pi_2 + \pi_3}{2} \quad (\text{or } 2\pi_1 - \pi_2 - \pi_3 \neq 0)$$

we use the statistic

$$\chi'^2 = \sum_{i=1}^{4} \frac{(f_i - h_i)^2}{h_i}$$

which is distributed approximately as the chi-square distribution with 1 degree of freedom. The critical region is

$$CR \doteq \{\chi^2 \mid \chi^2 > 3.84\}$$

The hypothetical frequencies are computed in the usual way and are shown in Table 8.6 with the observed frequencies. So

$$\chi'^2 = \frac{(180-153.8)^2}{153.8} + \frac{(230-256.2)^2}{256.2} + \frac{(120-146.2)^2}{146.2} + \frac{(270-243.8)^2}{243.8}$$
$$= 14.65$$

Table 8.6. Number of Responses* to Draft Question

Response	Liberal Arts College Men	Liberal Arts College Women	Total
Yes	180 (153.8)	230 (256.2)	410
No	120 (146.2)	270 (243.8)	390
Total	300	500	800

*The numbers in parentheses are hypothetical frequencies.

Since 14.65 falls in the critical region, we conclude that the proportion of yes responses depends on whether the college is for men or for women.

We may omit the frequencies for one or more columns and still test the independence of response and college (see Exercise 8.41). A sample (i.e., column) may be omitted because it is not part of the hypothesis or because one or two of the two hypothetical frequencies is less than 5.

The concepts illustrated in this section apply generally. If the responses are replaced by successes and failures and the colleges by c samples, where c is any positive integer greater than one, we would use the approximate chi-square statistic

(8.32) $$\chi'^2 = \sum_{i=1}^{2c} \frac{(f_i - h_i)^2}{h_i}$$

with $c - 1$ degrees of freedom to determine whether the true proportion of successes varies from population to population. Also, in the resulting $2 \times c$ contingency table, we could combine or omit any number of columns (keeping at least two) and still use the methods illustrated to test independence of the row and the column variables.

8.8.3. Test of Independence in $k \times c$ Contingency Tables

In Sections 8.8.1 and 8.8.2 we discussed procedures for testing the independence of the probabilities of success of two or more binomial populations. Since, in a binomial distribution, the probability of failure is one minus the probability of success, we also tested the independence of the probabilities of failure in two or more binomial populations. Put another way, we tested the **homogeneity of a sequence of successes (π) and failures $(1 - \pi)$**.

The concepts of these two sections may be extended to a comparison of c multinomial populations described on page 272. That is, if $\pi_1, \pi_2, \cdots, \pi_k$ denote the true probabilities of values falling in the k mutually exclusive categories of a population, then we may test to determine whether c samples could have been drawn from populations with the same set of probabilities. (Of course, it is assumed that the k categories are defined in the same way for each population.) If we fail to reject the hypothesis, we say the samples could have come from populations with the same set of probabilities $\pi_1, \pi_2, \cdots, \pi_k$. That is, the set of k proportions $(p_1, p_2, \cdots, p_k)$ does not differ significantly from sample to sample. In case the hypothesis is rejected we say some of the samples were drawn from different populations. Such a test may be termed a test of homogeneity of a sequence of c sets of probabilities $\pi_1, \pi_2, \cdots, \pi_k$. It is also called a test of independence of the set of k true probabilities and c populations.

The test statistic is

(8.33)
$$\chi'^2 = \sum_i^c \sum_j^k \frac{(f_{ij} - h_{ij})^2}{h_{ij}}$$

where f_{ij} and h_{ij} denote the observed and hypothetical frequencies, respectively, in the jth category of the ith population. The hypothetical frequencies are computed in the usual way. It can be shown that χ'^2 is approximately distributed as a chi-square distribution with $(k - 1)(c - 1)$ degrees of freedom.

In the $k \times c$ contingency table discussed above the column totals are fixed, but the row totals are left to chance (i.e., they are values of random variables). There is a second type of $k \times c$ contingency table in which the grand total is fixed, but both the row and column totals are chance variables. As an illustration of the second type, suppose 400 random students at a university responded to two separate questions as indicated in Table 8.7. We use this sample information to determine whether the university student stand on issues is related to political preference.

We test the null hypothesis that "the university student stand on issues is independent of political preference" against the alternative hypothesis that

Table 8.7. Sample of 400 Students at a University

		Political Preference			Total
		Democrat	Republican	Independent	
Stand on Issues	Conservative	25	20	5	50
	Moderate	75	100	25	200
	Liberal	70	30	50	150
	Total	170	150	80	400

"stand on issues is related to political preference" by the same chi-square procedure we have used with other contingency tables of Section 8.8. However, a different interpretation is used in computing the hypothetical frequencies. For example, the hypothetical frequency of conservative Democrats h_{11} is the product of the proportion of conservative students times the proportion of student Democrats times the total number of students. That is

$$h_{11} = \left(\frac{50}{400}\right)\left(\frac{170}{400}\right)(400) = \frac{(50)(170)}{400} \doteq 21.2$$

Similarly, the hypothetical frequency of moderate Democrats is

$$h_{12} = \left(\frac{200}{400}\right)\left(\frac{170}{400}\right)(400) = \frac{(200)(170)}{400} = 85.0$$

We could compute the 9 hypothetical frequencies in this way, but it should be apparent that we need compute only 4 this way and then obtain the remaining 5 by subtraction using row totals or column totals. The other hypothetical frequencies are

$$h_{13} = 63.8 \quad h_{21} = 18.8 \quad h_{22} = 75.0 \quad h_{23} = 56.2$$
$$h_{31} = 10.0 \quad h_{32} = 40.0 \quad h_{33} = 30.0$$

Thus

$$\chi'^2 = \frac{(25 - 21.2)^2}{21.2} + \frac{(75 - 85.0)^2}{85.0} + \cdots + \frac{(50 - 30.0)^2}{30.0}$$
$$\doteq 44.54$$

For a chi-square distribution with $(3 - 1)(3 - 1) = 4$ degrees of freedom and $\alpha = 0.05$ the critical region is

$$CR \doteq \{\chi^2 | \chi^2 > 9.49\}$$

Since 44.54 falls in the critical region we conclude that "stand on issues" is related to "political preferences" of students at this university. However, we do not know precisely to what to attribute the relation.

In the computation of the hypothetical frequencies for Table 8.7 we observed the following general rule for any $k \times c$ contingency table:

> The hypothetical frequency of any cell in a contingency table may be obtained by dividing the grand total into the product of the row total and column total to which the cell belongs.

This rule works equally well for the contingency tables of Sections 8.8.1 and 8.8.2 even though the hypothetical frequencies have a different interpretation. For special contingency tables there are special computational rules (see Exercises 8.36, 8.40, and 8.44).

In a contingency table with more than four cells, when we conclude two variables are related it is difficult to know the "direction" of the relationship. This is because the differences $f_{ij} - h_{ij}$ are squared and the alternative hypothesis is always a two-sided alternative. However, we may determine to some extent the nature of the relationship between two variables by testing a 2×2 part of the contingency table or by combining rows or columns (see Example 8.11) so that a 2×2 table is obtained. Then we can use the standard normal z statistic and the methods of Section 8.8.1 to test the significance of one-sided alternative hypotheses. The reader should be cautioned about interpreting a relationship between two variables as a *causal relationship*. The statistical analysis alone is not enough to claim that one variable causes another variable to behave in a certain way.

The total sample size n in a contingency table must be large enough for the discrete χ'^2 statistic to be a satisfactory approximation of the continuous χ^2 statistic. The size of n actually depends on the number of cells and the sizes of the row and column totals. Generally, as the number of cells increases, the total sample size must increase. The rule usually applied is an extension of the one given in Section 8.5. That is,

> We may use the χ'^2 statistic to approximate χ^2 whenever each cell in a $k \times c$ contingency table has a hypothetical frequency of five or larger.

So, when any hypothetical frequency is less than five it is advisable to make some kind of modification in the contingency table. The usual method is to ignore rows or columns containing cells with hypothetical frequencies below five or to combine such rows or columns with others. Of course, rows or columns should be combined only when it "makes sense" to do so; otherwise, it would be better to exclude those rows or columns with small cell totals from the analysis.

The two-variable contingency table can be extended to any number of variables. When the variables are not independent it is much more difficult to describe the relationship (or relationships) which do exist. For example, since three two-variable contingency tables can be constructed from a three-variable contingency table, an investigator would want to know whether the significant chi-square for a three-variable contingency table was due to all three variables or some one or more of the two-variable contingency tables.

8.9. EXERCISES

8.30. In a poll taken among university students, 23 out of 100 men favored a certain proposition, and 17 out of 100 women favored it.
  (a) Use Formula (8.28) to test the hypothesis that the true proportions favoring the proposition are equal.
  (b) Show that the values obtained by Formulas (8.29) and (8.30) are the same.

8.31. In a poll of the television viewers in a city, 46 out of 200 men disliked a specific program and 51 out of 300 women disliked it. Use Formula (8.30) to determine whether men and women in the city differ in their opinions of the program.

C8.32. (a) When $\pi = 0.5$ the value of $\sqrt{\pi(1-\pi)}$ is 0.5. Make a table to show how $\sqrt{p(1-p)}$ varies about 0.5 as p changes from 0.25 to 0.75.
(b) When $\pi = 0.4$ the value of $\sqrt{\pi(1-\pi)}$ is 0.490. Show how $\sqrt{p(1-p)}$ varies about 0.490 as p changes from 0.20 to 0.70.
(c) An estimator of the standard error of $p_1 - p_2$ is

$$\sqrt{p(1-p)\left[\frac{1}{n_1} + \frac{1}{n_2}\right]}$$

Make a table to show how this estimator varies as p changes by increments of 0.05 from 0.20 to 0.50 and $n_1 = n_2 = n$ changes from 50 to 200 to 450 to 800 to 1,250.

8.33. In an experiment with bluegrass it was found for Grower A that 900 out of 1,000 seeds germinated and for Grower B that 880 out of 1,000 seeds germinated.
(a) Determine whether Grower A produced seeds with greater germination rate than Grower B.
(b) Find a 99 percent confidence interval for the difference in germination rates of the two growers.

8.34. Manufacturer A claims that his insecticide kills at least 3 percent more ants than the comparable insecticide made by Manufacturer B. Suppose the insecticide of Manufacturer A kills 776 out of 800 ants and the other insecticide kills 1,488 out of 1,600 ants. Do these data confirm the claim of Manufacturer A?

8.35. Use the data of Exercise 8.30 to find a .95 confidence interval for the difference in the proportion of all men and all women in the university who favor the proposition.

E8.36. (**Formulas for 2 × 2 Contingency Tables**). Table 8.8 is a general 2 × 2 contingency table for observed frequencies a, b, c, d with totals $n_1 = a + c$, $n_2 = b + d$, $T_1 = a + b$, $T_2 = c + d$, $n = n_1 + n_2$ (or $T_1 + T_2$)
(a) It is possible to find the value of χ'^2 directly from the observed frequencies without doing any intermediate calculations as with Formulas (8.29) and (8.30). The value is obtained from

(8.34) $$\chi'^2 = \frac{n(ad - bc)^2}{n_1 n_2 T_1 T_2}$$

Use this formula to find the value of χ'^2 for Table 8.4.

Table 8.8.

	Sample 1	Sample 2	Total
Success	a	b	T_1
Failure	c	d	T_2
Total	n_1	n_2	n

(b) Show that Formula (8.34) is algebraically equivalent to Formula (8.29).
(c) Formulas (8.29), (8.30), and (8.34) give alternate but equivalent ways to approximate χ^2. By using Yates' correction for continuity we may adjust these formulas so that the chi-square distribution with one degree of freedom better approximates their common distribution. Applying Yates' correction Formula (8.34) becomes

(8.35) $$\text{adj } \chi'^2 = \frac{n\left(|ad - bc| - \frac{n}{2}\right)^2}{n_1 n_2 T_1 T_2}$$

Compute adj χ'^2 for Table 8.4 and compare its value with 16, the value of χ'^2. *Note:* Since adj χ'^2 is less that χ'^2 we tend to wrongly reject the null hypothesis more often than 100 α percent of the time when using χ'^2.

(d) Compute adj χ'^2 for Exercise 8.30 and compare its value with 1.125, the value of χ'^2.

(e) If possible change Formulas (8.29) and (8.30) so that their adjusted formulas are equivalent to Formula (8.35).

E8.37. **(Probabilities for 2 × 2 Contingency Tables).** Using the hypergeometric distribution (see Exercise 5.53) it can be shown, for fixed marginal totals, that the probability of a pattern of observed frequencies a, b, c, d in the 2 × 2 contingency table of Exercise 8.36 is

(8.36) $$P(a, b, c, d) = \frac{n_1! n_2! T_1! T_2!}{n! a! b! c! d!}$$

(a) If $n_1 = 10$, $n_2 = 15$, $T_1 = 7$, and $T_2 = 18$, find values of the 8 probabilities $P(a, b, c, d)$ and verify that

$$P(0, 7, 10, 8) + P(1, 6, 9, 9) + \cdots + P(7, 0, 3, 15) \doteq 1.0000$$

(b) Use the probabilities of part (a) to determine a critical region of patterns for a two-sided alternative hypothesis and a significance level equal to or less than 0.05; 0.10.

At best, the application of χ'^2 or adj χ'^2 leads to approximate results. So in doubtful cases (especially where one or more of the expected frequencies is less than 5) an exact procedure should be used even though it requires considerably more calculations.

(c) For the pattern $\begin{array}{|cc|} 0 & 7 \\ 10 & 8 \end{array}$ compute $\sqrt{\chi'^2}$ and $\sqrt{\text{adj } \chi'^2}$ and use the normal distribution to find $P(z > \sqrt{\chi'^2})$ and $P(z > \sqrt{\text{adj } \chi'^2})$. Compare both of these probabilities with the probability $P(0, 7, 10, 8) \doteq 0.0134$.

Note: The distribution of patterns is not symmetric. $P(a, b, c, d)$ may be used to test the usual null hypothesis against either a one-sided or a two-sided alternative hypothesis.

8.38. Use the observed frequencies of Table 8.9 to determine whether the defective rates of the common product made by three manufacturers are different:

Table 8.9.

Category of Product	Manufacturer		
	A	B	C
Defective	21	22	11
Nondefective	379	278	289

8.39. Four groups of students were asked the same question: Table 8.10 shows frequencies of "yes" and "no" responses.

Table 8.10.

Response to Question	High School Boys	High School Girls	College Boys	College Girls
Yes	36	18	65	30
No	84	102	55	90
Total	120	120	120	120

(a) Determine whether the proportions of affirmative responses differ significantly among the four groups.
(b) Determine whether there is a significant difference in the proportions of affirmative responses for boys and girls; for high school and college.

E8.40. Table 8.11 shows how a random sample of 200 university students is classified by sex and time devoted to study of courses each week.
(a) Use these data to decide whether male and female students have different study habits in terms of hours per week.

Table 8.11.

Sex	Number of Hours in Study per Week			Total
	Less than 20	20 to 40	More than 40	
Female	15	70	15	100
Male	30	50	20	100

(b) Verify that χ'^2 may be calculated by the formula

(8.37) $$\chi'^2 = \sum_{i=1}^{c} \frac{(f_{i1} - f_{i2})^2}{f_{i1} + f_{i2}}$$

when $f_{11} + f_{21} + \cdots + f_{c1} = f_{12} + f_{22} + \cdots + f_{c2}$ and f_{ij} denotes the observed frequency in the jth category of the ith population.

(c) Suppose each of two populations has k categories. Write a formula similar to Formula (8.37). Let the populations be "freshmen" and "seniors," and illustrate how you apply your formula for four categories.

8.41. Omit column 3 from Table 8.7 and test the hypothesis that "the university student stand on issues is independent of political preference" where political preference is restricted to Democrat and Republican.

^c**8.42.** Table 8.12 shows the grade distribution (by school in which enrolled) of 500 students in a freshman mathematics class.

(a) Use these data to determine whether grade is independent of school.
(b) Test two other hypotheses of interest to you.

Table 8.12.

School	Grade		
	A or B	C	D or F
Agriculture	10	10	10
Engineering	75	50	25
Home Economics	5	10	5
Liberal Arts	45	30	25
Science	80	80	40

^c**8.43.** Two days after a Democratic candidate for federal office disclosed that he had been hospitalized for nervous exhaustion the following question was asked of a national cross-section of 500 voters: Do you feel that the candidate should or should not resign from the Democratic ticket? The results are shown in Table 8.13.

Table 8.13.

Opinion on Question	Political Preference			Total
	Democrat	Republican	Independent	
Should	57	63	35	155
Should Not	128	63	84	275
Undecided	25	14	31	70
Total	210	140	150	500

(a) Test the hypothesis that "opinion on the question is independent of political preference."
(b) Test the hypothesis that the Democrats do not differ from the "others" in their opinion on the question.

(c) Omit the undecided row and the independent column from Table 8.13 and use the remaining entries to decide whether the Democrats and Republicans differ in their opinions.

E8.44. **(Binomial Index of Dispersion).** Suppose c samples, each of size n, are drawn from dichotomous distributions. Let $n_{i1}(i = 1, \cdots, c)$ denote the number of successes in the n independent trials from the ith distribution. Let $n_{.1} = n_{11} + \cdots + n_{c1}$ denote the total number of successes in c samples. Then $\bar{n}_{.1} = n_{.1}/c$ denotes the mean number of successes per sample, cn the total number of trials in c samples, and $cn - n_1$ the total number of failures in c samples. Then, for sufficiently large n, it can be shown that

(8.38) $$\sum_{i=1}^{c} (n_{i1} - \bar{n}_{.1})^2 \Big/ \left[\bar{n}_{.1}\left(1 - \frac{\bar{n}_{.1}}{n}\right) \right]$$

is distributed approximately as chi-square with $c - 1$ degrees of freedom and is identical to χ'^2 defined by Formulas (8.32) and (8.33). The statistic of Formula (8.38) is sometimes called the **binomial index of dispersion**.

(a) Use Formula (8.38) to compute the value of χ'^2 for Exercise 8.39(a).
(b) Use Formula (8.38) to compute the values of χ'^2 for the two parts of Exercise 8.39(b).

9
DISTRIBUTION-FREE METHODS

9.1. INTRODUCTION

A distribution-free statistic, also termed **nonparametric statistic**, does not require the assumption of the precise form (or shape) of any population sampled. For some distribution-free tests it is assumed that the sampled distribution (or distributions) be continuous or that distributions be symmetric about the same point or that two or more distributions have the same shape or some other mild restriction. But the assumptions are never elaborate enough to specify the populations being sampled, as is the case when normality is specified.

Since distribution-free procedures do not require that we know the shapes of populations, we do not usually deal with parameters. That is, distributions are often compared without the use of parameters. For this reason, distribution-free tests are also called nonparametric tests, even though neither term is entirely satisfactory in some cases. In Chapter 7 and the beginning of Chapter 8 the tests were *parametric*, and they required that we assume the samples be drawn from normal populations. (In other tests we might assume the samples to be drawn from other specified populations.) At the end of Chapter 8, the chi-square test procedures are distribution-free.

In Section 9.7 we give a list of advantages and disadvantages of the distribution-free test procedures. However, at this time it should be noted that the distribution-free tests require fewer assumptions and less accuracy of measurements than the parametric tests, but that generally such tests are not as *powerful*.

In this chapter we describe some nonparametric tests for one and two samples. There are nonparametric tests which correspond to a wide variety of parametric tests. In all tests the assumption of randomness is basic. However, there are times when an experimenter has reason to question this assumption. Hence, in Section 9.4.2 we describe a test which has been developed to determine whether the sample values are nonrandom.

9.2. TESTS OF TWO CORRELATED SAMPLES

Two samples are often used to determine whether two treatments (or conditions) are different, or whether one treatment is better than the other. In Section 7.11 we illustrated a test procedure for the difference between two means for samples drawn from normal populations, and in Section 8.8.1 we compared two proportions (means) for binomial populations. In both cases the samples were independently drawn.

There are many situations in which observations of two samples are not independent. However, the two treatments may be compared if the observations are obtained from matched (or paired) objects or from the same objects before and after some treatment. Some typical illustrations of experiments with paired observations are the reactions of hospital patients before and after the use of a drug, the attitudes of husband and wife toward a political candidate, the effect of nitrogen fertilizer on the yield of sugar beets, or the growth of matched animals fed different diets.

The purpose in pairing objects is to overcome the difficulty of dealing with extraneous variables which might otherwise be introduced. For example, if we wanted to compare two teaching methods and failed to match students by ability (and any other variable which might affect performance), we might say two teaching methods differed when the real difference was due to ability of students or we might fail to recognize a real difference between teaching methods due to the presence of other factors.

Pairing objects can be difficult, even for an experienced investigator. If matching is not done properly, any difference due to matching is combined with whatever difference exists in treatments to mislead the experimenter. Thus, whenever feasible, it is desirable that each object be used as its own control. Certainly, no better matching is possible than that of an object with itself.

When a paired-comparison test is used to compare two observations on the same object (or matched objects), the experimental design consists in taking n random objects (or matched objects), making an observation y_1 on each one, and then after some treatment making a second observation y_2 in the same unit as the first. The objective is to determine whether any real difference, on the average, occurs as a result of the treatment. Thus, the difference in observations $d = y_1 - y_2$ for each object is obtained, and these differences constitute a random sample which is used to test the hypothesis that there is no real difference in the populations.

In this section we illustrate and compare three distribution-free tests which are useful for paired observations with restricted information. In Section 10.4.2 we discuss the parametric paired-t test and make further comparisons.

9.2.1. The Sign Test

The sign test uses plus and minus signs of the difference d. All that is required of the data is that each pair of objects be randomly selected and that we be able to *order their measurements*. Example 9.1 indicates that this is a very simple test in which the binomial statistic is applied.

EXAMPLE 9.1. A random sample of 18 male students was selected from a university. Each student was asked to determine both his left- and right-hand grip on an instrument designed to measure grip in pounds. The results are shown in Table 9.1. We wish to know if males at the university have more grip in their right hands.

Table 9.1. Grip of 18 University Students

Grip in Pounds		Sign of Difference	Grip in Pounds		Sign of Difference
Left Hand y_1	Right Hand y_2	$(y_1 - y_2)$	Left Hand y_1	Right Hand y_2	$y_1 - y_2$
42	45	−	29	37	−
37	39	−	25	26	−
36	35	+	33	33	0
41	42	−	35	37	−
40	40	0	40	38	+
29	30	−	48	51	−
45	46	−	45	43	+
39	37	+	38	41	−
32	34	−	28	31	−

We first illustrate the test procedure by the general method, then we discuss some other points of interest:

1. *Hypotheses*. H_o: the median of the population of differences is zero. That is, there are as many students who have stronger grip in their left hands as there are students who have stronger grip in their right hands.
H_a: the median of the population of differences is negative. That is, more students have stronger grip in their right hands.

2. *Assumptions and specifications*. Assume 18 students are randomly selected from the university, that the observations are accurate enough that we can determine which is the larger measurement for each student, and that the variable of differences is a continuous random variable. Let $\alpha = 0.05$.

3. *Statistic*. Let n denote the total number of positive or negative signs, ignoring the zeros. In this case $n = 16$ since there are two zeros. Let T denote the total number of positive signs. Under the assumption that the null hypothesis is true, T has a binomial distribution $b(T)$ with $\pi = 0.5$.

4. Critical Region. The critical region is the set of all values of T equal to or less than an integer k for which

$$P(T \leq k) < 0.05$$

Thus, by using Table 9.2 or any standard table of cumulative binomial probabilities we find that

$$CR = \{T | T \leq 4\}$$

since

$$\sum_{T=0}^{4} b(T) = 0.0383 < 0.0500$$

and

$$\sum_{T=0}^{5} b(T) = 0.1050 > 0.0500.$$

Table 9.2. Critical Values of k for the Sign Test
[Table gives largest integral values of k such that $P(T \leq k) < \alpha$ where T has a binomial distribution with $\pi = 0.5$]

n \ α	0.005	0.01	0.025	0.05	0.125
6	—	—	0	0	1
7	—	0	0	0	1
8	0	0	0	1	1
9	0	0	1	1	2
10	0	0	1	1	2
11	0	1	1	2	3
12	1	1	2	2	3
13	1	1	2	3	3
14	1	2	2	3	4
15	2	2	3	3	4
16	2	3	3	4	5
17	2	3	4	4	5
18	3	3	4	5	6
19	3	4	4	5	6
20	3	4	5	5	6
21	4	4	5	6	7
22	4	5	5	6	7
23	4	5	6	7	8
24	5	5	6	7	8
25	5	6	7	7	9
30	7	8	9	10	11
35	9	10	11	12	13
40	11	12	13	14	15
45	13	14	15	16	18
50	15	16	17	18	20

5. Computations. With the aid of Table 9.1 we find that the number of positive signs is $T_c = 4$.

6. Conclusion. Since 4 is one of the five critical region values, we accept the alternative hypothesis. Thus, we conclude that male students at the selected university do have stronger grips in their right hands.

In Table 9.2, n is always the total number of plus and minus signs. If there are ties, n is less than the sample size. By assuming the random variable for the difference to be continuous, we know there will not be many ties, and that the ones which do occur are probably due to rounding-off errors.

If in Example 9.1 the alternative hypothesis had been two sided with $\alpha = 0.05$, we would have used the column with $\alpha = 0.025$. Then the critical region would have been

$$CR = \{T \mid T \leq 3 \quad \text{or} \quad T \geq 13\}$$

since a binomial distribution is always symmetric when $\pi = 0.5$.

In general, if we assume any pair of values (y_1, y_2) to be randomly drawn from the same continuous distribution, we expect $d = y_1 - y_2$ to be positive half the time and negative half the time in repeated trials. For this reason, the null hypothesis states that the difference d has a distribution with median zero. An equivalent statement is that the true proportion of positive (or negative) signs is $\pi = \frac{1}{2}$. Thus, regardless of the shape of the distribution from which the jth pair (y_{1j}, y_{2j}) is drawn, we expect the $+$ and $-$ signs to have a dichotomous distribution with $\pi = \frac{1}{2}$. Hence, in n independent trials for which the probability of a positive (or negative) sign on each trial is $\pi = \frac{1}{2}$, the probability of T positive (or negative) signs is given by the binomial probability distribution

$$(9.1) \qquad b(T) = \frac{n!}{T!(n-T)!\, 2^n} \quad \text{for} \quad T = 0, 1, \cdots, n$$

When n is larger than 25, say, we may use the normal approximation to the binomial distribution. The mean and standard deviation of the normal distribution are taken as

$$\mu = n\pi = 0.5n$$

and

$$\sigma = \sqrt{n\pi(1-\pi)} = 0.5\sqrt{n}$$

Hence

$$(9.2) \qquad \frac{T - 0.5n}{0.5\sqrt{n}}$$

is approximately normally distributed with zero mean and unit variance. When $T < 0.5n$ use $T + 0.5$ in place of T in Formula (9.2), and when $T > 0.5n$ use $T - 0.5$.

For the sign test described above we have actually hypothesized that $\mu_1 = \mu_2$ or $\delta \equiv \mu_1 - \mu_2 = 0$. In Example 9.1 the statement would be "the mean grip in the left hand is equal to the mean grip in the right hand." However, there are experiments in which we would be interested in hypothesizing that one method is a fixed number of units, δ_0, better than the other. In this case, we would require a test of the hypothesis $\mu_1 = \mu_2 + \delta_0$ or $\delta = \delta_0$, where δ_0 is some specified real number.

The procedure is simply to apply the sign test to the signs of the differences

(9.3) $\quad y_{11} - (y_{21} + \delta_0), y_{12} - (y_{22} + \delta_0), \cdots, y_{1n} - (y_{2n} + \delta_0)$

If the number of positive signs, say, is significantly different from $n/2$ at the α level, we reject the null hypothesis that

$$\mu_1 = \mu_2 + \delta_0$$

Otherwise, we fail to reject this hypothesis.

9.2.2. Wilcoxon's Signed Rank Test

The sign test requires only information about the *order* of the differences for pairs of observations. When the *magnitude* as well as the *order* of the difference of a pair is known, Wilcoxon has described a more powerful test procedure. It gives more weight to a pair with large difference than to a pair with small difference. Neither the sign test nor Wilcoxon's signed rank test requires the normality (or any other specific distribution) assumption. An illustration should clarify the procedure.

EXAMPLE 9.2. Illustrate Wilcoxon's signed rank test with the actual differences, $d_j = y_{1j} - y_{2j}$, between the left and right hand grips of the 18 university students listed in Table 9.1.

Ignoring the ties, the remaining 16 differences in grip are

$$-3, -2, 1, -1, -1, -1, 2, -2, -8, -1, -2, 2, -3, 2, -3, -3$$

Take the absolute value of each difference and arrange the resulting differences in increasing order of magnitude to obtain

(9.4) $\quad +1, 1, 1, 1, 1, 2, +2, 2, 2, +2, +2, 3, 3, 3, 3, 8$

where a plus sign is attached to each difference which was originally positive. Next, assign ranks and then attach the sign of the original differences. Whenever the absolute value of two or more differences are equal, assign to each the mean of the ranks they would have had if all were different. For example, since the first five values in Sequence (9.4) are the same, we assign the rank

$$\frac{1+2+3+4+5}{5} = 3$$

to each, attaching a minus sign to ranks which originated with -1. Following this procedure we find the signed ranks for the sequence in (9.4) to be

$$3, -3, -3, -3, -3, -8.5, 8.5, -8.5, -8.5, 8.5, 8.5, -13.5, -13.5,$$
$$-13.5, -13.5, -16$$

These last signed ranked numbers are the ones on which we base our test procedure.

We wish to test the null hypothesis that "grip in the left hand of male students does not differ from grip in the right hand" against the alternative hypothesis that "grip in the right hand is greater than in the left hand." The test statistic T is the sum of the positive or negative ranks, whichever is smaller. For a one-sided 0.025 level test we find from Table 9.3 that the critical region is

$$CR = \{T \mid T < 30\}$$

If the null hypothesis is true, we would expect the sum of the positive ranks and the sum of the negative ranks to be about the same. But if the sum of the positive ranks is quite different from the sum of the negative ranks, we would infer that the two treatments are not the same. Thus, for our one-sided test we

Table 9.3. Critical Values of T for Wilcoxon's Signed Rank Two-sided Test*

(Absolute values of T less than the tabulated values occur with indicated probability. In a two-sided test the probabilities are 0.05, 0.02, and 0.01, respectively.)

n Pairs	Probability			n Pairs	Probability		
	0.025	0.01	0.005		0.025	0.01	0.005
6	1	–	–	16	30	24	19
7	2	0	–	17	35	28	23
8	4	2	0	18	40	33	28
9	6	3	2	19	46	38	32
10	8	5	3	20	52	43	37
11	11	7	5	21	59	49	43
12	14	10	7	22	66	56	49
13	17	13	10	23	73	62	55
14	21	16	13	24	81	69	61
15	25	20	16	25	90	77	68

*This table is abridged from Table 2 of Frank Wilcoxon and Roberta A. Wilcox, *Some Rapid Approximate Statistical Procedures*, Lederle Laboratories, Pearl River, New York (1964), with permission of the Lederle Laboratories, Division, American Cyanamid Company.

reject H_o if the sum of the positive ranks is very small (or the sum of the negative ranks is very large). For a two-sided test we reject H_o when the sum of positive ranks is very small or very large (or the sum of negative ranks is very large or very small).

Since the sum of positive ranks in our problem is

$$T_c = 3 + 8.5 + 8.5 + 8.5 = 28.5$$

and since 28.5 falls in the critical region, we conclude that "grip in the right hand of male students at the selected university is actually greater than in the left hand."

For a 0.05 significance level, both the sign test and Wilcoxon's signed rank test would have led to the same conclusion for the grip data of Examples 9.1 and 9.2, but this is not always true. Whenever both tests are appropriate, as for the grip data, Wilcoxon's signed rank test should be used since it is more powerful.

In an experiment with more than 25 pairs with non-zero differences, Table 9.3 cannot be used. In such a case it can be shown that T is approximately normally distributed with mean

(9.5) $$\mu_T = \frac{n(n+1)}{4}$$

and variance

(9.6) $$\sigma_T^2 = \frac{n(n+1)(2n+1)}{24}$$

Therefore

(9.7) $$z' = \frac{T - \mu_T}{\sigma_T}$$

is approximately normally distributed with zero mean and unit variance. Thus, z' is used in experiments with more than 25 ranked differences.

We may also use the signed rank test for the null hypothesis that the true difference δ in means is some specified δ_o. In fact, the same test procedure may be used to test the null hypothesis that the median of a single group of observations is equal to some specified value, say, μ_o. The general test procedure for both Wilcoxon's signed rank test and the median test is as follows:

1. Subtract δ_o (or μ_o) from each difference $d = y_1 - y_2$ (or value y) and disregard all zero values

2. Rank the resulting adjusted differences (values) in order of size, ignoring sign. In case of ties in adjusted differences (or values), assign the mean rank to each tied adjusted difference (value)

3. Attach the sign of the adjusted difference (or value) to the corresponding rank

4. Find the sum, T, of the positive or negative ranks, whichever is smaller in absolute value. The total of both the positive and negative ranks is $n(n+1)/2$

5. Compare T with the critical value in Table 9.3. If T falls in the critical region, reject the null hypothesis that the mean difference (value) is δ_o (μ_o); otherwise, fail to reject the null hypothesis

In the procedure listed above, the value of δ_o (or μ_o) can be zero as in Example 9.2.

9.2.3. Comparison of Tests of Two Correlated Samples

Other test procedures are based on differences $d_j = y_{1j} - y_{2j}$ obtained from two matched samples. For purposes of comparison we introduce two of these.

Walsh describes a very powerful nonparametric test based on ranking the differences of two related samples which are drawn from symmetrical populations. For symmetrical populations the mean and median are the same. The populations may or may not be normal and the d_j's do not need to be from the same populations.

In Chapter 10 we describe the *parametric t-test for paired observations.* This test requires that the differences be normally distributed. The *t*-test is more powerful than any of the nonparametric test procedures. In fact, the parametric *t*-test is the most powerful test procedure for testing differences of paired observations, but it requires the normality assumption.

For all of these tests, the sign, Wilcoxon's, Walsh's, and the paired-*t*, we assume the observation variables have continuous distributions. The measurements of the sign test may be of the crudest kind as long as *order of pairs of measurements* can be determined. Wilcoxon's signed rank test requires that measurements be such that we can not only *order each pair of observations,* as with the sign test, but that we be able to *order* (i.e., rank) *the n differences.* The other two tests require that the magnitude of the observations be on the same scale and that the populations be symmetric (for the *t*-test they must also be normal).

We may compare these four tests provided we make the assumptions required of all four tests, i.e., make the assumptions underlying the paired *t*-test. The sign test is about 95 percent as powerful as the *t*-test when $n = 6$, but its power reduces to about 65 percent of that of the *t*-test for large samples. The Wilcoxon test is about 95 percent of that of the *t*-test for large samples and is not much less for smaller samples. Walsh's test is about 95 percent as powerful as the *t*-test for most values of n and α.

When the assumptions of the *t*-test are satisfied we should normally use the *t*-test. However, if observations are easy to obtain or saving of time is crucial, we might prefer to use a nonparametric test. Of course, when the assumptions of the *t*-test are not met we have no alternative except to use the appropriate nonparametric test.

9.3. EXERCISES

9.1. In a statistics course 13 students made the following scores on two tests, each marked on the basis of a maximum score of 100:

Test I	86	64	78	93	73	81	77	65	74	79	85	58	82
Test II	79	76	75	90	68	88	71	64	72	71	82	71	73

Use the .025 level sign test to determine whether Test II was more difficult than Test I.

9.2. During a 15-month period two athletes, Bob and Joe, along with others, competed 16 times in the 100 yard dash. Let W and L denote "win" and "lose," respectively. The results of the competition were

Bob	W	L	W	L	W	W	L	W	W	L	W	L	W	L	W	W
Joe	L	L	L	W	L	L	L	L	L	L	W	L	W	L	W	L

Determine whether one of these two atheletes is superior in the 100 yard dash.

9.3. Four flavors of Jello were ranked by each of 14 taste testers. Each tester ranked the flavors, 1, 2, 3, or 4 with 1 being the most preferred. The results are shown in the following table:

Taste Tester	A	B	C	D	Taste Tester	A	B	C	D
1	2	3	1	4	8	1	3	2	4
2	1	3	2	4	9	2	4	1	3
3	3	4	2	1	10	1	2	4	3
4	1	3	2	4	11	1	2	4	3
5	3	1	2	4	12	3	1	2	4
6	4	1	3	2	13	1	2	4	3
7	1	2	3	4	14	3	4	1	2

(a) Use a 0.5 level test to determine whether Flavor A or Flavor D is preferred.
(b) Write a statement in which you compare the four flavors.

9.4. Use the normal approximation to verify all entries in Table 9.2 corresponding to $n = 50$.

Hint: Suppose y is normally distributed. For each α show that the k specified in Table 9.2 is such that $P(T \leq k) \doteq P(y \leq k + \frac{1}{2}) < \alpha$ but $P(T \leq k + 1) \doteq P(y \leq k + 1 + \frac{1}{2}) \geq \alpha$.

^C**9.5.** Use the method of Exercise 9.4 to
(a) decide whether the normal approximation gives the values in Table 9.2 corresponding to $n = 25$
(b) find values of k corresponding to $n = 56$

9.6. Two experimental animals were taken from each of 12 litters. One member of each litter was fed diet I and the other was fed diet II. The amount of weight (in grams) gained during a specified period of time was as follows:

Litter	1	2	3	4	5	6	7	8	9	10	11	12
Diet I	52	44	43	47	59	43	45	56	59	44	40	74
Diet II	53	51	47	54	60	45	48	62	58	49	55	65

(a) Use the .025 level sign test to decide whether greater gain is associated with Diet II.
(b) Test H_o of part (a) by Wilcoxon's signed rank test and compare the results of the two tests.

9.7. Use Wilcoxon's signed rank test to determine whether Test II in Exercise 9.1 was more difficult than Test I.

9.8. Two methods of analysis were used to determine the following octane ratings of 10 types of gasoline:

Type	a	b	c	d	e	f	g	h	i	j
Analysis I	97	85	106	92	75	88	99	101	90	82
Analysis II	95	83	102	91	76	82	104	96	84	75

(a) Use the .05 level sign test to decide whether the two methods of analysis differ.
(b) Test H_o in part (a) by Wilcoxon's signed rank test.

9.9. A claim was made that stimulus A increases systolic blood pressure by at least five units. The blood pressure of 20 subjects immediately before and m minutes after the application of A was as follows:

Subject	Before Stimulus	After Stimulus	Subject	Before Stimulus	After Stimulus
1	114	119	11	116	117
2	106	114	12	128	134
3	113	126	13	108	113
4	97	102	14	109	115
5	109	120	15	115	125
6	122	125	16	119	125
7	105	114	17	112	121
8	102	107	18	111	119
9	113	124	19	123	130
10	120	127	20	110	118

(a) Use the sign test to determine if the claim is justified.
(b) Apply Wilcoxon's signed rank test to H_o of part (a).

9.10. Thirty-six students in the same course in high school mathematics were divided into 18 pairs of nearly equal IQ scores. One member of each pair was randomly assigned to Group I and the others were assigned to Group II. The same instructor taught both groups the same mathematics, but different methods of instruction were used with the two groups. At the end of the semester the 36 students took the same examination with the following resulting grades:

Pair No.	1	2	3	4	5	6	7	8	9	10
Group I	91	79	66	86	72	69	74	83	88	79
Group II	93	83	72	87	80	68	83	90	88	84

Pair No.	11	12	13	14	15	16	17	18
Group I	74	84	76	68	79	78	49	76
Group II	84	85	87	73	80	86	60	82

Test the hypothesis that the method of instruction used with Group II improves the examination score by 3 points, on the average. Test by two methods.

E9.11. The median of all differences $d = y_1 - y_2$ from two populations is not necessarily equal to the difference in the medians of these two populations, i.e., $M_d \neq M_1 - M_2$ sometimes. Thus, equality of two population medians does not necessarily imply that the median of all differences from two populations will be zero. However, when the two populations have the same distribution it follows that $M_d = M_1 - M_2$. (Recall that $\mu_d = \mu_1 - \mu_2$ regardless of the nature of the two populations.)

Suppose Population 1 has five equally likely values 1, 2, 5, 6, 7 and Population 2 has five equally likely values 3, 4, 5, 8, 9. Find the theoretical sampling distribution of $d = y_1 - y_2$ and verify the statements (about medians) made in the above paragraph. *Note:* For any symmetric population it can be shown that $M = \mu$.

C9.12. Use the normal approximation for Wilcoxon's signed rank T to
(a) check the values in Table 9.3 corresponding to $n = 25$
(b) find critical values of T corresponding to $n = 30; 35$

9.13. The observations 42, 55, 46, 63, 38, 30, 48, 47, 22, 37, 58, 44, 36 were randomly drawn from a normal population with unknown mean and standard deviation 10. Test the hypothesis that $M = 50$ using (a) the sign test; (b) Wilcoxon's signed rank test. (c) Test the hypothesis $\mu = M = 50$ using the fact that $\bar{y}$ is normally distributed.

9.14. It is claimed that the median running time for rats in a certain type maze is 30 seconds. Suppose 30 rats had the following running times (in seconds):

26 18 36 23 90 27 30 28 25 16
90 29 12 25 30 15 25 19 75 17
21 20 25 33 14 18 90 22 26 31

(a) Use the sign test on these data to decide whether the claim is justified. *Hint:* In practice, sample values equal to the hypothesized value would usually be eliminated.

(b) Test the hypothesis of part (a) with Wilcoxon's signed rank T statistic.

E9.15. (McNemar Test). This test is designed to determine whether responses of a group of subjects changes after experiencing a treatment. For example, an investigator might wish to know whether persons change their attitudes or prejudices after seeing a particular film.

Suppose each subject may be classified as belonging to one of two mutually exclusive categories "before" and "after" some treatment has been applied. The results, in terms of observed frequencies, may be displayed as in the following 2 × 2 table:

		Before Treatment	
		Category 1	Category 2
After Treatment	Category 1	a	c
	Category 2	b	d

Note that $a + d$ subjects do not show change after treatment, but b subjects change from category 1 before treatment to category 2 after treatment, and c subjects change from category 2 to category 1. Thus, after the treatment is applied a total of $b + c$ subjects change in one direction or the other. If the treatment has no effect we would expect $(b + c)/2$ to change from category 1 to category 2 and $(b + c)/2$ to change from category 2 to category 1.

We are interested in the hypothesis that in a population the proportion of subjects π to change in one direction (say, from category 1 before treatment to category 2 after treatment) is equal to the proportion π' to change in the opposite direction. It can be shown that this hypothesis may be tested with the statistic

$$\chi'^2 = \frac{(b-c)^2}{b+c}$$

which is approximately distributed as chi-square with one degree of freedom.

(a) Suppose 40 students in a politics course were asked "whether the television newsmen of a specific network were biased in reporting events of the Vietnam war. After members of the class viewed an authentic hour-long documentary film of the war they were asked the same question. The results are shown in the following table:

After Film	Before Film	
	Reporting Biased	Reporting Not Biased
Reporting Biased	14	16
Reporting Not Biased	6	4

Can you conclude that the film is effective in changing student attitudes toward bias in reporting news (on the specific network)?

(b) When Yate's correction for continuity is used the formula for χ'^2 is

$$\text{adj } \chi'^2 = \frac{(|b-c|-1)^2}{b+c}$$

Use the formula for adj χ'^2 to test H_o in part (a).

(c) Prove that

$$\frac{\left(b - \frac{b+c}{2}\right)^2}{\frac{b+c}{2}} + \frac{\left(c - \frac{b+c}{2}\right)^2}{\frac{b+c}{2}} = \frac{(b-c)^2}{b+c}$$

E9.16. Use adj χ'^2 of Exercise 9.15(b) to test the appropriate hypothesis in each of the following:

(a) Exercise 9.9 where category 1 denotes the number is less than 115.
(b) Exercise 9.10 where category 1 denotes the number is less than 80.

9.4. ONE-SAMPLE TESTS

We have already used a single random sample to test whether

    1. a sample mean is significantly different from a hypothesized population mean (see Example 7.15)

    2. a sample proportion is significantly different from a hypothesized population proportion (see Examples 7.13 and 8.5)

    3. several sample proportions are significantly different from corresponding population proportions or hypothesized population proportions (see Examples 8.6 and 8.8); and

    4. a sample of 125 test scores could have come from a normal curve (see Example 8.7)

All of these may be considered goodness-of-fit tests with the first being a parametric test and the others being nonparametric tests. The normal distribution was used for the parametric test, and in Chapter 10 we illustrate how the t-test may be used for the same purpose. The binomial and chi-square distributions were applied in the nonparametric tests. In Section 9.4.1 we introduce one more nonparametric test, the Kolmogorov-Smirnov test, which may be used to examine goodness-of-fit. At the end of the section we compare these three tests.

A single sample may also be used in another kind of test. In Section 9.4.2 we describe a nonparametric procedure which is designed to answer the following question: Is it reasonable to believe that a particular sample is a random sample from some known population?

9.4.1. The Kolmogorov–Smirnov Test of Goodness-of-Fit

The Kolmogorov-Smirnov test procedure specifies the cumulative distribution in the null hypothesis and compares it with an observed (or empirical) cumulative relative frequency distribution. The maximum absolute difference D in the hypothetical cumulative distribution and empirical cumulative distribution is used as a test statistic. If D is larger than an α level value, we reject the hypothetical cumulative distribution and accept one that is closer to the observed cumulative distribution; otherwise we conclude that the observed cumulative distribution is not significantly different from that hypothesized.

Before illustrating the test procedure in Example 9.3 we introduce some symbolism. Let $F_o(y)$ denote the probability cumulative distribution of a continuously distributed random variable y which is completely specified by the null hypothesis. Let $S_n(y)$ denote the observed cumulative relative frequency distribution of a random sample of n observations. That is, for any possible value of y, $S_n(y)$ is a step function

(9.8) $$S_n(y) = \frac{k}{n}$$

where k is the number of observations equal to or less than y. Under the null hypothesis that the sample has been drawn from the specified distribution, step function $S_n(y)$ should be fairly close to the continuous function $F_o(y)$. Thus, we define

(9.9) $$D = \max |S_n(y) - F_o(y)|$$

as the test statistic. Assuming the hypothesized continuous cumulative distribution to be true, then the sampling distribution of D is known and critical values of D for a two-sided test are as shown in Table 9.5.

EXAMPLE 9.3. Suppose we wish to test the hypothesis that the six numbers .096, .552, 1.968, 1.344, 1.272, and .336 were chosen at random from numbers uniformly distributed between 0 and 2.4.

By the general test procedure we have:

1. *Hypotheses.* H_o: The variable y is uniformly distributed between 0 and 2.4.
   H_a: The variable y has a distribution different from that specified by H_o.

2. *Assumptions and Specifications.* Assume the six observations are randomly drawn from the same distribution. Let $\alpha = 0.05$.

3. *Statistic.* Use the statistic D defined by Equation (9.9).

4. *Critical Region.* Entering Table 9.5 with $\alpha = 0.05$ and $n = 6$ we find that

$$CR = \{D | D > .519\}$$

5. *Computations.* From Table 9.4 we find that the sample value of D is 0.273.

Table 9.4. Comparison of the Observed Distribution with the Hypothetical Distribution

Observed Value y	Obs. Cum. Distribution $S_6(y)$	Hyp. Cum. Distribution $F_o(y)$	Difference in Distributions $\|S_6(y) - F_o(y)\|$
.096	1/6 ≐ .167	.096/2.4 = .040	.127
.336	2/6 ≐ .333	.336/2.4 = .140	.193
.552	3/6 ≐ .500	.552/2.4 = .230	.270
1.272	4/6 ≐ .667	1.272/2.4 = .530	.137
1.344	5/6 ≐ .833	1.344/2.4 = .560	.273
1.968	6/6 ≐ 1.000	1.968/2.4 = .820	.180

6. *Conclusion.* Since the computed value of $D = 0.273$ does not fall in the critical region, we fail to reject H_o. That is, we conclude that the sample could have come from the uniform distribution specified by H_o.

Note: If a one-sided test is required the statistic to use is

$$D^* = \text{maximum} |S_n(y) - F_o(y)|$$

where only negative values or only positive values of $S_n(y) - F_o(y)$ are used to compute D^*, but not both. Thus, an α level test with D^* is equivalent to one side of a 2α level test when D is applied.

The chi-square test can be applied to qualitative measurements, but the Kolmogorov–Smirnov test cannot. Further, strictly speaking, the Kolmogorov–Smirnov test does not apply to discrete populations. However, the Kolmogorov–Smirnov test may be used with a discrete population provided the user knows that the actual significance level for the test can be expected to be less than that indicated in the test procedure (see Exercises 9.20 and 9.21). On the other hand, when a small sample of quantitative measurements is analyzed the Kolmogorov-Smirnov test can be used, and the chi-square test may not be applied (see Exercises 9.17, 9.18, and 9.20). In situations where both tests can be applied it appears that the Kolmogorov–Smirnov test is more powerful, for this test uses every value in the analysis while the chi-square test loses information when values are combined into categories.

The binomial test may be applied when there are only two categories and the sample size is small. If the sample size is large and there are more than two categories, the chi-square test is useful. Both tests can be applied for qualitative data.

Table 9.5. Critical Values for the Two-sided Kolmogorov–Smirnov Test of Goodness of Fit*

Sample Size n	Level of Significance for $D = \max\lvert S_n(y) - F_0(y) \rvert$				
	0.20	0.10	0.05	0.02	0.01
1	.900	.950	.975	.990	.995
2	.684	.776	.842	.900	.929
3	.565	.636	.708	.785	.829
4	.493	.565	.624	.689	.734
5	.447	.509	.563	.627	.669
6	.410	.468	.519	.577	.617
7	.381	.436	.483	.538	.576
8	.358	.410	.454	.507	.542
9	.339	.387	.430	.480	.513
10	.323	.369	.409	.457	.489
11	.308	.352	.391	.437	.468
12	.296	.338	.375	.419	.449
13	.285	.325	.361	.404	.432
14	.275	.314	.349	.390	.418
15	.266	.304	.338	.377	.404
16	.258	.295	.327	.366	.392
17	.250	.286	.318	.355	.381
18	.244	.279	.309	.346	.371
19	.237	.271	.301	.337	.361
20	.232	.265	.294	.329	.352
21	.226	.259	.287	.321	.344
22	.221	.253	.281	.314	.337
23	.216	.247	.275	.307	.330
24	.212	.242	.269	.301	.323
25	.208	.238	.264	.295	.317
30	.190	.218	.242	.270	.290
35	.177	.202	.224	.251	.269
40	.165	.189	.210	.235	.252
45	.156	.179	.198	.222	.238
50	.148	.170	.188	.211	.226
Asymptotic Formula	$\dfrac{1.07}{\sqrt{n}}$	$\dfrac{1.22}{\sqrt{n}}$	$\dfrac{1.36}{\sqrt{n}}$	$\dfrac{1.52}{\sqrt{n}}$	$\dfrac{1.63}{\sqrt{n}}$

*This table is adopted from Leslie H. Miller, "Table of Percentage Points of Kolmogorov Statistics," *Journal of American Statistical Association*, Vol. 51 (1956), pp. 113–121, with the permission of the editor of the journal.

We have described three goodness-of-fit tests for a single population. In choosing among them an investigator should take into account the (1) type of variable used, (2) size of the sample, (3) number of categories, and (4) power of the test.

9.4.2. Runs Test to Detect Nonrandomness

The assumption of randomness is necessary in all problems of statistical inference. Whenever sample values are obtained by random number tables or some recognized probability method, the assumption is not likely to be questioned. But when the investigator has little or no control over the selection of sample values, the randomness assumption is subject to question. For example, a prediction of sales of some manufactured product must be based on whatever records are available. So we might wish to test to see if such observations could be considered to be random. Also, in an investigation of statistical quality control, we might wish to test the randomness assumption in an effort to detect the presence (or absence) of assignable causes.

So far we have not generally taken into account the effect of time on the experiments presented. Still, knowledge of the *order in time* of each observation in a sample might be valuable information in detecting nonrandomness. The first two patterns of observations shown in Figure 9.1 may indicate how nonrandomness can be detected when order of observations is known.

In Figure 9.1 the horizontal axes are for "order in which the observations were made," and the vertical axes are for "the magnitude of the observations." The pattern of dots in Figure 9.1(a) shows a cyclic pattern and that in Figure 9.1(b) shows a linear trend. Both patterns illustrate nonrandomness. Randomness is not as easy to recognize. Figure 9.1(c) indicates possible randomness. Certainly "order of occurrence" of observations is likely to be useful in constructing tests of nonrandomness.

Several tests have been developed for detecting nonrandomness, and they are typically based on the *order* in which observations occur. Some such common tests are based on *number of runs, length of runs, runs up and down,* and *control charts*. We describe the test procedure based on *number of runs*.

Distribution of Total Number of Runs of Two Kinds of Observations. Suppose the n observations in a sample are classified as two types, called a and b. Let n_1 of the observations be of type a and n_2 of type b. For example, the two types (classes) might be "observations above the median" and "observations equal to or below the median." The following is a sequence of 17 such observations:

(9.10) $\qquad\qquad a\ b\ b\ a\ b\ a\ a\ a\ b\ b\ b\ a\ a\ a\ a\ b\ b$

In the sequence there are 9 a's and 8 b's with four runs of a's and four runs of b's. The lengths of the runs of a's are 1, 1, 3, and 4, respectively; the lengths of the runs of b's are 2, 1, 3, and 2, respectively. So the particular arrangement of two kinds of observations in Sequence (9.10) contains a total of 8 runs.

According to Equation (5.9) there are

$$\frac{17!}{9!\,8!} = \frac{(10)(11)\cdots(17)}{(1)(2)\cdots(8)} = 24{,}310$$

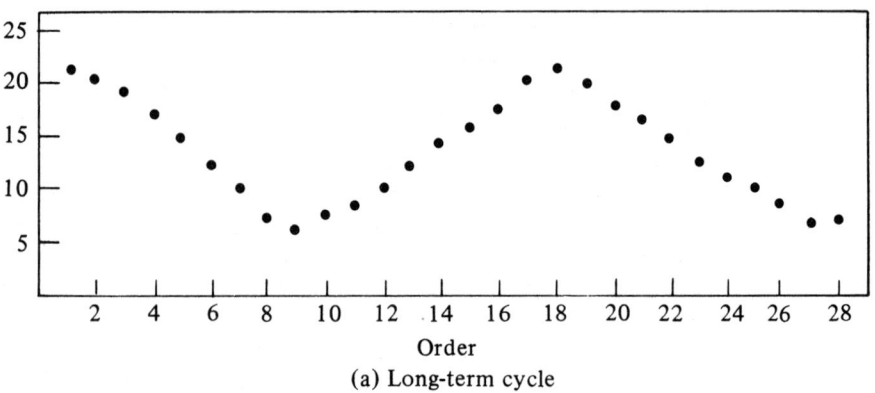
(a) Long-term cycle

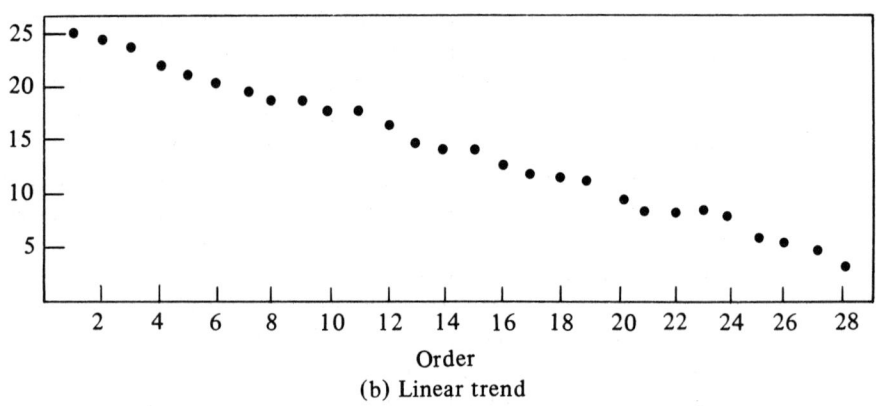
(b) Linear trend

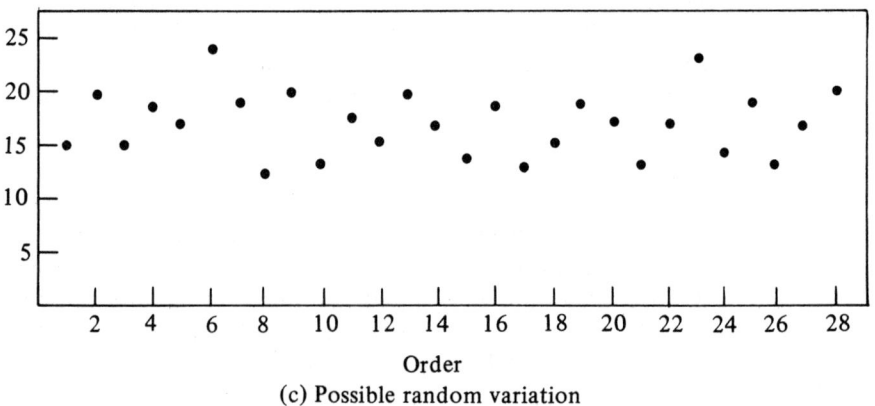
(c) Possible random variation

Figure 9.1. Dot Diagrams Indicating Order in which Observations were Drawn

possible arrangements of 9 objects ot one kind and 8 objects of a second kind. As illustrations of other arrangements, the sequence

(9.11) $$a\ b\ a\ b\ a\ b\ a\ b\ a\ b\ a\ b\ a\ b\ a\ b\ a$$

has 9 runs of a's and 8 runs of b's, each of length one, and the sequence

(9.12) $$a\ a\ a\ a\ a\ a\ a\ a\ a\ b\ b\ b\ b\ b\ b\ b\ b$$

has one run of a's of length 9 and one run of b's of length 8. So we observe that the 9 a's and 8 b's can be arranged so that the total number of runs varies from 2 to 17.

In an experiment with instruction "toss a well-balanced coin 17 times" and where a denotes "heads" and b denotes "tails," we might expect Sequence (9.10) to occur but we probably would not expect the other two sequences to occur. Sequences (9.11) and (9.12) would indicate nonrandomness and Sequence (9.10) could well indicate randomness. This leads us to believe that sequences with too many or too few runs could indicate the absence of randomness. Thus, we introduce the total number of runs, r, in a sequence as a statistic which could be useful in tests of randomness.

As we have suggested, sequences may result from sampling a dichotomous population or they may result by using values on both sides of a median. Such sequence may also be formed from continuous populations by letting the observation values above and below a given value y_0 be indicated by a and b, respectively. The value y_0 may be any quantile value, but the median is most frequently used.

For fixed n_1 and n_2 it is not difficult to find the probability of the total number of runs, $f(r)$. Then $f(r)$ can be used to compute lower, r', and upper, r'', critical values of r. For $n_1 \leq n_2 \leq 20$, Table 9.6 shows the upper and lower 0.025 critical values. The lower critical value, r', is the largest value of r for which

(9.13) $$\sum_{r=2}^{r'} f(r) = P[r \leq r'] \leq 0.025$$

and the upper critical value, r'', is the smallest value r for which

(9.14) $$\sum_{r=r''}^{n} f(r) = P[r \geq r''] \leq 0.025$$

Table 9.6 can be used for a 0.025 one-sided test or for a 0.05 two-sided test.

EXAMPLE 9.4. During an hour, 25 travel trailers, T, and campers, C, entered a national park in the following order:

$$T\ T\ T\ C\ C\ T\ C\ C\ C\ T\ T\ C\ T\ C\ T\ T\ T\ C\ C\ T\ C\ C\ T\ T\ T$$

Did these 25 vehicles enter the park in nonrandom order?

Table 9.6. Lower 0.025 Critical Value for Number of Runs*
(The Probability is 0.025 or Less that r is Equal to or Less than the Value Tabled.)

n_1 \ n_2	5	6	7	8	9	10	11	12	13	14	15	16	17	18	19	20
2								2	2	2	2	2	2	2	2	2
3		2	2	2	2	2	2	2	2	2	3	3	3	3	3	3
4	2	2	2	3	3	3	3	3	3	3	3	4	4	4	4	4
5	2	3	3	3	3	3	4	4	4	4	4	4	4	5	5	5
6		3	3	3	4	4	4	4	5	5	5	5	5	5	6	6
7			3	4	4	5	5	5	5	5	6	6	6	6	6	6
8				4	5	5	5	6	6	6	6	6	7	7	7	7
9					5	5	5	6	6	6	7	7	7	7	8	8
10						6	6	7	7	7	7	8	8	8	8	9
11							7	7	7	7	8	8	8	9	9	9
12								7	8	8	8	9	9	9	10	10
13									8	9	9	9	10	10	10	10
14										9	9	10	10	10	11	11
15											10	10	11	11	11	12
16												11	11	11	12	12
17													11	12	12	13
18														12	13	13
19															13	13
20																14

Table 9.6 (continued). Upper 0.025 Critical Value for Number of Runs*
(The Probability is 0.025 or Less that r is Equal to or Greater than the Value Tabled.)

n_1 \ n_2	5	6	7	8	9	10	11	12	13	14	15	16	17	18	19	20
4	9	9														
5	10	10	11	11												
6		11	12	12	13	13	13	13								
7			13	13	14	14	14	14	15	15	15					
8				14	14	15	15	16	16	16	16	17	17	17	17	17
9					15	16	16	16	17	17	18	18	18	18	18	18
10						16	17	17	18	18	18	19	19	19	20	20
11							17	18	19	19	19	20	20	20	21	21
12								19	19	20	20	21	21	21	22	22
13									20	20	21	21	22	22	23	23
14										21	22	22	23	23	23	24
15											22	23	23	24	24	25
16												23	24	25	25	25
17													25	25	26	26
18														26	26	27
19															27	27
20																28

*This table is abridged from Frieda S. Swed and C. Eisenhart, "Tables for Testing Randomness of Grouping in a Sequence of Alternatives," *Annals of Mathematical Statistics*, Vol. 14 (1943), pp. 66–87, Tables II and III, with permission of the editor of the journal.

We use a two-sided 0.05 level test of the null hypothesis
H_o: the T's and C's occur in random order
against the alternative hypothesis
H_a: the T's and C's occur in nonrandom order
Since there are fewer C's we let $n_1 = 11$, the number of C's, and let $n_2 = 14$, the number of T's. According to Table 9.6 the critical region is

$$CR = \{r \mid r = 2, 3, \cdots, 8 \quad \text{or} \quad r = 19, 20, \cdots, 23\}$$

The total number of runs is 13. Thus, since 13 does not fall in the critical region we fail to reject H_o. That is, we conclude that the sample of 25 trailers and campers may have entered the park in random order. Certainly, we do not have enough evidence to say they entered in nonrandom order.

If both n_1 and n_2 are large, then r is approximately normally distributed with mean

(9.15) $$\mu_r = \frac{2n_1 n_2}{n} + 1$$

and variance

(9.16) $$\sigma_r^2 = \frac{2n_1 n_2 (2n_1 n_2 - n)}{n^2(n-1)}$$

Hence the statistic

(9.17) $$\frac{r - \mu_r}{\sigma_r}$$

is approximately normally distributed with mean zero and variance one. When both n_1 and n_2 exceed 20, the normal approximation is quite close. When $n_1 < 5$ and $n_2 > 20$, the probability of an extreme number of runs should be computed directly from the probability function given in Exercise 9.28. In other cases where Table 9.6 is not applicable, the normal approximation is satisfactory.

When an analysis is based on *order of the observations*, as in the runs test, we use information which is not taken into account by simply counting the *frequency of* observations, as in the chi-square test, or by testing a statistic relating to a *parameter of the population* from which the sample was drawn. For example, if in 25 tosses of a coin we obtain 12 heads and 13 tails, the binomial (or chi-square) test would not reveal any lack of "fairness" of the coin tosses. But in the runs test where order of observations is applied we might detect a striking lack of randomness (and thus lack of "fairness") if the tails and heads alternated throughout the sequence.

Other Tests for Randomness. When the number of runs is great as compared with $n_1 + n_2$, we expect each one to be short, but when the number of runs is small at least one run can be expected to be long. Thus, it would be natural to use the *length of the largest run* to test randomness. There is such a test. Even though a *control chart* is normally used to detect and eliminate

318   Distribution-Free Methods

assignable causes of variation, it is also useful in determining whether a sequence of observations can be considered a random or a controlled sequence. Such a chart is useful in assessing extreme variation of individual measurements. Other tests of randomness may be found in Exercises 9.31, 9.32, and 9.33. Actually, when the null hypothesis is true, any test may be considered a test for nonrandomness of a special type of discrepancy between the population and the sample, as compared with the tests for a general lack of randomness.

9.5. EXERCISES

9.17. Test the hypothesis that the five numbers 0.22, 0.10, 0.23, 0.62, and 0.65 were randomly drawn from numbers uniformly distributed between zero and one.

9.18. Test the hypothesis that the seven observations 62, 54, 57, 48, 34, 42, 37, were randomly drawn from a normal distribution with mean 50 and variance 100.

9.19. Use the formulas at the bottom of Table 9.5 to compute critical values for $n = 50$; 60; 80.

9.20. In 18 tosses of a die the numbers 1, 2, $\cdots$, 6 occurred with frequencies 2, 1, 0, 5, 6, 4, respectively. Could the die be considered well-balanced?

9.21. Use a 0.05 level Kolmogorov-Smirnov test with the data of Exercise 8.26.

9.22. Suppose y has a normal distribution with standard deviation 10 and mean equal to or greater than 50. Use the eight sample observations 55, 65, 48, 54, 46, 57, 56, 76 to test the hypothesis that the normal distribution has mean 50 against the alternative that the mean is greater than 50.
(a) Use a 0.05 level Kolmogorov-Smirnov test.
(b) Use the sample mean $\bar{y}$. Compare the conclusions obtained by these two methods.

E9.23. A well-balanced coin is tossed five times. Let y denote the number of heads on a single toss. Find the theoretical sampling distribution of D, the Kolmogorov-Smirnov statistic, and construct a probability histogram of the distribution of D. Compare the five critical values of Table 9.5 with those obtained by using your histogram.

E9.24. Suppose the probability of obtaining a head on a single toss of a coin is 0.2 (instead of 0.5). Do Exercise 9.23.

9.25. In an auditorium the seating arrangement of 12 females (F) and 15 males (M) in the fifth row was as follows:

M F M F F F F M M M M M F F F M M F F F F M M M M M

Determine whether the order of the sequence could be considered nonrandom.

9.26. The sequence of heads (H) and tails (T) in 30 tosses of a coin was

H T T T H H T H T H T H T H T H T H T T H T T T H H T H T T

Does the order of heads and tails in the sample appear to be nonrandom?

9.27. Four a's and five b's may be arranged in 126 sequences.
  (a) List all sequences with only two runs; three runs.
  (b) List all sequences with at least eight runs.
  (c) Find the probability of two runs, of at most three runs, and of at least eight runs.
  (d) Show that the two critical values of Table 9.6 are correct.

E9.28. (**Probability of r Runs**). Under the assumption that every sequence of n_1 a's and n_2 b's is equally likely, it can be shown that the probability of r runs is given by

$$f(r) = \begin{cases} \dfrac{2\binom{n_1-1}{\frac{r}{2}-1}\binom{n_2-1}{\frac{r}{2}-1}}{\binom{n_1+n_2}{n_1}} & \text{when } r \text{ is even} \\[2em] \dfrac{\binom{n_1-1}{r-1}\binom{n_2-1}{r-3} + \binom{n_1-1}{r-3}\binom{n_2-1}{r-1}}{\binom{n_1+n_2}{n_1}} & \text{when } r \text{ is odd} \end{cases}$$

  (a) Use this formula to find the probabilities of Exercise 9.27(c).
  (b) Use $f(r)$ to verify the entries in Table 9.6 corresponding to $n_1 = 5$ and $n_2 = 8$.

9.29. Use the normal approximation to test the hypothesis in Exercise 9.26 and compare with the exact test of Exercise 9.26.

9.30. The following digits are from a table of random sampling numbers:

03 27 48 97 41 75 86 52 01 17 48 25 81 65 24
69 23 38 09 16 08 64 71 62 76 82 13 42 19 83

  (a) Determine whether the odd digits (O) and even digits (E) appear in nonrandom order.
  (b) Does the sequence of sixes (S) and non-sixes (F) appear to be nonrandom?

9.31. The sample median (or any central value of location) may be used to decide whether a sequence of observations is nonrandom. For such a test let each observation less than the sample median be denoted by a and each observation equal to or greater than the median be denoted by b. Then to detect nonrandomness the runs test of Section 9.4.2 may be applied. Use the sample median to determine whether the following 12 observations were randomly drawn:

22 79 4 11 28 29 31 48 91 82 35 43

9.32. Use the method of Exercise 9.31 to decide whether the following 20 observations were nonrandomly drawn:

$$49\ \ 47\ \ 58\ \ 61\ \ 43\ \ 52\ \ 51\ \ 47\ \ 65\ \ 45$$
$$31\ \ 40\ \ 26\ \ 61\ \ 47\ \ 49\ \ 54\ \ 41\ \ 40\ \ 34$$

E9.33. For a sample $y_1, y_2, \ldots, y_n$, the statistic

$$d' = \frac{\sum_{i=1}^{n-1}(y_{i+1} - y_i)^2}{\sum_i (y_i - \bar{y})^2}$$

termed *mean square successive difference*, may be used to detect nonrandomness. In particular, when observations are from a normal population and the sample size n is moderately large, say 20 or greater, then

$$\frac{1 - \dfrac{d'}{2}}{\sqrt{\dfrac{n-2}{n^2-1}}}$$

is approximately normally distributed with mean zero and variance one. Large positive values of this approximate z statistic are associated with long trends and large negative values with many short trends.

The observations in Exercise 9.32 were drawn from a normal population with mean 50.
(a) Use d' to decide whether the sample is nonrandom.
(b) Replace "sample median" by "population mean" in Exercise 9.31 and use the runs test to decide whether the sample in Exercise 9.32 was randomly drawn.

E9.34. Compute the value of each of the two statistics defined in Exercise 9.33 for the following sequence:
(a) 1   2   3   4   5   9
(b) 1   9   2   5   3   4
(c) 1   2   9   5   3   4
(d) Compare the values of "approximate z" for parts (a), (b), and (c).

9.6. TESTS FOR TWO INDEPENDENT SAMPLES

We have already used two independent samples to compare similar characteristics of two populations. In Section 7.11 we presented a *parametric test* to compare the means of two normal populations, and in Section 8.8.1 we compared two proportions from binomial populations. Again in Chapter 10 we illustrate how the parametric *t*-test may be used to compare two or more means of normal populations.

In the four distribution-free tests to be considered in this section we reduce the requirements made in the above test procedures. We assume that the measurements are accurate enough to be ordered (or ranked) and that the two independent random samples are drawn from two continuous populations which have the same form. (The continuity assumption is made to avoid ties, even though we recognize that tied scores may still result due to the nature of measuring instruments.)

The tests described in Section 7.11 and Chapter 10 assume the populations to be *normal*. In the tests introduced in this section we specify, under the assumption that the null hypothesis is true, that the *form* of the two populations be the same, but we do not require that their form be specified. They may be normal or they may not. Thus, when the null hypothesis is true, the random and independent samples

$$y_{11}, y_{12}, \ldots, y_{1n_1} \quad \text{and} \quad y_{21}, y_{22}, \ldots y_{2n_2}$$

come from a single population. The alternative hypothesis that the two populations are different may be *one-sided* or *two-sided*.

With the *parametric tests* our interest is centered on differences in measures of central tendency rather than differences in dispersion or differences in shape of distributions. With distribution-free procedures we test simultaneously for differences in two or three of these. The test statistics of this section are *not* expressed in terms of sample means (or proportions), but in terms of sample sizes and order of sample values. The first two tests are extensions of tests already described (see Sections 8.8.2 and 9.4.2).

9.6.1. The Median Test

The median test is designed to determine whether two independent samples of sizes n_1 and n_2 were drawn from two populations with the same median. The null hypothesis is that the two populations have the same median; usually the alternative hypothesis is that the two populations have different medians.

In the analysis we first find the median of the $n = n_1 + n_2$ values of the combined sample. Then we divide both samples into two groups, depending on whether a sample value is greater than the combined median or is equal to or less than the median. (If those sample values equal to the median are dropped, the combined sample size is less and the second class is made up of those values less than the median. An investigator may wish to have values in the analysis which are either less than or greater than the combined median even though information is lost.) Suppose m_1 and m_2 of the y_1 and y_2 values, respectively, are greater than the combined median. Then, as is illustrated in Table 9.7, $m = m_1 + m_2$ values are larger than the combined median and $n - m$ values are equal to or less than the combined median.

Table 9.7. Grouped Data for the Median Test
(In Terms of Number of Observations)

Number of Sample Observations	Sample of y_1-values	Sample of y_2-values	Total
Greater than the combined median	m_1	m_2	m
Equal to or less than the combined median	$n_1 - m_1$	$n_2 - m_2$	$n - m$
Total	n_1	n_2	n

If both $n_1 m/n$ and $n_2 m/n$ are equal to or greater than 5 we may use the χ''^2 statistic, corrected for continuity. That is, the statistic distributed approximately as chi-square with one degree of freedom is

$$(9.18) \quad \chi''^2 = \frac{n\left(|m_1(n_2 - m_2) - m_2(n_1 - m_1)| - \frac{n}{2}\right)^2}{n_1 n_2 m(n - m)}$$

EXAMPLE 9.5. Suppose the individual annual incomes (to the closest thousand dollars) of a random sample of 12 residents of City A are

9 4 11 23 8 13 11 2 10 6 15 9

and the individual annual incomes of a random sample of 15 residents of City B are

8 5 7 4 6 9 35 7 6 3 2 9 28 7 8

Is there a difference in the median incomes of individuals in the two cities?
We test, at the 0.05 level, the null hypothesis
H_o: the median annual income of individuals in City A
is equal to the median annual income of individuals in City B
against the alternative hypothesis
H_a: the median annual income of individuals in Cities
A and B are different.
Using the chi-square statistic with one degree of freedom, the critical region is

$$CR \doteq \{\chi''^2 \mid \chi''^2 > 3.84\}$$

To get the median of the combined sample of $12 + 15 = 27$ residents we arrange the sample values in the following order, placing a "bar" over the values for City A:

$\bar{2}$, 2, 3, $\bar{4}$, 4, 5, $\bar{6}$, 6, 6, 7, 7, 7, $\bar{8}$, 8, 8, $\bar{9}$, $\bar{9}$, 9, 9, $\overline{10}$, $\overline{11}$, $\overline{11}$, $\overline{13}$, $\overline{15}$, $\overline{23}$, 28, 35

The fourteenth ordered value of the combined sample is the median of 8 thousand dollars. Thus, it follows that

$$m = 12 \quad m_1 = 8 \quad m_2 = 4$$

$$\chi''^2 = \frac{27\left(|(8)(11) - (4)(4)| - \frac{27}{2}\right)^2}{12 \times 15 \times 12 \times 15} \doteq 2.85$$

Since 2.85 falls in the noncritical region, we fail to reject H_o. That is, on the basis of our samples, we cannot say the median annual income of individuals in City A differs from that of City B.

Note 1: If we had failed to make the correction for continuity, the value of the statistic would have been

$$\chi'^2 = \frac{27(8 \times 11 - 4 \times 4)^2}{12 \times 15 \times 12 \times 15} = 4.32$$

Since χ'^2 falls in the critical region we would have accepted the alternative hypothesis that the median annual income of individuals in the two cities differs.

Note 2: Recall that the upper tail of a chi-square distribution with one degree of freedom corresponds to two symmetric tails of the standard normal distribution. Thus, for $\alpha = 0.10$, the critical region $\{\chi^2 | \chi^2 > 2.71\}$ is the same as the critical region $\{z | z < -1.65 \text{ or } z > 1.65\}$, and both may be used in a 0.10 level test of Example 9.5. However, for the one-sided alternative hypothesis that the median annual income of individuals in City A is greater than in City B, we should use only the right tail of the standard normal distribution. That is, the 0.05 critical region is

$$CR = \{z | z > 1.65\}$$

Due to the special relationship between z and χ^2 with one degree of freedom, we may take the positive square root of χ''^2 as an approximation of z. In particular,

$$z \doteq \sqrt{2.85} \doteq 1.69$$

Thus, since 1.69 falls in the critical region we accept the alternative hypothesis that the median income in City A is greater than in City B. This illustrates again how important it is to state the appropriate alternative hypothesis.

When both n_1 and n_2 are quite large we may use the statistic of Expression (8.28), i.e., we may use the statistic

(9.19) $$z' = \frac{\frac{m_1}{n_1} - \frac{m_2}{n_2}}{\sqrt{\frac{m}{n}\left(1 - \frac{m}{n}\right)\left(\frac{1}{n_1} + \frac{1}{n_2}\right)}}$$

which approximates the standard normal statistic z. We may use z with either one-sided or two-sided alternative hypotheses. Since the square of z' is the uncorrected χ'^2 statistic mentioned in Note 1 above, we may also use the chi-square distribution with one degree of freedom when a two-sided alternative hypothesis is specified.

If either $n_1 m/n$ or $n_2 m/n$ is less than 5, the hypergeometric distribution should be used. The hypergeometric distribution gives the probability that exactly m_1 of the y_1 values and exactly m_2 of the y_2 values are above the combined median, i.e., the probability is

$$\binom{n_1}{m_1}\binom{n_2}{m_2} \bigg/ \binom{n}{m}$$

Thus, to test the null hypothesis of equal medians against the alternative hypothesis that the median of the distribution of y_2 values is larger than the median of the distribution of y_1 values, we need to know the probability that m_1 is equal to or less than the specified value m_1'; i.e., we require

$$(9.20) \qquad P(m_1 \leq m_1') = \sum_{m_1=0}^{m_1'} \frac{\binom{n_1}{m_1}\binom{n_2}{m_2}}{\binom{n}{m}}$$

Such values have been tabled by Lieberman and Owen [13] and may be used in the test indicated. These tables may also be applied for two-sided alternative hypotheses.

Median Test for c Samples. We may test the null hypothesis that c populations with identical distributions have equal medians against the alternative hypothesis that some of the medians are different. Following the procedure described for two medians, the data are arranged in a $2 \times c$ contingency table and tested with an approximate χ'^2 statistic with $c - 1$ degrees of freedom [see Formula (8.32)]. The null hypothesis is rejected if the computed χ'^2 is significantly large. For an illustration see Exercise 9.39.

Extension of the Median Test. In place of the median we may use any percentile to divide a sample into two parts. Or we may use two or more percentiles to group the sample values into three or more categories. For example, we may wish to use the quartiles to group students from different classrooms into four categories. Then we would test to determine whether the proportions falling in the four categories differ significantly from classroom to classroom. A χ'^2 statistic with $(4 - 1)(c - 1)$ degrees of freedom should be used [see Formula (8.33)]. For an illustration see Exercise 9.39.

9.6.2. The Wald-Wolfowitz Runs Test

The Wald-Wolfowitz runs procedure is designed to test the null hypothesis that *two distributions are the same* against the alternative hypothesis that the *two distributions are different in some respect*. The alternative hypothesis includes any type of difference, difference in location or in variability or in skewness or in any combination of these or other characteristics of the shape of distributions.

To apply this test we assume random samples of sizes n_1 and n_2 are independently drawn from two continuous populations and that the measurements are accurate enough that they can be ordered. Then the two samples are combined and ordered as in the runs test to detect nonrandomness (see Section 9.4.2). If the values from the first sample are designated by a's and the values from the second sample are designated by b's, there will be runs of a's and runs of b's. We let r denote the total number of runs in the two samples.

If the null hypothesis is true (i.e., if the samples come from the same distribution) the a's and b's should be well mixed and the total number of runs will be relatively large. If the alternative hypothesis is true, r will be relatively small. For example, if two populations have the same shape, the median of the first population being much smaller than that of the second population, there will likely be a long run of a's at the lower end of the ordered sequence and a long run of b's at the upper end. Thus, r will be relatively small. Further, suppose two populations have the same shape and same mean, but the first population has a much smaller dispersion. Then we would expect a long run of b's at each end of the ordered sequence of the combined sample and thus a relatively small value of r. Similar arguments can be presented to show that we expect r to get smaller as the differences in two populations get greater. Thus, in general we *fail to reject* the null hypothesis when r is large, but *reject* the null hypothesis that the distributions are the same when r is "too small."

When both n_1 and n_2 are less than 21, the first part of Table 9.6 may be used for a 0.025 level test, and Table 9.8 for a 0.05 level test. An illustration should make the test procedure clear.

EXAMPLE 9.6. Suppose the audible sounds made by 9 month-old-infants are taped under standard control conditions and that the number of distinct sounds per unit of time is computed for each infant. The sounds are taped while each baby is alone, and comparable conditions are used with each child. Suppose the number of audible sounds for a random sample of 10 boys and an independent random sample of 12 girls are as follows:

Boys: 23 38 27 40 21 17 33 26 30 35
Girls: 29 46 36 39 45 57 75 38 55 51 47 42

Are the distributions of the number of audible sounds the same for boys and girls who are 9 months of age?

We use a 0.05 level test of the null hypothesis that "the distribution of the number of audible sounds is the same for boys and girls" against the alternative hypothesis that "the distributions are different." The number of boys is $n_1 = 10$ and the number of girls is $n_2 = 12$. From Table 9.8 we find that the critical region is

$$CR = \{r \,|\, r = 2, 3, \cdots, 7\}$$

Table 9.8. Lower 0.05 Critical Value for Number of Runs*
(The Probability is 0.05 or Less that r is Equal to or Less than the Value Tabled.)

n_1 \ n_2	4	5	6	7	8	9	10	11	12	13	14	15	16	17	18	19	20
2					2	2	2	2	2	2	2	2	2	2	2	2	2
3		2	2	2	2	2	3	3	3	3	3	3	3	3	3	3	3
4	2	2	3	3	3	3	3	3	4	4	4	4	4	4	4	4	4
5		3	3	3	3	4	4	4	4	4	5	5	5	5	5	5	5
6		3	3	4	4	4	4	5	5	5	5	5	6	6	6	6	6
7			4	4	4	5	5	5	5	6	6	6	6	6	7	7	7
8				4	5	5	5	6	6	6	6	7	7	7	7	8	8
9					5	5	6	6	6	7	7	7	8	8	8	8	9
10						6	6	6	7	7	7	8	8	8	9	9	9
11							6	7	7	8	8	8	9	9	9	10	10
12								7	7	8	8	9	9	9	10	10	10
13									8	8	9	9	9	10	10	10	11
14										9	9	9	10	10	10	11	11
15											10	10	11	11	11	11	12
16												11	11	11	12	12	12
17													11	12	12	13	13
18														12	13	13	13
19															13	14	14
20																14	15

*This table is abridged from Frieda S. Swed and C. Eisenhart, "Tables for Testing Randomness of Grouping in a Sequence of Alternatives," *Annals of Mathematical Statistics*, Vol. 14 (1943), pp. 66–87, Table II, with permission of the editor of the journal.

For the analysis, arrange the combined samples as follows, where a "bar" indicates the sample value is for boy:

(9.21)
$\overline{17}$ $\overline{21}$ $\overline{23}$ $\overline{26}$ $\overline{27}$ 29 $\overline{30}$ 33 $\overline{35}$ 36 $\overline{38}$
38 39 $\overline{40}$ 42 45 46 47 51 55 57 75

Thus, there are 8 runs. Since $r = 8$ falls in the noncritical region, we conclude that the distributions of number of audible sounds are not different for boys and girls.

If either n_1 or n_2 is larger than 20, Tables 9.6 and 9.8 cannot be used. In this case, the statistic of Expression (9.17) may be used to approximate the normal distribution. If $n_1 + n_2$ is not very large, a correction for continuity should be used since r is a discrete variable. The corrected statistic

(9.22)
$$z' = \frac{\left| r - \frac{2n_1 n_2}{n} - 1 \right| - 0.5}{\sqrt{\frac{2n_1 n_2 (2n_1 n_2 - n)}{n^2 (n-1)}}}$$

more nearly approximates the normal distribution and should usually be applied when Tables 9.6 and 9.8 cannot be used.

Ties. The continuity assumption is made to avoid ties, but inaccurate measurements in rounding-off errors occasionally result in ties. When ties occur in either the sample of a's or in the sample of b's, the number of runs, r, is not affected and thus the test conclusion is not affected. However, if observations from one sample are tied with observations from the other sample, we cannot expect to obtain a unique number of runs in the ordered sequence. In Example 9.6 we see a fairly rare exception where the measurement 38 occurred in both samples, but the number of runs remains unaffected by interchanging the tied measurements. This occurs only when one measurement from each sample is a member of the tied pair *and* the ordered values immediately on both sides of the tied pair are from the same sample.

When the arrangement of tied observations for two samples does affect the number of runs, the test procedure should be applied for each number of runs. If all arrangements lead to the same decision (either to reject or not to reject the null hypothesis), the conclusion is obvious. If different arrangements lead to different decisions, we recommend that no final decision be made. Clearly, the Wald-Wolfowitz runs test should not be used when there are several tied ranks.

9.6.3. The Rank-Sum Test

This test requires the same assumptions as the Wald-Wolfowitz runs test and is designed to test the same hypothesis. However, the rank-sum test is a more powerful test when the primary difference in distributions is in their measures of centrality.

Let the random and independent samples

$$y_{11}, y_{12}, \cdots, y_{1n_1} \quad \text{and} \quad y_{21}, y_{22}, \cdots, y_{2n_2}$$

be drawn from two continuous distributions. If the null hypothesis is true, these samples are drawn from the same distribution. In this case, the samples may be combined to give a single random sample of size $n = n_1 + n_2$. We get our test statistic by arranging these n observations in order of increasing size and assigning rank 1 to the smallest value, rank 2 to the second smallest, $\cdots$, rank n to the largest, preserving the identities of the original samples. Suppose $n_1 \leq n_2$. Let T denote the total of the ranks of the sample with n_1 observations, i.e., the sample with fewer observations.

It can be shown that T has a symmetric distribution (see Exercise 9.46) and that the probability of obtaining a particular value of T is the same as the probability of $n_1(n_1 + n_2 + 1) - T$. If the two random samples are drawn from two different distributions with quite different centrality parameters, we expect T to be relatively small or relatively large. Significantly small values to T can be found in Table 9.9, and either one-sided (0.005 or 0.025 level) or two-sided (0.01 or 0.05 level) tests may be applied, depending on the statement of the alternative hypothesis.

EXAMPLE 9.7. Use the data of Example 9.6 and the rank-sum test to determine whether the distributions of audible sounds are the same for boys and girls.

We use a 0.05 level rank-sum test of the null hypothesis against the two-sided alternative hypothesis stated for Example 9.6. Since $n_1 = 10$ and $n_2 = 12$, we find from Table 9.9 that the critical region is

$$CR = \{T \mid T \leq 85 \text{ or } T \geq 145\}$$

where the upper critical value is given by

$$n_1(n_1 + n_2 + 1) - T = 10(23) - 85 = 145$$

Table 9.9. Critical Values for Rank Sum*
For a Two-sided Test They are 0.05 Level Values (or Less).
For a One-sided Test They are 0.025 Level Values (or Less).

n_2 \ n_1	2	3	4	5	6	7	8	9	10	11	12	13	14	15
4			10											
5		6	11	17										
6		7	12	18	26									
7		7	13	20	27	36								
8	3	8	14	21	29	38	49							
9	3	8	15	22	31	40	51	63						
10	3	9	15	23	32	42	53	65	78					
11	4	9	16	24	34	44	55	68	81	96				
12	4	10	17	26	35	46	58	71	85	99	115			
13	4	10	18	27	37	48	60	73	88	103	119	137		
14	4	11	19	28	38	50	63	76	91	106	123	141	160	
15	4	11	20	29	40	52	65	79	94	110	127	145	164	185
16	4	12	21	31	42	54	67	82	97	114	131	150	169	
17	5	12	21	32	43	56	70	84	100	117	135	154		
18	5	13	22	33	45	58	72	87	103	121	139			
19	5	13	23	34	46	60	74	90	107	124				
20	5	14	24	35	48	62	77	93	110					
21	6	14	25	37	50	64	79	95						
22	6	15	26	38	51	66	82							
23	6	15	27	39	53	68								
24	6	16	28	40	55									
25	6	16	28	42										
26	7	17	29											
27	7	17												
28	7													

Table 9.9 (cont.). Critical Values for Rank Sum*

For a Two-sided Test They are 0.01 Level Values (or Less).
For a One-sided Test They are 0.005 Level Values (or Less).

n_1 \ n_2	2	3	4	5	6	7	8	9	10	11	12	13	14	15
5				15										
6			10	16	23									
7			10	17	24	32								
8			11	17	25	34	43							
9		6	11	18	26	35	45	56						
10		6	12	19	27	37	47	58	71					
11		6	12	20	28	38	49	61	74	87				
12		7	13	21	30	40	51	63	76	90	106			
13		7	14	22	31	41	53	65	79	93	109	125		
14		7	14	22	32	43	54	67	81	96	112	129	147	
15		8	15	23	33	44	56	70	84	99	115	133	151	171
16		8	15	24	34	46	58	72	86	102	119	137	155	
17		8	16	25	36	47	60	74	89	105	122	140		
18		8	16	26	37	49	62	76	92	108	125			
19	3	9	17	27	38	50	64	78	94	111				
20	3	9	18	28	39	52	66	81	97					
21	3	9	18	29	40	53	68	83						
22	3	10	19	29	42	55	70							
23	3	10	19	30	43	57								
24	3	10	20	31	44									
25	3	11	20	32										
26	3	11	21											
27	4	11												
28	4													

*"The Use of Ranks in a Test of Significance for Comparing Two Treatments" by C. White; *Biometrics* Vol. 8 (1952), Number 1, pages 37–38, reprinted by permission of the Editor of *Biometrics*.

For the analysis we find, using the arrangement in the solution of Example 9.6, that the ranks are as follows:

Boys: 1 2 3 4 5 7 8 9 11.5 14
Girls: 6 10 11.5 13 15 16 17 18 19 20 21 22

For the tied ranks we found the mean of the ranks that the tied observations would have if they were distinguishable and then assigned each this mean rank. That is, both measurements of 38 received ranks of 11.5 since they fell in the 11th and 12th position of the ordered sequence.

The total rank for the 10 boys is $T = 64.5$, and it falls in the lower part of the critical region. Thus, we conclude that the distributions of audible sounds are different for boys and girls. In fact, the distribution of audible sounds for boys has smaller measurements than the distribution for girls. In other words, the hypothetical data indicate that "girls seem to make more sounds than boys at a very early age." If we had used a one-sided 0.025 level test the conclusions would have been the same. (Note we do not have a chance to use a one-sided test with the Wald-Wolfowitz test.)

In many problems, as in Examples 9.6 and 9.7, the Wald-Wolfowitz runs and the rank-sum tests lead to different conclusions. Even though they are used to test the same hypothesis, it can be shown that the rank-sum test is more powerful than the runs test whenever the primary difference between the two distributions is with respect to measure of location (or centrality), *but* the runs test is more powerful whenever the distributions are very similar relative to location and quite different in dispersion or shape. In our illustration with audible sounds, the bulk of the distribution for boys is below that for girls. Thus, the rank-sum test would appear to be the more appropriate test.

Ties. In case of ties, the usual procedure is to do what we did in the illustration although there are good alternatives. When the tied observations fall in one sample, the test is not affected; neither is the test seriously affected when they fall in different samples.

Normal Approximation. It can be shown that the random variable T has a mean of

(9.23) $$\mu_T = \frac{n_1(n_1 + n_2 + 1)}{2}$$

and a variance of

(9.24) $$\sigma_T^2 = \frac{n_1 n_2 (n_1 + n_2 + 1)}{12}$$

As n_1 and n_2 increase, the distribution of T approaches the normal distribution. Thus, for values of n_1 or n_2 outside the range of Table 9.9, the statistic

(9.25) $$z' = \frac{T - \frac{n_1(n_1 + n_2 + 1)}{2}}{\sqrt{\frac{n_1 n_2 (n_1 + n_2 + 1)}{12}}}$$

is a very good approximation of the standard normal statistic z unless n_1 is small. As an example of the nearness of the approximation, it can be shown (see Exercise 9.47) that when $n_1 + n_2 = 30$ and $\alpha = 0.05$, the approximate critical values are the same as those in Table 9.9 in 12 of 14 cases and only one unit larger in the other two.

Extension to Test for Several Distributions. Suppose c random samples of sizes $n_1, \cdots, n_c$ are independently drawn from c distributions. If the null hypothesis that "the distributions are identical" is true, we may combine the $n = \sum_i n_i$ sample values and treat the combined samples as one random sample. Arrange the n observations in order of increasing size and assign ranks as we did in the two-sample rank-sum test. Let T_i denote the sum of ranks of the ith sample and let the test statistic be

(9.26) $$\chi'^2 = \frac{12}{n(n+1)} \sum_i \frac{T_i^2}{n_i} - 3(n+1)$$

If each n_i is greater than 5 and the null hypothesis is true, it can be shown that χ'^2 is approximately distributed as chi-square with $c - 1$ degrees of freedom. Large values of χ'^2 lead to the rejection of the null hypothesis. Any ties in observation are treated as in the two-sample case.

9.6.4. The U Test

This test, also known as the Wilcoxon or the Mann-Whitney test, requires exactly the same assumptions as the Wald-Wolfowitz runs test and the rank-sum test, and it is designed to test the same hypothesis. In fact, it can be shown that the rank-sum test and the U test lead to identical results even though the U statistic seems quite different from the T statistic of the rank-sum test. (Since the U test is commonly applied in a number of areas, it seems desirable that we describe the test procedure and compare the U statistic with the T statistic of the rank-sum test.)

Suppose the two random and independent samples of sizes n_1 and n_2 are combined and arranged in ascending order of magnitude, preserving the identity of the original samples. Let each value in the sample of n_1 observations be designated by a and each value in the sample of n_2 observations be designated by b. Then the U statistic is defined to be *the number of times a b precedes an a.* For example, suppose the heights (in inches) of 4 boys and 4 girls are arranged and identified as follows:

Order of value	62	64	65	66	68	69	71	74
Identity of sample	a	a	b	a	b	a	b	b

Since the b (boy) height of 65 inches precedes each of the a (girl) heights of 66 and 69 inches, the b height of 68 inches precedes the a height of 69 inches, and the b heights of 71 and 74 inches precede no a heights, the sample value of the U statistic is

$$U = 2 + 1 + 0 + 0 = 3$$

The U test, in its original form, was designed to test the null hypothesis
H_o: the y_1 distribution is identical to the y_2 distribution
against the alternative hypothesis
H_a: the location parameter of the y_2 distribution is *larger* than the corresponding location parameter of the y_1 distribution (or the bulk of the y_2 distribution is to the right of the bulk of the y_1 distribution).

When H_a is true, we expect U to be small. For the one-sided alternative hypothesis that the location parameter of the y_2 distribution is *smaller* than the location parameter of the y_1 distribution, we compute the statistic U' defined to be *the number of times an a precedes a b*. For a one-sided test, Auble [1] gives tables of critical values of U (or of U') for significance levels of 0.001, 0.01, 0.025, and 0.05. For a two-sided test, compute U or U', whichever is smaller, and use Auble's tables, remembering that the significance levels are now 0.002, 0.02, 0.05, and 0.10. (The reader may wish to refer to Exercise 9.52 for more information on the computation of probabilities of U.)

It is not necessary to give tables of the U statistic because of a simple relationship between U and the rank-sum statistic T. If T_i denotes the sum of the ranks assigned to the sample of size n_i ($i = 1, 2$), it can be shown that

(9.27) $$U = n_1 n_2 + \frac{n_2(n_2 + 1)}{2} - T_2$$

Similarly U' is found by

(9.27a) $$U' = n_1 n_2 + \frac{n_1(n_1 + 1)}{2} - T_1$$

Usually in practice U or U' is computed by Equation (9.27) or Equation (9.27a) since it is tedious to find either by the definition.

When the null hypothesis is true, the random variable U has a mean of

(9.28) $$\mu_U = \frac{n_1 n_2}{2}$$

and a variance of

(9.29) $$\sigma_U^2 = \frac{n_1 n_2 (n_1 + n_2 + 1)}{12}$$

When both n_1 and n_2 are larger than eight, the U statistic is approximately normally distributed, and the approximation gets better as the sample sizes increase.

9.7. SUMMARY AND COMPARISONS

In Chapter 8 we showed how the binomial and the chi-square distributions can be used to make a variety of distribution-free tests of "independence" and of "goodness-of-fit." In this chapter we have presented distributions of other statistics (based on such things as number of runs, rank, and total rank) which are useful in such distribution-free tests as "goodness-of-fit," "nonrandomness," "difference in population medians," "difference in measure of centrality of two populations," and "any difference in two populations" tests. We found that the *statistic* used for the runs test of nonrandomness of a single sample is the same one used in the Wald-Wolfowitz runs test to compare two distributions. This multiple use applies to other statistics. For example, the statistic of the Kolmogorov-Smirnov test of goodness-of-fit can also be used to compare two distributions (see Exercise 9.51). Unfortunately, in a general text, it is possible to give only a very brief description of a few of the simplest procedures.

Six of the tests of this chapter are for two samples. For paired observations we presented the sign test and Wilcoxon's signed rank test. If the measurements are only good enough to determine which member of each pair is larger, then the sign test should be applied since the conditions required for Wilcoxon's signed test and the paired t-test are not satisfied. However, if enough is known of paired measurements to rank their differences, Wilcoxon's signed rank test should be used because it is more powerful than the sign test. Finally, if we know the paired differences come from a normal population then the paired t-test, presented in Chapter 10, should normally be preferred since it is the most powerful test. (For more details see Section 9.2.3).

Four tests for two-independent samples were presented. The rank sum and U tests give identical results. The measurements of the median test are not restricted as much as those of the other three tests. In case all measurements cannot be ordered, the median test should be used to compare the centrality of two distributions. Whenever the measurements can be ordered (or ranked) the other three tests may be applied. If the primary interest is in the difference of the location of two distributions, then either the rank-sum or U test is best. However, if the primary interest in the difference of two distributions is in their dispersions or in their shapes, the Wald-Wolfowitz runs test is more powerful.

Samples must be randomly drawn in order for any of these tests to be valid. In all tests which require that the measurements be ordered (or ranked), the observations are assumed to be from continuous distributions. In no case do we assume the shape of the distributions to be known or specified. In particular, we do not require that the samples be from normal distributions or any other specific family of distributions.

Generally, we have described test procedures which apply to one or two distributions. These test procedures can be generalized to distribution-free methods which are useful in comparing more than two distributions. Also, there are simple distribution-free tests relating to other concepts in statistics which we have not considered so far.

Sometimes a parametric test should be applied and sometimes a nonparametric test should be preferred. Some comments which might be useful in comparing parametric and nonparametric tests are

1. The calculations for distribution-free tests are generally simpler. This is especially true with small samples.

2. The assumptions for distribution-free tests are few and not very restrictive. The form of distributions is not required and knowledge of the mean or variance is not needed.

3. The distribution-free methods are relatively easy to understand and to apply. Usually all that is required by way of analysis is ranking, counting, adding, and subtracting.

4. The distribution-free tests may be applied to a much larger class of distributions since fewer assumptions are required.

5. Real differences in location parameters may be more easily detected with distribution-free tests. This is particularly true when the distributions deviate considerably from normality or the location parameters are not means.

6. If the assumptions for a parametric test are satisfied, it is likely to be more powerful than any distribution-free test. Some information may be lost when the quantitative nature of the data is replaced by ranks.

The reader may want more detailed information on the tests presented in this chapter or wish to read about other more specialized procedures. A number of useful distribution-free tests are presented in a book by Siegel [21]. An extensive bibliography of nonparametric methods has been compiled by Savage [19] and Walsh [25].

9.8. EXERCISES

9.35. The following random samples were independently drawn from populations A and B, respectively:

Sample A	24	36	24	8	44	77	9	57	7	54	41	4
Sample B	30	56	44	25	45	37	56	55	97	11	47	79	72

Test the hypothesis that the two populations have the same median.

9.36. One typist prepared the first 400 pages and a second typist the last 400 pages of a manuscript. A random sample of 15 pages from each of the typists' work revealed that the number of errors per page was as follows:

Typist I	0	5	1	2	0	4	2	0	7	0	3	1	3	0	1
Typist II	2	0	4	3	5	4	3	6	0	3	8	5	4	3	1

In terms of medians, does Typist II make more errors per page than Typist I?

9.37. Independent random samples of sizes 40 and 60 have a common median of 30. If each sample has 25 observations greater than the median, can we conclude that the median of the population from which 40 observations was drawn is greater than the median of the other population?

*9.38. On a given day the maximum temperature (in degrees fahrenheit), as recorded by 11 instruments of the same type, were as follows:

County A	79	74	88	68	70	
County B	76	80	85	73	85	86

Use the hypergeometric distribution to decide whether the median for County A is smaller than the median for County B.

9.39. The three quartiles of IQ's of all students from three classrooms are used to classify each room of students into four categories as follows:

Number of Students	Classroom I	Classroom II	Classroom III
Highest Quarter of all IQ	5	7	13
Second Highest Quarter	6	8	11
Third Highest Quarter	11	5	9
Fourth Highest Quarter	8	10	7

(a) Do the proportions of students falling in the four categories differ significantly from classroom to classroom?
(b) What can you say about the medians of the three populations from which the groups of students came?
(c) Are the third quartiles of the first and third populations unequal?

9.40. Use the Wald-Wolfowitz runs test on the data of Exercise 9.38 to decide whether the distribution of maximum temperatures (on a given day) in Counties A and B are different.

9.41. Use the data of Exercise 9.35 to decide whether the two populations A and B have different distributions.
(a) Apply the Wald-Wolfowitz runs test.
(b) Apply the rank-sum test.

9.42. In Example 9.6 suppose the number of audible sounds of an additional 12 nine-month-old girls are as follows:

$$42 \quad 44 \quad 61 \quad 31 \quad 54 \quad 37 \quad 38 \quad 59 \quad 43 \quad 52 \quad 32 \quad 39$$

(a) For the sample of 10 boys and 24 girls use Equation (9.22) to decide whether the distributions of the number of audible sounds are the same for boys and girls.
(b) Use Equation (9.25) to test the hypothesis in part (a).

9.43. Along two stretches of highway 55-miles-per-hour speed limit signs are regularly posted. The speeds of drivers who were caught exceeding the limit are as follows:

Highway I	65	64	100	83	67	68	70	87
Highway II	65	77	69	73	72	66	74	71	69

Decide whether the distributions of the two populations of excessive speeds are different by applying the (a) Wald-Wolfowitz runs test; (b) rank-sum test.

9.44. Use the rank-sum test on the data of Exercise 9.38 to decide whether the median for County A is smaller than the median for County B.

9.45. The following data were obtained from a study in which 12 four-year-old boys and 12 four-year-old girls were observed during two 15-minute play sessions, and each child's play during the 30 minutes was scored for incidence of and degree of aggression:

Girl's Score	58	16	7	26	9	16	36	20	15	55	40	22
Boy's Score	86	69	72	65	113	65	118	45	141	104	41	50

Apply both the Wald-Wolfowitz runs and the rank-sum tests to these scores to decide whether boys and girls differ in amount of aggression under the conditions indicated. Scores from Siegel [20].

*9.46. Let T denote the total of the ranks of the smaller sample in the rank-sum test.
(a) If $n_1 = 2$ and $n_2 = 3$, show that T has a symmetric distribution provided each arrangement of the $n_1 + n_2$ observations is equally likely.
(b) In general, show that T can be as small as $n_1(n_1 + 1)/2$ or as large as $n_1(n_1 + 2n_2 + 1)/2$. Then show that T has a symmetric distribution.

c9.47. Show that when $n_1 + n_2 = 30$ and the significance level for a two-sided test is 0.05, then critical values T_0, obtained by Formula (9.25), are the same as those in Table 9.9 in 12 of 14 cases and only one unit larger in the other two cases. *Hint:* Correcting for continuity, use the standard normal tables to find z' such that $P(z < z') = 0.025$ and then find T_0 from the equation

$$z' = \frac{(T_0 - \mu_T) + \frac{1}{2}}{\sigma_T}$$

c9.48. The following random samples were independently drawn from four distributions:

Sample 1	58	13	6	36	20	17	35	38	9	41	1	9	18
Sample 2	25	21	30	26	34	27	29	35	36	22	25	21	32
Sample 3	33	43	42	3	45	43	71	53	17	34	75	54	27
Sample 4	62	57	58	92	56	96	61	74	80	75	71	50	67

Use Formula (9.26) to test the hypothesis that the four distributions are identical.

9.49. Use the data of Exercise 9.45 to compute the value of the U statistic by
(a) the definition
(b) Formula (9.27)

9.50. Use the standard normal approximation for the U statistic and the data of Samples 2 and 3 from Exercise 9.48 to compare the two distributions from which these samples were drawn.

E9.51. (**Two-sample Kolmogorov-Smirnov Test**). Let $S_{n_1}(y)$ and $S_{n_2}(y)$ denote observed cumulative relative frequency distributions based on independent random samples of sizes n_1 and n_2, respectively. The two-sample Kolmogorov-Smirnov statistic is defined as

$$D = \text{maximum} |S_{n_1}(y) - S_{n_2}(y)|$$

where the same intervals for y are used for both $S_{n_1}(y)$ and $S_{n_2}(y)$. This D statistic is used to test the null hypothesis that the two samples have come from the same distribution against either a two-sided or one-sided alternative hypothesis that the samples are from different distributions.

We consider a restricted application. If $n_1 = n_2 = n$ it follows that $D = k/n$ where k is the maximum difference in the numerators of $S_{n_1}(y)$ and $S_{n_2}(y)$ for an interval of y. Critical values k' of k for a two-sided alternative hypothesis when $\alpha \leq 0.05$ are as follows:

n	5 or 6	7–9	10–13	14–17	18–22	23–27	28–33	34–39	40
k'	5	6	7	8	9	10	11	12	13

Any value of k equal to or greater than k' falls in the critical region of an $\alpha = .05$ level two-sided test. (Values of k' for other α values of D for more general cases can be found in statistical tables.)

(a) Use Samples 1 and 2 of Exercise 9.48 to compare the two distributions from which these samples were drawn. Let intervals of y be -0.5 to 4.5, 4.5 to 9.5, and so forth.

(b) In Exercise 9.48, suppose Samples 1 and 2 came from from Distribution A and Samples 3 and 4 came from Distribution B. Use intervals -0.5 to 9.5, 9.5 to 19.5, and so forth to decide whether Distributions A and B are different.

E9.52. (**Probability of U**). In Section 9.6.4 the U statistic was defined as the number of times a b precedes an a. Suppose there are m a-values and n b-values. Let the probability of a specific U be denoted by $P_{m,n}(U)$. For example, if $m = n = 1$, then the only arrangements of sample values are ab and ba with corresponding U-values of 0 and 1, respectively. Since the arrangements are assumed to be equally likely it follows that

$$P_{1,1}(0) = P_{1,1}(1) = \frac{1}{2}$$

Further, if $m = 3$ and $n = 1$, then the arrangements $aaab$, $aaba$, $abaa$, $baaa$ have values

0, 1, 2, 3, respectively. Thus

$$P_{3,1}(0) = P_{3,1}(1) = P_{3,1}(2) = P_{3,1}(3) = \frac{1}{4}$$

Values of $P_{m,n}(U)$ are easily found in this way when m and n are small. However, for larger m and n it is often better to use the formula

$$P_{m,n}(U) = \frac{m}{m+n} P_{m-1,n}(U-n) + \frac{n}{m+n} P_{m,n-1}(U)$$

where it is understood that $P_{m,n}(U) = 0$ when U is negative and that $P_{0,n}(0) = P_{m,0}(0) = 1$.

(a) Find values of $P_{3,3}(U)$ and $P_{3,4}(U)$.
(b) Use the formula to find values of $P_{3,5}(U)$.

10

APPLICATIONS OF STUDENT t DISTRIBUTIONS

10.1. PROPERTIES OF STUDENT t DISTRIBUTIONS

When, in Chapter 7, the sample mean was used to obtain a confidence interval or to test a hypothesis about a population mean, it was necessary that the population variance be known *or* that the sample size be large. However, in most experiments it is unrealistic to require the population variance to be known or to expect the sample size to be large enough so that the sample variance may replace the population variance. Fortunately, due to efforts of W.S. Gosset, who used the pseudonym *Student* in his statistical writings, the t distributions were introduced to remove the restrictive conditions mentioned above.

When the population variance σ^2 is replaced by a sample variance s^2 in the standard normal statistic

(10.1) $$z = \frac{\bar{y} - \mu}{\sqrt{\frac{\sigma^2}{n}}}$$

we obtain the Student t statistic, or simply t statistic,

(10.2) $$t = \frac{\bar{y} - \mu}{\sqrt{\frac{s^2}{n}}}$$

We would expect the distribution of t to be similar to that of z because the only change is from σ^2 to s^2. Since σ^2 is a fixed unknown constant and s^2 is a random variable which depends on the sample size, we would expect t to fluctuate more from sample to sample than z. Indeed, it can be shown that the mean of t is zero and that the variance is greater than one, the variance of z.

The distribution of the statistic t depends on the number of degrees of freedom of s^2. There is a different distribution (or curve) for each different degree of freedom of s^2. Thus, there is really a family of t distributions, one corresponding to each positive integer. Further, as the number of degrees of freedom increases, the random variable s^2 gets closer to σ^2 and t approaches z. So z may be considered a special case of t when the number of degrees of freedom is infinite.

We summarize these properties in the following two statements (which are proved in more mathematical books):

Theorem 10.1. If a random sample $y_1, \cdots, y_n$ is drawn from a normal distribution with mean μ and variance σ^2, then the statistic

$$t = \frac{\bar{y} - \mu}{\sqrt{s^2/n}}$$

is distributed as the Student t distribution with $n-1$ degrees of freedom, y and s^2 being defined in the usual way.

Theorem 10.2. As the number of degrees of freedom of s^2 approaches infinity, the Student t distribution approaches the normal distribution with mean 0 and variance 1.

Graphs of three t distributions are shown in Figure 10.1. Clearly, t distribution curves are bell-shaped and look like a standard normal curve. However, they are more spread out, and this becomes more noticeable as the degrees of freedom get smaller. Thus, for example, the 95th percentile (or upper 0.05 point) becomes larger with decreasing degrees of freedom.

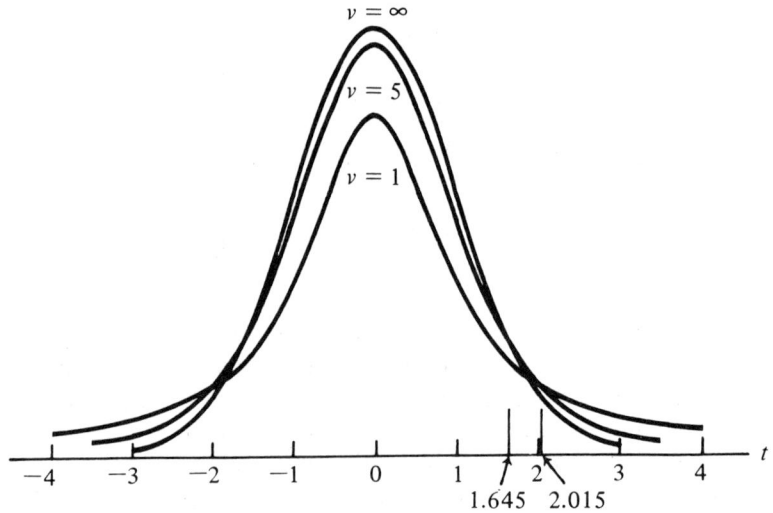

Figure 10.1. Student t Distributions for $v = 1, 5,$ and ∞. Two Upper 0.05 Points ($t_{.05}$) are Indicated

We often require percentiles in the tails of a t distribution with ν degrees of freedom. Let t_α be that value of t (the random variable of a Student t distribution with ν degrees of freedom) for which the area under the curve and to its right is α. That is, for any α between 0 and 0.5, t_α is that value of t for which

(10.3) $$P(t > t_\alpha) = \alpha$$

Figure 10.1 shows two different values of $t_{.05}$ (the third is 6.314). Values of t_α for selected values of α and ν are shown in Table V in the Appendix. Since a t distribution is symmetric, it is not necessary that we table lower tail values. The probability that t is less than $-t_\alpha$, a value in the left tail of a t distribution with ν degrees of freedom, is given by

(10.4) $$P(t < -t_\alpha) = P(t > t_\alpha)$$

The reader should note that the last row in Table V gives values for the standard normal distribution.

The statistic of Theorem 10.1 (along with others to be introduced later) is a special case of the more general statistic in the following statement:

> **Theorem 10.3.** If z has a normal distribution with mean 0 and variance 1, χ^2 has a chi-square distribution with ν degrees of freedom, and z and $\sqrt{\chi^2}$ are independently distributed, then the random variable
>
> $$t = \frac{z}{\sqrt{\chi^2/\nu}}$$
>
> has the Student t distribution with ν degrees of freedom with mean 0 and variance $\nu/(\nu - 2)$ for $\nu > 2$.

To understand how Theorem 10.1 is actually a corollary of Theorem 10.3 we first state that "the mean and variance of a random sample are independent random variables provided the sample is from a normal distribution." By Theorem 10.1 the random sample is from a normal distribution. Thus, $\bar{y}$ and s^2 are independently distributed. So

(10.5) $$\frac{\bar{y} - \mu}{\sqrt{\sigma^2/n}} \quad \text{and} \quad \frac{(n-1)s^2}{\sigma^2}$$

are independently distributed, μ, σ, and n being constants. The two random variables of Expression (10.5) are distributed as z and χ^2 of Theorem 10.3 where $\nu = n - 1$. Hence

$$t = \frac{z}{\sqrt{\frac{\chi^2}{\nu}}} = \frac{\frac{\bar{y} - \mu}{\sqrt{\sigma^2/n}}}{\sqrt{\frac{(n-1)s^2}{\sigma^2}/(n-1)}} = \frac{\bar{y} - \mu}{\sqrt{\frac{s^2}{n}}}$$

is distributed as the Student t variable with $n - 1$ degrees of freedom.

The two statistics $\bar{y}$ and s^2 used to compute t may or may not be determined from the same sample. If they are computed from samples of different sizes, the number of degrees of freedom for the t statistic is the same as that of s^2.

10.2. APPLICATIONS OF t DISTRIBUTIONS: SINGLE SAMPLE

The t statistic is applied in the same way as the standard normal z statistic, except the sample variance is required instead of the population variance. This leads to more computations and more elaborate tables, but one assumption is removed—the population variance does not need to be specified. Example 10.1 brings out the differences.

EXAMPLE 10.1. The national mean score on a verbal standardized test for high school seniors is known to be 500. Suppose a random sample of 13 seniors at Gamma High School had scores of

$$560 \quad 470 \quad 600 \quad 440 \quad 520 \quad 650 \quad 500$$
$$610 \quad 600 \quad 550 \quad 580 \quad 400 \quad 540$$

(a) Is there reason to believe that seniors at Gamma High School have higher verbal scores than those in the nation? (b) Find a symmetric 0.95 confidence interval for the mean score of all seniors at Gamma High School.

For the answer to part (a) we use the following general test procedure:

1. *Hypotheses.* $H_o : \mu = 500$
   $H_a : \mu > 500$

2. *Assumptions and Specifications.* The 13 scores are randomly drawn from a normal population with unknown variance. Let $\alpha = 0.05$.

3. *Statistic.* Under the assumption that the null hypothesis is true

$$t = \frac{\bar{y} - 500}{\sqrt{s^2/13}}$$

has the Student t distribution with 12 degrees of freedom.

4. *Critical Region.* For a one-sided 0.05 level test

$$CR = \{t \mid t > 1.782\}$$

Note: If we knew the population variance, the critical region would be the set $\{z \mid z > 1.645\}$.

5. *Computations.* Since

$$\sum_i y_i = 7{,}020 \quad \text{and} \quad \sum_i y_i^2 = 3{,}853{,}600$$

it follows that

$$\bar{y} = \frac{7{,}020}{13} = 540$$

$$s^2 = \frac{13(3{,}853{,}600) - (7{,}020)^2}{(12)(13)} = 5{,}233.3$$

$$t = \frac{540 - 500}{\sqrt{\frac{5{,}233.3}{13}}} = \frac{40}{20.1} \doteq 1.99$$

Notice that the computations are the same as for the z statistic, except we must also find $\sum_i y_i^2$ and s^2.

6. *Conclusion.* Since $t = 1.99$ falls in the critical region, we reject the null hypothesis and conclude that seniors at Gamma High School do have higher mean verbal scores than those in the nation. Of course, the conclusion is based on a specific standardized test.

Note how important it is to specify the correct alternative hypothesis. For a two-sided 0.05 level test we would have had $CR = \{t \mid t < -2.179 \text{ or } t > 2.179\}$. Therefore, we would have failed to reject the null hypothesis.

To find a confidence interval for μ we use the Student t statistic and the method illustrated in Examples 7.11 and 7.12, replacing $\sigma_{\bar{y}} = \sqrt{\sigma^2/n}$ by its estimate $s_{\bar{y}} = \sqrt{s^2/n}$. Thus, the symmetric 0.95 confidence limits are given by

(10.6) $$\bar{y} \pm t_{.025} \times s_{\bar{y}} \quad \text{or} \quad \bar{y} \pm t_{.025} \times \sqrt{s^2/n}$$

From the solution to part (a) we know that

$$\bar{y} = 540 \qquad t_{.025} = 2.18 \qquad s_{\bar{y}} = 20.1$$

Thus $t_{.025} \times s_{\bar{y}} = 43.8$ and the limits for the mean verbal scores at Gamma High School are 540 ± 43.8, or 496.2 and 583.8. So the symmetric 0.95 confidence interval for μ is

(10.7) $$496.2 < \mu < 583.8$$

Power. In Section 9.4 we noted that, of the three goodness-of-fit distribution-free tests presented, the Kolmogorov–Smirnov test is the most powerful. When it is assumed that a random sample is drawn from a normal distribution, it can be shown that the Student t test is the most powerful test of the null hypothesis that the population mean is some specified value. Put another way, for a specified α the probability of a type II error (i.e., of failing to reject the null hypothesis when we should) is smallest with the Student t test. So in Example 10.1 if the alternative hypothesis is actually true the Student t test is least

10.3. EXERCISES

10.1. Find the standard deviation of the Student t distribution with v degrees of freedom for $v = 3, 4, \cdots, 10$.

10.2. Random samples of sizes 5 and 10 are independently drawn from a normal distribution with mean 50. The sample of size 5 has a mean of 48.2 and the sample of size 10 has a variance of 8.52.
(a) Does the value of the t statistic belong to the set $\{t \mid t < -t_{.05}\}$? Explain.
(b) Find a 0.95 confidence interval of the population mean. Does the true mean fall in this interval?

10.3. A sample of size 12, drawn from a normal distribution, has a mean of 22.2 and a standard deviation of 3.66. Use a 0.05 level test of the hypothesis that $\mu = 20$.

10.4. The heights (in inches) of a random sample of 10 adult males from state university were

$$69 \quad 72 \quad 65 \quad 76 \quad 66 \quad 70 \quad 72 \quad 73 \quad 68 \quad 75$$

Assuming the distribution of heights of all males at the university to be approximately normally distributed, test the hypothesis that $\mu = 68$ inches.

10.5. Use the data of Exercise 10.4 to find a 0.99 confidence interval of μ.

10.6. A salesman of fishing lines claims that his 5-pound test line will average at least 7 pounds test. Is his claim justified if a sample of 8 pieces of line break at the following pounds?

$$7.2 \quad 8.5 \quad 6.8 \quad 6.5 \quad 7.7 \quad 8.3 \quad 7.6 \quad 8.0$$

10.7. Use the data in Exercise 10.6 to find a 95 percent confidence interval of the mean test strength of the fishing line.

E10.8. Suppose t has the Student t distribution with v degrees of freedom. In the special case where $\mu = 0$, show the following:

(a) $t^2 = \dfrac{n y^{-2}}{s^2}$

(b) $t^2 = \dfrac{n-1}{nA - 1}$ where $A = \dfrac{\sum y_i^2}{(\sum y_i)^2}$

(c) $A = \dfrac{1}{n}\left[\dfrac{n-1}{t^2} + 1\right]$

Applications of t Distributions: Two Samples

C10.9. Substitute values from Table V in the equation of Exercise 10.8(c) to show that 0.05 level critical values of A for $\nu = 1, 2, \cdots, 12$ are 0.503, 0.369, 0.324, 0.304, 0.293, 0.286, 0.281, 0.278, 0.276, 0.274, 0.273, 0.271, respectively. Note that values in the two tails of a t distribution correspond to small values of an A distribution. For example, when $\nu = 5$ the sets $\{t \mid t < -2.571 \text{ or } t > 2.571\}$ and $\{A \mid A < 0.293\}$ are equivalent critical regions for a 0.05 level test. (Clearly, the alternative hypothesis must be of the form $\mu \neq 0$ before the A statistic can be applied. For more on the A-test see Sandler [18].)

E10.10. In Exercise 10.4 subtract 68 from each observation and use the A test of Exercise 10.9 to test $H_o: \mu - 68 = 0$ against $H_a: \mu - 68 \neq 0$. *Hint:* Compute $A = \sum y_i^2/(\sum y_i)^2$.

10.4. APPLICATIONS OF t DISTRIBUTIONS: TWO SAMPLES

Whenever two population means are compared, Student t distributions are most commonly used. These distributions are applied to both tests of hypotheses and confidence interval estimation.

In Chapter 9 we noted that t-tests for both paired and independent samples are very powerful tests. Generally, when the parent populations can be assumed to be normal (or when the sample sizes are large enough that we may assume the sample means to be approximately normally distributed), a t-test is the most powerful test an experimenter can use. Actually, it turns out that when dealing with means, t-tests are very *robust* even when the normality assumption fails; i.e., the probabilities of type I and type II errors do not change much unless the populations are drastically nonnormal and at least one sample size is small.

10.4.1. Independent Samples

In Section 6.7 we learned that when random samples $y_{11}, \cdots, y_{1n_1}$ and $y_{21}, \cdots, y_{2n_2}$ of sizes n_1 and n_2 are independently drawn from normal populations with means μ_1 and μ_2 and variances σ_1^2 and σ_2^2, respectively, then the difference in sample means $\bar{d} = \bar{y}_1 - \bar{y}_2$ is normally distributed with mean $\mu_1 - \mu_2$ and variance

$$(10.8) \qquad \frac{\sigma_1^2}{n_1} + \frac{\sigma_2^2}{n_2}$$

In Section 7.11 we illustrated how $\bar{d}$ could be used to test hypotheses about $\delta = \mu_1 - \mu_2$.

When the two normal populations have common unknown variance $\sigma_1^2 = \sigma_2^2 = \sigma^2$, the variance of $\bar{d}$ is

$$(10.9) \qquad \sigma_{\bar{d}}^2 = \sigma^2 \left(\frac{1}{n_1} + \frac{1}{n_2} \right)$$

From Section 8.4 we know that s_1^2, s_2^2, and

(10.10) $$s_p^2 = \frac{(n_1-1)s_1^2 + (n_2-1)s_2^2}{n_1+n_2-2} = \frac{SS_1 + SS_2}{n_1+n_2-2}$$

are all unbiased estimators of the common variance σ^2 and that they have n_1-1, n_2-1, and n_1+n_2-2 degrees of freedom, respectively. Since s_p^2 has the largest number of degrees of freedom we use

(10.11) $$s_{\bar{d}}^2 = s_p^2\left(\frac{1}{n_1}+\frac{1}{n_2}\right)$$

as the unbiased estimator of $\sigma_{\bar{d}}^2$. It can be shown that $\bar{d}$ and $s_{\bar{d}}^2$ are independently distributed. Hence, with the aid of Theorem 10.3, we obtain the following very important statement:

Theorem 10.4. If random samples $y_{11}, \ldots, y_{1n_1}$ and $y_{21}, \ldots, y_{2n_2}$ of sizes n_1 and n_2 are independently drawn from two normal populations with common variance σ^2, then the statistic

$$\frac{\bar{d}-\delta}{\sqrt{s_{\bar{d}}^2}} = \frac{\bar{y}_1 - \bar{y}_2 - (\mu_1-\mu_2)}{\sqrt{s_p^2\left(\frac{1}{n_1}+\frac{1}{n_2}\right)}}$$

has the Student t distribution with $n_1 + n_2 - 2$ degrees of freedom.

As we shall soon see, this theorem can be applied in obtaining confidence intervals for and in testing hypotheses about $\delta = \mu_1 - \mu_2$ when the population variances are unknown.

EXAMPLE 10.2. The mean tensile (or tear) strength of paper bags made under standard conditions is to be compared to that of the same product made with a new chemical additive. The experimenter claims that the additive will increase tensile strength by at least two pounds. However, there is reason to believe that the application of the new chemical additive will not affect the variability of tensile strength. (Such an assumption should come from someone who understands the process.) Suppose a random sample of 8 standard paper bags has tensile strengths (in pounds) of

39.7 41.1 40.3 39.2 40.1 38.5 39.6 39.3

and a random sample of 10 "new" paper bags, independently drawn, has tensile strengths of

42.1 40.2 43.6 42.4 42.5 41.4 41.7 43.4 41.5 41.8

(a) Test the experimenter's claim. (b) Obtain a symmetric 0.95 confidence interval for the difference in mean tensile strengths.

Let the subscripts 1 and 2 refer to the "standard" and "new" conditions, respectively. For the solution to part (a) we use the general test procedure to obtain:

1. H_o: The mean tensile strength μ_2 of paper bags with the new chemical treatment is equal to or less than the mean tensile strength μ_1 for standard bags plus two pounds. That is, $\mu_2 \leq \mu_1 + 2$ or $\mu_2 - \mu_1 \leq 2$ or $\mu_1 - \mu_2 \geq -2$.
H_a: $\mu_2 - \mu_1 > 2$ or $\mu_1 - \mu_2 < -2$, since we are interested in knowing if the new chemical additive increases tensile strength *at least* two pounds.

2. Assume the two random samples of sizes 8 and 10 are independently drawn from two normal populations with common unknown variance σ^2. Let the significance level be $\alpha = 0.05$.

3. The test statistic, under the temporary assumption that the null hypothesis is true, is

$$t = \frac{\bar{y}_1 - \bar{y}_2 - (-2)}{\sqrt{s_p^2 \left(\frac{1}{8} + \frac{1}{10}\right)}}$$

with $8 + 10 - 2 = 16$ degrees of freedom.

4. The left tail 0.05 critical region is

$$CR = \{t \,|\, t < -1.746\}$$

5. From

$$\sum_j^8 y_{1j} = 317.8 \qquad \sum_j^{10} y_{2j} = 420.6$$

$$\sum_j^8 y_{1j}^2 = 12{,}628.94 \qquad \sum_j^{10} y_{2j}^2 = 17{,}699.32$$

we obtain

$$\bar{y}_1 = 39.72 \qquad \bar{y}_2 = 42.06 \qquad \bar{d} = \bar{y}_1 - \bar{y}_2 = -2.34$$

$$SS_1 = \frac{34.68}{8} = 4.335 \qquad SS_2 = \frac{88.84}{10} = 8.884$$

$$s_p^2 = \frac{4.335 + 8.884}{8 + 10 - 2} = 0.8262$$

$$s_{\bar{d}} = \sqrt{0.8262 \left(\frac{1}{8} + \frac{1}{10}\right)} = \sqrt{.1859} \doteq .431$$

$$t = \frac{-2.34 - (-2)}{.431} = -.789$$

6. Since $t = -.789$ does not fall in the critical region, we fail to reject the null hypothesis. Thus, we *cannot accept* the experimenter's claim that "the chemical additive actually does increase tensile strength at least two pounds."

Next, by the usual method we obtain, from Theorem 10.4, the following symmetric 0.95 confidence limits:

(10.12) $\quad \bar{y}_1 - \bar{y}_2 \pm t_{.025} \times s_{\bar{d}} \quad$ or $\quad \bar{y}_1 - \bar{y}_2 \pm t_{.025} \times \sqrt{s_p^2\left(\dfrac{1}{n_1} + \dfrac{1}{n_2}\right)}$

Thus, to obtain the solution of part (b) we find $t_{.025} = 2.120$ and $t_{.025} \times s_{\bar{d}} = (2.12)(.431) = .914$. Finally, the 0.95 confidence limits are $-2.34 \pm .91$ or -3.25 and -1.43. So, we estimate with 0.95 confidence that the true difference in the means is in the interval

$$-3.25 < \mu_1 - \mu_2 < -1.43$$

or

$$1.43 < \mu_2 - \mu_1 < 3.25$$

That is, the mean tensile strength of paper bags treated with the new chemical additive are from 1.43 pounds to 3.25 pounds stronger than the mean tensile strength of paper bags prepared by the standard process.

When the population variances are unknown, it is seldom known whether they are equal or unequal. As a consequence, the method illustrated for equal variances is typically applied without knowing whether the variances are equal. However, this is not necessarily troublesome, for it has been found that there will be no serious consequence if the population variances are only moderately different *and* the sample sizes are equal. Thus, by using equal sample sizes the experimenter can use the method described unless the population variances are greatly different.

An approximate solution to the problem of unequal variances, known as the Behrens–Fisher solution, is described below. Let

$$s_{\bar{y}_1}^2 = \dfrac{s_1^2}{n_1} \quad \text{and} \quad s_{\bar{y}_2}^2 = \dfrac{s_2^2}{n_2}$$

Then it can be shown that

(10.13) $\quad t' = \dfrac{\bar{d} - \delta}{\sqrt{s_{\bar{y}_1}^2 + s_{\bar{y}_2}^2}}$

is approximately distributed as Student t with ν degrees of freedom where

(10.14) $\quad \nu = \dfrac{(s_{\bar{y}_1}^2 + s_{\bar{y}_2}^2)^2}{\dfrac{(s_{\bar{y}_1}^2)^2}{n_1 - 1} + \dfrac{(s_{\bar{y}_2}^2)^2}{n_2 - 1}}$

Since ν is not likely to be a whole number it should be rounded-off to the nearest integer. The statistic t' is applied like the t statistic of Theorem 10.4, but should be used sparingly. (For problems see Exercises 10.16 and 10.17.)

Linear Combination of Two Means. There are many other applications involving two independent sample means. For example, we may want to use a weighted sample mean (see Section 3.4) to make a statement about the corresponding population mean. This would lead us to consider the distribution and application of a more general statistic like

(10.15) $$c = a_1 \bar{y}_1 + a_2 \bar{y}_2$$

where a_1 and a_2 are real numbers with at least one being different from zero.

If random samples $y_{11}, \cdots, y_{1n_1}$ and $y_{21}, \cdots, y_{2n_2}$ are independently drawn from normal populations with means μ_1 and μ_2 and variances σ_1^2 and σ_2^2, respectively, it can be shown that c is distributed normally with mean

(10.16) $$\gamma = a_1 \mu_1 + a_2 \mu_2$$

and variance

$$\sigma_c^2 = a_1^2 \frac{\sigma_1^2}{n_1} + a_2^2 \frac{\sigma_2^2}{n_2}$$

In case the populations have common variance $\sigma_1^2 = \sigma_2^2 = \sigma^2$, the variance of c is

(10.17) $$\sigma_c^2 = \sigma^2 \left(\frac{a_1^2}{n_1} + \frac{a_2^2}{n_2} \right)$$

When the common variance σ^2 is estimated by s_p^2

(10.18) $$s_c^2 = s_p^2 \left(\frac{a_1^2}{n_1} + \frac{a_2^2}{n_2} \right)$$

is an unbiased estimator of σ_c^2 and we have the following useful general statement:

Theorem 10.5. If random samples of sizes n_1 and n_2 are independently drawn from two normal populations with means μ_1 and μ_2, respectively, and common variance σ^2, then the statistic

$$t = \frac{c - \gamma}{\sqrt{s_c^2}} = \frac{a_1 \bar{y}_1 + a_2 \bar{y}_2 - (a_1 \mu_1 + a_2 \mu_2)}{\sqrt{s_p^2 \left(\frac{a_1^2}{n_1} + \frac{a_2^2}{n_2} \right)}}$$

has the Student t distribution with $n_1 + n_2 - 2$ degrees of freedom.

Note that Theorem 10.4 is a special case of Theorem 10.5 when $a_1 = 1$ and $a_2 = -1$.

Applications are similar to those of Theorem 10.4. For example, suppose we know the mean annual incomes for heads of households in twin cities 1 and 2 are $9,400 and $9,700 and that the sample sizes are 20 and 25, respectively.

350 Applications of Student t Distributions

If 40 percent of the households are in city 1 and 60 percent are in city 2, we might wish to find a 0.95 confidence interval for the mean annual income $(0.40)(\mu_1) + (0.60)(\mu_2)$ in the combined city. The limits of such an interval are

$$a_1 \bar{y}_1 + a_2 \bar{y}_2 \pm t_{.025} \sqrt{s_p^2 \left(\frac{a_1^2}{n_1} + \frac{a_2^2}{n_2} \right)}$$

or, in particular

$$(0.40)(9400) + (0.60)(9700) \pm 2.02 \sqrt{s_p^2 \left(\frac{(0.4)^2}{20} + \frac{(0.6)^2}{25} \right)} = 9580 \pm .302 s_p$$

where s_p^2 is the pooled variance with 43 degrees of freedom. As another example, we might wish to use the mean yield of a new variety of corn on two quite different sample plots to test the hypothesis that the new variety actually gives better yield than the older variety.

10.4.2. Correlated Samples

The reasons for pairing (or matching) objects in two samples are given in Section 9.2. In Section 9.2.1 we illustrated how the sign test may be used to compare differences in paired objects when measurements for each pair can be ordered. In Section 9.2.2 Wilcoxon's signed rank test was presented for cases where the differences in paired observations may be ranked. We have already indicated that the t-test for the comparison of observations from two correlated samples is more powerful when we can assume the differences are normally distributed.

Suppose y_{1j} and y_{2j} denote the observed values of the jth pair of a random sample of n paired objects. Assume the differences

(10.19) $\quad d_1 = y_{11} - y_{21}, \ d_2 = y_{12} - y_{22}, \ \cdots, \ d_n = y_{1n} - y_{2n}$

are a sample of n random observations from a normal population with mean δ and variance σ_δ^2. Denote the sample mean and variance by $\bar{d}$ and s_d^2, respectively, and compute them by the formulas

(10.20) $\quad \bar{d} = \dfrac{\sum_j d_j}{n} \quad \text{and} \quad s_d^2 = \dfrac{n \sum_j d_j^2 - \left(\sum_j d_j \right)^2}{n(n-1)}$

If there are extraneous factors affecting the jth pair, we assume that such factors affect both observations of the pair in exactly the same way and that subtraction removes the effect. With these assumptions it follows that

(10.21) $\quad t = \dfrac{\bar{d} - \delta}{\sqrt{\dfrac{s_d^2}{n}}}$

Applications of t Distributions: Two Samples

is distributed as Student t with $n - 1$ degrees of freedom. Example 10.3 illustrates the test procedure as well as the method for finding a confidence interval for the mean difference δ.

EXAMPLE 10.3. Use the data shown in Table 9.1 (a) to determine whether mean grip in the right hand is greater than mean grip in the left hand, and (b) to find a symmetric 0.95 confidence interval for the difference in mean grip in the two hands.

We use the same data so that comparisons among the three procedures are easier. (It should be realized that for a specific collection of data only one of the three procedures should be applied; the experimenter should be able to determine which one from the kind of measurements and the assumptions he is willing to make.)

As in Table 9.1, y_{1j} and y_{2j} denote grip in left and right hand, respectively, of the jth student. The test procedure for part (a) is as follows:

1. H_o: The mean of the population of differences is zero or $\delta = 0$. That is, the mean grip in the left hand is the same as the mean grip in the right hand.

H_a: The mean of the population of differences is negative or $\delta < 0$.

2. Assume 18 students are randomly selected from the university and that the differences $d_j = y_{1j} - y_{2j}$ are normally distributed with unknown variance σ_δ^2. Let $\alpha = 0.05$.

3. Under the assumption that the null hypothesis holds, the test statistic is

$$t = \frac{\bar{d}}{\sqrt{s_d^2/n}}$$

and it has the Student t distribution with 17 degrees of freedom.

4. The critical region for the one-sided test is

$$CR = \{t \mid t < -1.740\}$$

5. From Table 9.1 we find the differences to be

$$\begin{array}{rrrrrrrrr} -3 & -2 & 1 & -1 & 0 & -1 & -1 & 2 & -2 \\ -8 & -1 & 0 & -2 & 2 & -3 & & 2 & -3 & -3 \end{array}$$

We find that

$$\sum_j^{18} d_j = -23 \qquad \sum_j^{18} d_j^2 = 129$$

$$\bar{d} = \frac{-23}{18} = -1.28$$

$$s_d^2 = \frac{129 - \frac{(-23)^2}{18}}{17} = 5.859$$

$$\sqrt{s_d^2/n} = \sqrt{5.859/18} = 0.571$$

$$t_c = \frac{-1.28}{0.571} = -2.24$$

6. Since $t_c = -2.24$ is in the critical region, we accept the alternative hypothesis. Thus, we conclude that the male students at the university do have stronger mean grip in their right hands.

For part (b) we find $t_{.025} = 2.110$ and

$$t_{.025}\sqrt{\frac{s_d^2}{n}} = (2.11)(0.571) = 1.20$$

Thus the symmetric 0.95 confidence limits of δ are

$$\bar{d} \pm t_{.025}\sqrt{\frac{s_d^2}{n}} = -1.28 \pm 1.20 \quad \text{or} \quad -2.48 \text{ and } -0.08.$$

So the required confidence interval is

$$-2.5 < \delta < -0.1 \quad \text{or} \quad 0.1 < \mu_2 - \mu_1 < 2.5$$

where $\delta = \mu_1 - \mu_2$. That is, we conclude with 0.95 confidence that the mean grip in the right hand is from 0.1 pounds to 2.5 pounds greater than in the left hand.

Note: If, in part (a) of Example 10.3, we had used a two-sided test, only the paired t-test would have rejected the null hypothesis; the sign test and Wilcoxon's signed rank test would have failed to reject. Put another way, for the three tests described, the value of the statistic for the paired t-test fell farthest out in the critical region, the value of the statistic for Wilcoxon's signed rank test fell next farthest out in the critical region, and the value of the statistic for the sign test was the same as the critical point. This is illustrative of what can be expected generally when the assumptions underlying the t-test are satisfied. That is, if the null hypothesis is not true, the paired t-test will most often lead to that conclusion.

The statements of the last paragraph should not be surprising, for we lost information in running the sign and Wilcoxon's tests; for example, in both we simply disregarded two values, which reduced the sample size. Further, in Wilcoxon's test we changed the magnitudes of measurements when we ranked differences, and in the sign test we not only changed the magnitudes of measurements, we simply ignored them.

Comparison of Independent and Paired t Tests. It might be informative to describe what our conclusion would have been in Example 10.3 if we had erroneously run the independent t-test on the data of Table 9.1. For such a t statistic with $18 + 18 - 2 = 34$ degrees of freedom we would find the critical region to be

$$CR = \{t \mid t < -1.69\}$$

Further, we would find

$$\sum_{j}^{18} y_{1j} = 662 \qquad \sum_{j}^{18} y_{2j} = 685$$

$$\sum_{j} y_{1j}^2 = 25{,}058 \qquad \sum_{j} y_{2j}^2 = 26{,}715$$

$$\bar{y}_1 = 36.78 \qquad \bar{y}_2 = 38.06 \qquad \bar{d} = -1.28$$

$$SS_1 = 711.11 \qquad SS_2 = 646.94$$

$$s_p^2 = \frac{711.11 + 646.94}{34} = 39.94$$

$$s_{\bar{d}} = \sqrt{39.94\left(\frac{1}{18} + \frac{1}{18}\right)} = \sqrt{4.438} = 2.11$$

$$t_c = \frac{-1.28}{2.11} = -0.61$$

Since $t_c = -0.61$ does not fall in the critical region, we fail to reject the null hypothesis.

Even though we have twice as many degrees of freedom in the independent t-test, we do not reject the null hypothesis as we did with the paired t-test. In other words, the increase in the number of degrees of freedom is not enough to compensate for the difference in hand strengths of the boys. Clearly, since $\bar{d}$ has the same value (-1.28) in both tests, the estimate of the standard error makes the difference. The standard error for the independent t-test is very inflated because of the addition of variation in strengths of boys; in the process of forming pairs, this variation was eliminated.

In summary, if there is some extraneous factor present which affects the observations in a systematic way, the paired t-test should usually be applied. However, if this extraneous factor is not present, the independent t-test is preferred, since it is then more powerful than the paired t-test. This is due to the larger number of degrees of freedom—for the independent t-test has twice as many as the corresponding paired t-test. The difference in power (or in probability of type II error) is considerable for small n, but is relatively small when n is moderately large, say $n \geq 12$. One the other hand, for the paired t-test we do not assume that y_{1j} and y_{2j} are independent nor that their variances are equal as in the case for the independent t-test. Actually, we do not even assume the variances σ_1^2 and σ_2^2 remain constant throughout the experiment. All we need assume is that the variance of $y_{1j} - y_{2j}$, i.e., $V(y_{1j} - y_{2j})$, remains constant for the n paired observations.

Selection of Experimental Units. In our three illustrations of paired observations we have used n experimental units (objects), and for each we made two observations. In this way we did not have the problem of matching experimental units (or objects) prior to running the experiment. However, in many experiments we must match $2n$ objects before the observations can be made.

For example, a supervisor might want to determine which of two methods of instruction is more effective. Suppose the usual instruction procedure is designated as method A and the new or experimental procedure is termed method B. Equal size groups of students are to be taught the same subject by the same teacher but with different methods. At the end of the term the same examination will be given to both groups, and comparisons made on the basis of differences in matched students.

Assume the $2n = 70$ children in the experiment represent all children available or represent a random sample from among all those eligible students in the population. Before the two methods of instruction may be applied the students must be separated into two groups so that each student in one group is matched (or paired) with one student in the second group. This matching should be done by someone familiar with variables which are (or may be) important to the study. For example, scores on standardized tests should be matched as well as possible. Also, environmental background, sex, past academic performance, and so forth, should probably be taken into account when students are paired. Once the $n = 35$ pairs have been formed, one member of each pair should be assigned randomly to group A, say, and the other one is automatically assigned to group B. Thus, there are 35 random choices to group assignments.

On the other hand, for an independent t-test, the same 70 students would not be matched. A table of random numbers could be used to assign each of the 70 students to a group. In the end, one group could be assigned 40 students and the other group would have 30 students. (If the classes were to be made the same size, then once one group was assigned 35 students, the remaining students would be placed in the second group.) In this case each of 70 students is assigned at random. Note that one of the objectives of this random assignment is to attempt to keep the extraneous factor from affecting one group to a greater extent than the other. While the pairing tries to control for, say, IQ by carefully matching, randomization of all students tries to "spread out" IQ over the two groups.

The use of paired observations is not limited to humans. Animals may be paired in a feeding experiment. Two plots in the same field may be used to compare the effect of fertilizer on the yield of a crop. Cars may be paired to compare similar parts. The list of possibilities is almost endless.

Objects are paired to minimize the magnitude of the estimate of the population variance. Any variation from one pair to another is eliminated by subtraction. So when an experimenter decides to use the paired t-test in place of the independent t-test, he or she should take into account the magnitude of the variation among pairs as compared to the loss in $n - 1$ degrees of freedom.

10.5. EXERCISES

10.11. Samples of sizes $n_1 = 11$ and $n_2 = 8$ are from normal populations with common variance σ^2. It is found that $\bar{y}_1 = 27.2$, $s_1 = 3.3$, $\bar{y}_2 = 24.1$, and $s_2 = 4.8$.

(a) Use s_p^2 to find a 0.95 confidence interval of $\mu_1 - \mu_2$. What assumptions are required?

(b) Use s_1^2 to find a 0.95 confidence interval of $\mu_1 - \mu_2$. What assumptions are required?

(c) Which of the two above confidence intervals is preferred? Why?

10.12. Random samples of sizes $n_1 = 10$ and $n_2 = 15$ are independently drawn from two normal populations with common standard deviation. Determine whether the mean of the first population is less than that of the second population if it is found that $\bar{y}_1 = 48.2$, $\bar{y}_2 = 55.1$, $s_1^2 = 65.50$, and $s_2^2 = 80.00$.

10.13. If sample values are $n_1 = n_2 = 6$, $\bar{y}_1 = 156.7$, $\bar{y}_2 = 172.3$, $s_1 = 13.6$, and $s_2 = 16.7$, test the hypothesis that $\mu_1 + 3 = \mu_2$. Write a complete report.

10.14. Assume the two samples of Exercise 9.38 are from normal populations with common variance. Use the t-test to decide whether the mean maximum temperature of County A was different from the mean maximum temperature of County B.

10.15. Assume Samples 1 and 3 of Exercise 9.48 are from normal populations with common variance. Find a 0.99 confidence interval of $\mu_3 - \mu_1$.

10.16. If sample values are $n_1 = 10$, $n_2 = 15$, $\bar{y}_1 = 187.2$, $\bar{y}_2 = 173.3$, $s_1^2 = 123.0$, and $s_2^2 = 415.5$, test the hypothesis that $\mu_1 = \mu_2$ by use of
(a) the statistic of Equation (10.13)
(b) theorem 10.4
(c) Compare results of the tests of parts (a) and (b)

10.17. Assume the two samples of Exercise 9.43 are from normal populations with unequal variances. Use the statistic of Equation (10.13) to obtain an approximate 0.95 confidence interval for the difference in the means of populations I and II.

10.18. Suppose two populations have a specific common variance σ^2. If $n_1 + n_2$ is a fixed constant, it can be shown (as in Exercise 6.33) that

$$V(\bar{y}_1 - \bar{y}_2) = \sigma^2 \left(\frac{1}{n_1} + \frac{1}{n_2} \right)$$

is smallest when $n_1 = n_2$.

(a) Let $n_1 + n_2 = 50$. Compute $V(\bar{y}_1 - \bar{y}_2)$ for $n_1 = 25$; for $n_1 = 20$; for $n_1 = 1$. Compare these three variances.

(b) Let $n_1 = 20$. Compute $V(\bar{y}_1 - \bar{y}_2)$ when $n_2 = 10, 100, 1{,}000$. Note that $V(\bar{y}_1 - \bar{y}_2)$ is greater than $0.05\sigma^2$ regardless of the size of n_2. Thus, the only way to further decrease $0.05\sigma^2$ is to increase the size of n_1. Find the smallest values of n_1 and n_2 such that $V(\bar{y}_1 - \bar{y}_2)$ is less than $0.05\sigma^2$.

10.19. Use the data of Exercise 10.12 to find a 0.95 confidence interval of $\mu_1 + 2\mu_2$; of $2\mu_1 + \mu_2$.

10.20. Use the data of Exercise 10.13 to find a 0.99 confidence interval of $\mu_1 + 3\mu_2$; of $(\mu_1 + 3\mu_2)/4$.

10.21. Find n_1 such that σ_c^2 of Equation (10.17) is a minimum when σ^2 is a constant, $a_1 = 2a_2$, and $n_1 + n_2 = 30$.

10.22. Use the paired t-test to do Exercise 9.1.

10.23. For Exercise 9.6 determine a 0.95 confidence interval of mean weight gained by Diet I minus mean weight gained by Diet II.

10.24. Apply the paired t-test to the data of Exercise 9.9 to determine whether the claim is justified.

10.25. For Exercise 9.8 erroneously apply the t-test for two independent samples and compare the resulting conclusion with that of the paired t-test. (For Exercise 9.8 the correct sample value is $t = 2.30$ with 9 degrees of freedom.)

10.26. Thirty students are to be divided into two groups each with 15 students. Suppose IQ score is associated with experimental results. Carefully explain how you would assign students to each group if you wish to use
(a) the paired t-test
(b) the t-test for independent groups

11

LINEAR REGRESSION

11.1. INTRODUCTION

There are many situations in which we need to study the relationships between two or more variables. In this chapter we present procedures to show how one variable, termed the **independent variable** and denoted by x, may be used to make inferences (or predictions) about a related variable, the **dependent variable**, which is denoted by y. Illustrations of this special type of relationship are numerous and can be found almost any place measurements occur in pairs. As examples we observe that weight (y) generally increases with height (x), yield of a crop (y) depends on the amount of fertilizer (x) applied, grade point average in college (y) is related to the grade point average in high school (x), and sale of a product (y) generally increases with the amount of advertising (x).

Such "statistical relationships" are quite different from the so-called "exact" functional relations of some areas of science in which y can be found with good accuracy when x is specified. For example, Boyle's law states that at a given temperature, the product of the volume (v) of a gas and the pressure (p) is constant (c), that is

$$vp = c$$

Thus, for a specific volume of a gas, say v_0, we find the corresponding pressure to be $p = c/v_0$, and the prediction is considered quite accurate for moderate pressures.

In most investigations, statistical or otherwise, y cannot be determined exactly once a value of x is known. For example, there is no functional relationship which gives (or predicts) the exact weight of a person whose exact height is known. Still, a tall man is likely to weigh more, *on the average*, than a short man. Thus, in problems of this type it is often assumed that a functional relationship does exist between x and the *mean* η of the array of y values associated with a specific x value. That is, it is assumed that

(11.1) $$\eta = g(x)$$

where $\eta = E(y|x)$ denotes the mean of all y values for a given x. [Sometimes

$\mu_{y \cdot x}$ is used to denote $E(y|x)$.] Clearly, for each value of x there is a single value of η. The functional relation of Equation (11.1) is known as a **regression equation** of y on x for the population of pairs (x, y).

In regression analysis one of the first problems is to specify the best functional form (or family of equations), and then to determine values of the parameters (or estimates of the parameters) which give the most appropriate equation. But how are we to know the nature (shape) of the particular regression equation for a population being studied? Two common methods used by an investigator are to (1) take into account the investigator's knowledge of the theoretical structure existing or seeming to exist among the variables involved, or to (2) plot sample points (or, if finite and small, population points) in a **scatter diagram**. Both methods can be important in experimentation. Method (1) should be considered first since it is also useful in selecting the appropriate variables before the observations are made. Method (2) should be used to supplement method (1) or, in case little is known about the theoretical structure, to make decisions about the form of the relationship. (Some principles of linear regression are introduced by using the simple but unrealistic hypothetical bivariate population of Example 11.1.)

EXAMPLE 11.1. Suppose a bivariate population contains the following five pairs of values:

x	1	1	4	4	6
y	7	5	4	2	1

(a) Plot the scatter diagram of the bivariate population, and (b) find the regression equation of the line that passes through the means of the y arrays corresponding to $x = 1, 4, 6$.

For part (a) we plot the five points as indicated by the dots in Figure 11.1(a). Even though these five points do not lie on a straight line, their pattern is along a straight line. That is, there is a **straight line trend**.

To solve part (b), we think of the bivariate population as being made up of three *subpopulations* or *arrays* corresponding to the three different values of x. That is, the subpopulations are (1, 5), (1, 7); (4, 2), (4, 4); and (6, 1). In other words, for $x = 1$ the y array is 5, 7; for $x = 4$ the y array is 2, 4; and for $x = 6$ the y array is 1. When $x = 1$, the mean of the y array is

$$\eta = \mu_{y \cdot 1} = E(y|x=1) = (5+7)/2 = 6$$

Likewise when $x = 4$, $\eta = 3$ and when $x = 6$, $\eta = 1$. The points for these three pairs, (1, 6), (4, 3), (6, 1), are indicated by small circles in Figure 11.1(b), and the **regression line** is the line segment from (1, 6) to (6, 1). It can be shown, by any of several methods from analytical geometry, that the equation of the

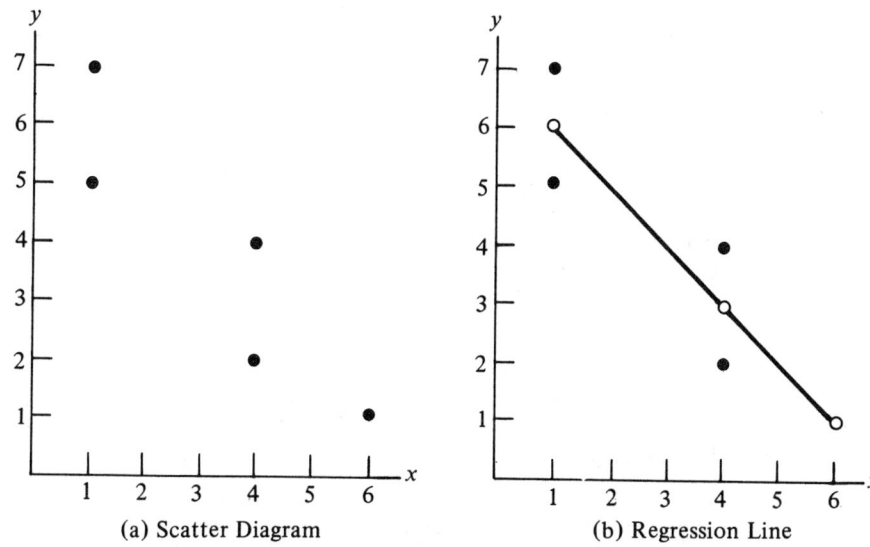

Figure 11.1. Scatter Diagram and Regression Line for Population of Example 11.1

regression line is

(11.2) $$\eta = 7 - x \qquad \text{for } x = 1, 4, 6$$

The coefficient of x, -1, is the **slope**, denoted by β, of the regression line, and it can be found by using the coordinates of any two points on the line. For example, for the points $(1, 6)$ and $(4, 3)$ we find the slope to be $\beta = (6 - 3)/(1 - 4) = 3/(-3) = -1$. A slope of -1 means that for a one-unit increase in the value of x there is a one-unit decrease in the mean value of y.

The means of the x and y values are

$$\mu_x = \frac{1 + 1 + 4 + 4 + 6}{5} = \frac{16}{5} = 3.2$$

and

$$\mu_y = \frac{7 + 5 + 4 + 2 + 1}{5} = \frac{19}{5} = 3.8$$

respectively. Clearly, the point (μ_x, μ_y) falls on the regression line for the values $x = \mu_x = 3.2$ and $\eta = \mu_y = 3.8$ satisfy Equation (11.2). That is, $3.8 = 7 - 3.2$. Since $(\mu_x, \mu_y) = (3.2, 3.8)$ does fall on the regression line, we may also write the regression equation of y on x as

(11.3) $$\eta = 3.8 - 1(x - 3.2) \qquad \text{for } x = 1, 4, 6$$

This slightly more cumbersome form turns out to be more useful than Equation (11.2) for many purposes of this chapter.

In Example 11.1 the regression of y on x is said to be **linear.** In this chapter we consider only **linear regression.** We write the general equation for linear regression of y on x as

(11.4) $\quad\quad\quad\quad \eta = \mu_y + \beta(x - \mu_x) \quad\quad$ for $c \leq x \leq d$

where β is the rate of change of the mean of y arrays with respect to x and is called the **regression coefficient** of y on x. In geometric terms, β is the slope of the regression line. The regression line is only a segment of a straight line. That is, it has finite length for practical problems and it starts at $x = c$, the smallest value of x, and ends at $x = d$, the largest value of x. The coefficient term μ_y is the mean of the y array means η. In Example 11.1, $\mu_y = (6 + 6 + 3 + 3 + 1)/5 = 19/5 = 3.8$.

One reason for working with linear regression is its simplicity. But the primary reason is that many relationships are actually of this form or closely approximate linear regression over a restricted set of values of x.

An advantage in having a regression equation is that every y-array mean η can be expressed in terms of three numbers, μ_y, β, and μ_x. In Example 11.1, the three y-array means are expressed in terms of $\mu_y = 3.8$, $\beta = -1$ and $\mu_x = -3.2$. Even when there are one hundred or an infinite number of y-array means, they can all be expressed in terms of only three parameters. That is, the linear regression of y on x reduces the number of parameters from an arbitrary number to only three if we use Equation (11.4). [An equation in the form of Equation (11.2) requires only two parameters.] As we shall learn in later sections, once estimates of μ_y and β are obtained, the estimate of any η can be obtained indirectly through an estimated regression equation.

Each y value can be expressed as the sum of an array mean η and a deviate ϵ, the amount y deviates from η, that is

$$y = \eta + \epsilon$$

or, substituting from Equation (11.4)

(11.5) $\quad\quad\quad\quad y = \mu_y + \beta(x - \mu_x) + \epsilon$

We call Equation (11.5) the **model equation** for pairs of values (x, y) with a bivariate distribution. The model equation for Example 11.1 is

$$y = 3.8 - 1(x - 3.2) + \epsilon$$

For $(x, y) = (1, 7)$, we have $\eta = 6$ and $\epsilon = 1$. For the points

$$(1, 5) \quad (4, 4) \quad (4, 2) \quad (6, 1)$$

the values of ϵ are

$$-1 \quad 1 \quad -1 \quad 0$$

respectively.

Assumptions for Linear Regression of y on x. All properties presented in later sections of this chapter are based on samples which come from a specific bivariate population with two or more of the following characteristics:

1. For every pair of values (x, y), x is fixed and y is a random variable with mean $\eta = \mu_{y \cdot x}$. That is, for each fixed value of x there is a corresponding array of y's with mean η. When we say x is fixed we mean that its values are controlled by the experimenter and assumed observable without error.

2. For each x, the mean y value, η, falls on a straight line segment called the linear regression of y on x.

3. All the y arrays have the same variance σ^2.

4. Each y array is a normal distribution.

At the appropriate place we indicate which of these assumptions are required for each specific procedure.

The model for linear regression of y on x which has all four characteristics listed above is illustrated by the three-dimensional graph shown in Figure 11.2. The three normal sections shown represent only three of many possible y arrays. (There are as many y arrays as there are values of x.) Since the normal sections

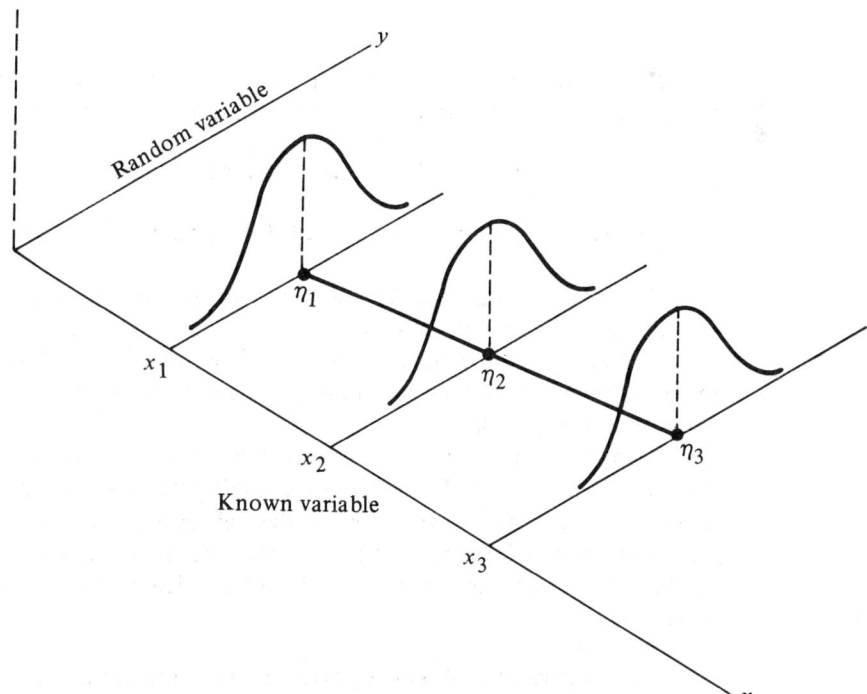

Figure 11.2. Model for Linear Regression of y on x with $\beta > 0$

have the same spread, this illustrates that the variances are equal. The heavy dark line is the regression line of y on x. Since the means of the y arrays increase with x, a positive regression coefficient β is indicated. For a negative β, the regression line would slope downward and the mean of the y array would decrease as x increases. For zero β, the regression line would be parallel to the x axis and the mean of the y array would remain the same regardless of the value of x.

11.2. FITTED REGRESSION LINE OF y ON x

We presented the model for linear regression in Section 11.1. The parameters of the "true" regression line are seldom known in practice. In which case they must be estimated from a sample of pairs

$$(x_1, y_1), (x_2, y_2), \cdots, (x_n, y_n)$$

where the x's are determined (or controlled) by the experimenter and the y's are random variables. The sample estimates of μ_y, β, μ_x, and η are denoted by a, b, $\bar{x}$, and $\hat{\eta}$, respectively. Then the estimated linear regression of y on x is

(11.6) $\qquad \hat{\eta} = a + b(x - \bar{x}) \qquad$ for $c \leq x \leq d$

We may use Equation (11.6) to estimate the mean of the y array for any x between c and d. In particular, an estimate of η for a sample value x_i is given by

(11.7) $\qquad \hat{\eta}_i = a + b(x_i - \bar{x}) \qquad$ for $i = 1, 2, \cdots, n$

We also refer to Equation (11.6) as the **fitted regression line of y on x** or, simply, **fitted line.**

The fitted line ought to have desirable properties and be as "close" to the true line as possible. But how do we go about finding such a line? There is a different line for every value of a and for every value of b, so how do we find a and b so that the fitted line, based on a particular sample, is as close as possible to the true unknown line?

Many principles have been presented, but currently one method is applied in almost all situations. The method for finding the fitted line which is usually considered "best" is known as the **method of least squares**. The least squares procedure gives estimators of the parameters μ_y, β, and η which are unbiased and have minimum variance of all estimators which are linear functions of the observations.

The **method of least squares** chooses values of a and b (which determine the fitted line) such that the sum of the squares of the vertical distances from the sample points to the fitted line is a minimum. In symbols, a and b must be values which make

(11.8)
$$Q = \sum_i [y_i - a - b(x_i - \bar{x})]^2$$

a minimum.

That is, the least squares line is the one which is, on the average, closest to all the points on the scatter diagram. The vertical distances from the points to the fitted line are, on the average, smallest when the least squares line is chosen. In this sense, the line is the "closest" one to all the points. Note that since the x values are assumed fixed (or controlled) and exact, the sample mean $\bar{x}$ is not a random variable.

Since x_i and y_i are data values and $\bar{x}$ is fixed, Q is a function of a and b. Thus, it can be shown by differential calculus that Q is a minimum when

(11.9)
$$a = \bar{y}$$

and

(11.10)
$$b = \frac{n \sum x_i y_i - (\sum x_i)(\sum y_i)}{n \sum x_i^2 - (\sum x_i)^2}$$

Other useful formulas for computing b are introduced later.

EXAMPLE 11.2. Use a sample of size three to find the least squares line of regression which "best fits" data from the bivariate population of five pairs in Example 11.1.

Of the ten possible samples of size three we use

(1, 5)   (4, 4)   (6, 1)

which is one of four samples with y values from each y array. From Figure 11.1(a) it is clear that these three points fail to fall on a straight line. Thus, we cannot apply the methods of analytical geometry to find the estimating equation. A typical procedure for computing the fitted regression line by the method of least squares is shown in Table 11.1. Only the entries in columns 2, 3, 4, and 5 are needed to find estimates of μ_y, β, η, and the population regression line. For the chosen sample, we observe that the estimate $\bar{y} = 3.33$ is less than $\mu_y = 3.8$, that $b = -0.763$ is numerically less than $\beta = -1$, and that

(11.11)
$$\hat{\eta} = 3.33 - 0.763(x - 3.67) \quad \text{or} \quad \hat{\eta} = 6.13 - 0.763x$$

estimates

$$\eta = 3.8 - 1(x - 3.2) \quad \text{or} \quad \eta = 7 - 1x$$

The three estimated values of η, as given by Equation (11.11), are shown in column 6 of Table 11.1. The amount $\hat{\eta}_i$ deviates from η_i is shown in column 7,

Table 11.1. Computations Leading to the Least Squares Fitted Regression Equation of y on x

Pair	x_i	y_i	x_i^2	$x_i y_i$	$\hat{\eta}_i$	$\hat{\eta}_i - \eta_i$	$y_i - \hat{\eta}_i = e_i$
1	1	5	1	5	5.37	−0.63	−0.37
2	4	4	16	16	3.08	0.08	0.92
3	6	1	36	6	1.55	0.55	−0.55
Total	11	10	53	27	10.00		0.00

$$\bar{x} = \frac{11}{3} \doteq 3.67 \qquad a = \bar{y} = \frac{10}{3} \doteq 3.33$$

$$n \sum x_i^2 - (\sum x_i)^2 = 3(53) - (11)^2 = 38$$

$$n \sum x_i y_i - (\sum x_i)(\sum y_i) = 3(27) - (11)(10) = -29$$

$$b = -\frac{29}{38} \doteq -0.763$$

$$\hat{\eta} = 3.33 - 0.763(x - 3.67)$$

and the amount y_i deviates from $\hat{\eta}_i$ is shown in column 8. These same comparisons can be made by referring to Figure 11.3 which shows both the population regression line and the fitted regression line. The true and estimated lines do not coincide, but the estimated line is the best representation we can get given the sample which we selected. Both Table 11.1 and Figure 11.3 are used to illustrate further general properties of linear regression analysis.

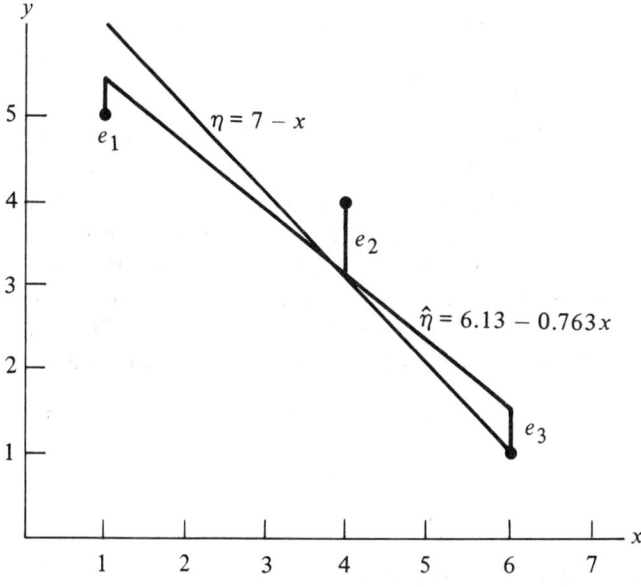

Figure 11.3. True and Fitted Regression Lines for Sample of Example 11.2

The estimated line of regression always passes through the point $(\bar{x}, \bar{y})$. For Example 11.2, when $x = \bar{x} = 3.67$ is substituted in Equation (11.11), we obtain $\hat{\eta} = 3.33$ which is $\bar{y}$.

Equation (11.6) may be used to find $\hat{\eta}$ for any value of x for which the true regression equation is a model. In Example 11.2, we have found the only estimates of η which are permissible since the population has only three values of x. [If we had used (1, 5), (1, 7) and (6, 1) to find a regression line, we could use this line to estimate η for $x = 4$ as well as for $x = 1$ and 6.] We have used a very restricted hypothetical population to illustrate principles. However, in most problems the line of best fit is assumed to hold for *all values* between the two extreme values of x in the experiment—this is the case when we use the four assumptions at the end of Section 11.1. But values of x outside the two extreme experimental values should not be used to find estimates of η unless it is known that the model extends beyond these values.

The slope of the estimated regression line may be expressed as

$$(11.12) \qquad b = \frac{[n \sum x_i y_i - (\sum x_i)(\sum y_i)]/n}{[n \sum x_i^2 - (\sum x_i)^2]/n}$$

by dividing both numerator and denominator of the right-hand side of Equation (11.10) by n. In this form, we recognize that the denominator is the computational form of

$$(11.13) \qquad SS_x = \sum_i (x_i - \bar{x})^2$$

which, in Chapter 8, was defined as the sum of squares about the mean. Similarly, it can be shown that the numerator of the right-hand side of Equation (11.12) is the computing form of the sum of products of paired values about their means, i.e.

$$(11.14) \qquad SP = \sum_i (x_i - \bar{x})(y_i - \bar{y}) = \frac{n \sum x_i y_i - (\sum x_i)(\sum y_i)}{n}$$

Hence, we may write

$$b = \frac{SP}{SSx} = \frac{\sum (x_i - \bar{x})(y_i - \bar{y})}{\sum (x_i - \bar{x})^2}$$

or

$$(11.15) \qquad b = \frac{(x_1 - \bar{x})^2 b_1' + \cdots + (x_n - \bar{x})^2 b_n'}{(x_1 - \bar{x})^2 + \cdots + (x_n - \bar{x})^2}$$

where

$$b_1' = \frac{y_i - \bar{y}}{x_i - \bar{x}} \qquad \text{for } i = 1, \cdots, n$$

is the slope of a line passing through (x_i, y_i) and $(\bar{x}, \bar{y})$. In this way we see that b is a weighted mean of slopes of individual lines passing through $(\bar{x}, \bar{y})$. The slope b'_i is the rate of change of y_i relative to x_i and the corresponding weight $(x_i - \bar{x})^2$ becomes larger as the horizontal distance of (x_i, y_i) from $(\bar{x}, \bar{y})$ increases. In Example 11.2, the points

$$(1, 5) \quad (4, 4) \quad (6, 1)$$

have slopes

$$b'_1 = \frac{5 - \frac{10}{3}}{1 - \frac{11}{3}} = \frac{5}{-8}, \quad b'_2 = \frac{4 - \frac{10}{3}}{4 - \frac{11}{3}} = \frac{2}{1}, \quad b'_3 = \frac{1 - \frac{10}{3}}{6 - \frac{11}{3}} = \frac{-7}{7}$$

with weights

$$(x_1 - \bar{x})^2 = \left(-\frac{8}{3}\right)^2 = \frac{64}{9}, \quad (x_2 - \bar{x})^2 = \frac{1}{9}, \quad (x_3 - \bar{x})^2 = \frac{49}{9}$$

respectively. Thus the slope of the estimated regression line is

$$b = \frac{\left(\frac{64}{9}\right)\left(-\frac{5}{8}\right) + \left(\frac{1}{9}\right)(2) + \left(\frac{49}{9}\right)(-1)}{\frac{64}{9} + \frac{1}{9} + \frac{49}{9}} = \frac{-87}{114} = -\frac{29}{38} \doteq -0.763$$

So, on the average, we find again that y decreases 29 units for every 38 units increase in x.

The fitted regression equation is expressed in the form of Equation (11.6) because the slope b is distributed independently of a, under the four assumptions of Section 11.1. This allows us to make inferences about the parameters μ_y and β independently of each other. (If the equation had been written in the conventional form as $\hat{\eta} = a' + bx$, then the two estimators a' and b would be correlated and inferences based on a' and b might be less meaningful.)

The entries in columns 3, 6, and 8 of Table 11.1 illustrate a very important property of all fitted regression lines. It is always true that

$$\Sigma y_i = \Sigma \hat{\eta}_i$$

or

(11.16) $$\sum_{i}^{n} (y_i - \hat{\eta}_i) = 0$$

The latter equation indicates that the sum of the distances of the n sample points from the estimated regression line is equal to zero. That is, the sum of the positive distances is equal to the sum of the negative distances. (A distance is termed positive or negative depending on whether the point is above or below the line. See Figure 11.3 for an illustration.)

The estimated mean $\hat{\eta}$ of a y array is obtained by using all values of y in the sample, not just those in the array above x. In Example 11.2 the mean of the y values in the array for $x = 1$ is 5 since there is only one value of y in the sample above $x = 1$. Clearly, this value 5 is different from $\hat{\eta} = 5.37$.

When values a and b are found which make Q a minimum, we call a and b least squares estimators of μ_y and β, respectively, and the value of Q, in this case, is called the **residual sum of squares** and denoted by $SSres$. The graph in Figure 11.3 may be used to illustrate why $SSres$ is as small as possible. The deviation $e_i = y_i - \hat{\eta}_i$ is the vertical distance between the ith point and the estimated regression line. The sum of squares of all these n deviations is the residual sum of squares. That is

(11.17) $$SSres = \sum_i e_i^2 = \sum (y_i - \hat{\eta}_i)^2$$

Clearly, the value of $SSres$ is affected by the position of the line relative to the points of a given sample. Since a and b are found so that $Q = SSres$ is a minimum, and since a and b uniquely determine a line, it follows that $SSres$ is as small as possible. (See Exercise 11.11 for more details.)

The **residual mean square**, $MSres$, is defined by

(11.18) $$MSres = \frac{SSres}{n - 2}$$

It can be shown that $MSres$ with $n - 2$ degrees of freedom is an unbiased estimator of σ^2, the common variance of the y arrays. There are $n - 2$ degrees of freedom since 2 parameters were estimated in finding the fitted regression line which was used to define $SSres$. We may interpret $MSres$ as the "average squared distance" from the n points to the fitted line. For Example 11.2

$$SSres = (-0.37)^2 + (0.92)^2 + (-0.55)^2 = 1.2858 \doteq 1.29$$

and

$$MSres = \frac{1.29}{3 - 2} = 1.29$$

Later we learn how $MSres$ may be used to make inferences about μ_y, β, and η.

When $b = 0$, the regression equation becomes $\hat{\eta} = \bar{y}$. In this case the sum of squares of residuals is equal to the sum of squares of y values about their mean, $\sum_i (y_i - \bar{y})^2$. That is

$$SSres = SSy = \sum_i (y_i - \bar{y})^2$$

When $b \neq 0$, it should be clear that $SSres < SSy$, and that the difference $SSy - SSres$ is due to the linear trend in the means of the y arrays. We illustrate this by referring to Table 11.1 and Figure 11.4 of Example 11.2. First we find that

$$SSy = (5 - 3.33)^2 + (4 - 3.33)^2 + (1 - 3.33)^2 \doteq 8.67$$

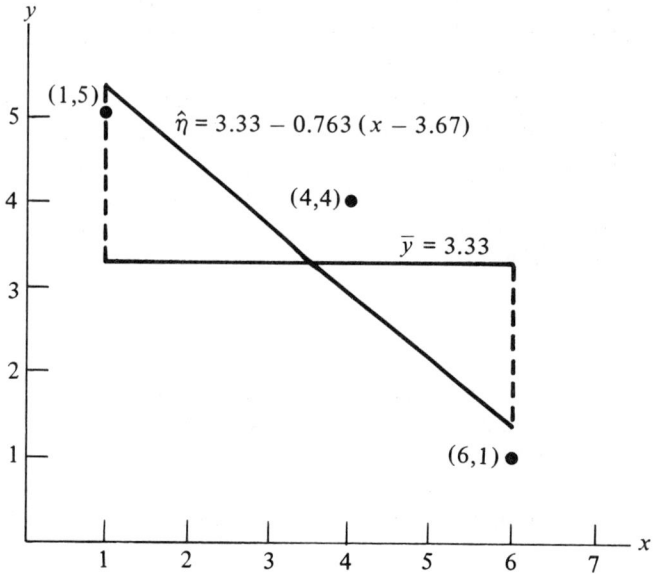

Figure 11.4. Deviation of Regression Line from $\bar{y} = 3.33$ for Three Sample Points of Example 11.2

The sum of squares for the deviation of the regression line from the horizontal line $\bar{y} = 3.33$, see Figure 11.4, is given by

$$\sum_{i=1}^{3} (\hat{\eta}_i - \bar{y})^2 = (5.37 - 3.33)^2 + (3.08 - 3.33)^2 + (1.55 - 3.33)^2 \doteq 7.39$$

We call this last value the **regression sum of squares** and denote it by *SSreg*. We note that

$$SSy - SSres = 8.67 - 1.29 = 7.38 = SSreg$$

except for rounding off errors. In general, if we let

(11.19) $$SSreg = \sum_{i=1}^{n} (\hat{\eta}_i - \bar{y})^2$$

it can be shown that

(11.20) $$SSy = SSreg + SSres$$

which is called the **sum of squares identity for regression** analysis.

When any two of the sums of squares in Equation (11.20) are known we may easily find the third sum of squares by substitution. Usually *SSy* and *SSreg* are found first by convenient computing formulas and then *SSres* is found by subtraction by the formula

(11.21) $$SSres = SSy - SSreg$$

where

(11.22a) $$SSreg = \frac{SP^2}{SSx}$$

or

(11.22b) $$SSreg = bSP$$

The derivation of Formulas (11.22a) and (11.22b) follows immediately from Equation (11.19) (see Exercise 11.15). Recall that the computing formula for the sum of squares of the y values about their mean is

$$SSy = \frac{n \sum y_i^2 - (\sum y_i)^2}{n}$$

We have illustrated some of the basic properties of linear regression by using only three points to estimate the regression for a very simple hypothetical bivariate population. But we have not always demonstrated the best computing procedure for doing the analysis. Thus, it might be useful to examine a more practical problem in which (1) the true regression equation is unknown and (2) desirable computing formulas are used.

EXAMPLE 11.3. Twenty similar plots in a controlled experiment are randomly divided into four groups of five plots each, and 50 pounds of fertilizer are applied to each plot in one group, 100 pounds to each plot in a second group, and so on. Then the same variety of some crop, say sugar beets, is planted in each of the 20 plots. Except for differences in amount of fertilizer, each plot is treated the same from planting to harvest. The hypothetical yields are shown in Table 11.2. Find

(a) the estimated linear regression of yield on amount of fertilizer;
(b) the residual mean square; and
(c) the symmetric 0.95 confidence interval for β.

Table 11.2. Yield of Crop in 10 Pounds

	Amount of Fertilizer per Plot (in pounds)			
	50	100	150	200
	151	166	159	186
	158	151	188	170
	139	168	170	184
	141	171	175	190
	146	154	168	175
Total	735	810	860	905
Mean	147	162	172	181

Let x denote amount of fertilizer in pounds and y the corresponding yield in 10 pounds. Thus, there are 20 pairs of numbers of the form

$$(50, 151) \quad (50, 158) \quad (50, 139) \ldots (200, 175)$$

from which we obtain the basic sums and sums of squares. The computations for parts (a) and (b) are conveniently summarized in Table 11.3.

Table 11.3. Computations for Linear Regression and Estimate of Variance for Data in Table 11.2

$n = 20$

$\sum x_i = 2{,}500$ $\qquad\qquad\qquad\qquad\qquad\sum y_i = 3{,}310$
$\sum x_i^2 = 375{,}000$ $\qquad\qquad\qquad\qquad\;\;\sum y_i^2 = 552{,}272$

$\qquad\qquad\qquad\sum x_i y_i = 427{,}750$

$\bar{x} = 125$ $\qquad\qquad\qquad\qquad\qquad\qquad a = \bar{y} = 165.5$
$SSx = 62{,}500$ $\qquad\qquad\qquad\qquad\qquad SSy = 4{,}467$

$\qquad\qquad\qquad SP = 14{,}000$

$b = \dfrac{SP}{SSx} = 0.224$ $\qquad\qquad\qquad\qquad SSreg = \dfrac{SP^2}{SSx} = 3{,}136$

$\boxed{\hat{\eta} = 165.5 + 0.224(x - 125)}$ $\qquad\quad SSres = SSy - SSreg = 1{,}331$

$\qquad\qquad\qquad\qquad\qquad\boxed{MSres = \dfrac{SSres}{n - 2} = 73.94}$

For part (a) the estimated regression line is

(11.23) $\qquad \hat{\eta} = 165.5 + 0.224(x - 125) \qquad$ for $50 \leq x \leq 200$

The slope of 0.224 indicates that we expect 2.24 pounds increase in yield for each pound increase in application of fertilizer from 50 to 200 pounds. Equation (11.23) may be used to find estimated mean yields, see Figure 11.5, of

$$1{,}487 \quad 1{,}599 \quad 1{,}711 \quad 1{,}823$$

pounds of sugar beets per plot corresponding to

$$50 \quad 100 \quad 150 \quad 200$$

pounds of fertilizer, respectively. Checking these estimated mean yields against the actual mean yields shown in Table 11.2, we find the estimated values are within 17, -21, -9, and 13 pounds, respectively, of the actual yields. On the assumption that the true means actually do fall on a straight line, the estimates are the values that should be used as predictions of future yields. For example, the mean yield of 1,470 pounds of the crop resulting from an application of 50 pounds of fertilizer is a value of a random variable based on only five observa-

Fitted Regression Line of y on x

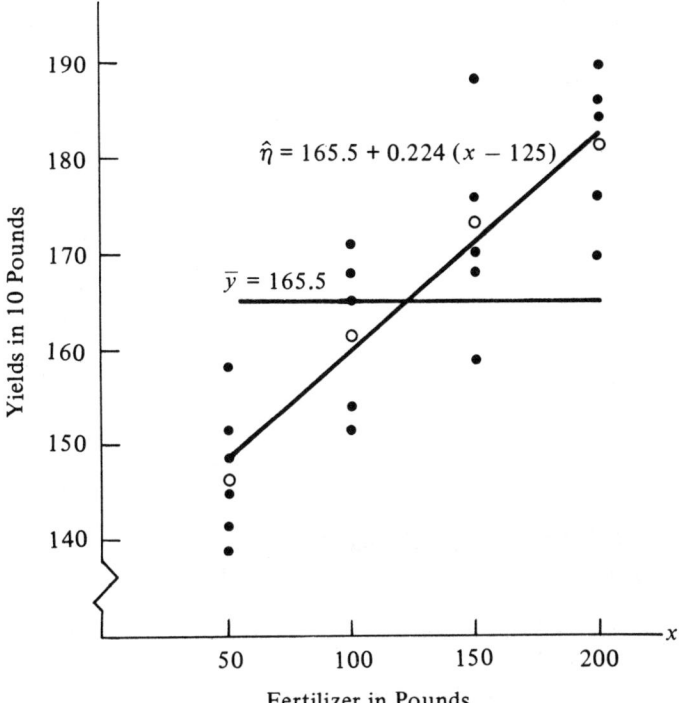

Figure 11.5. Scatter Diagram and Fitted Regression Line for Data of Example 11.3. Array Means are Denoted by Small Circles.

tions, but the regression value of 1,487 pounds is the value of a random variable based on 20 observations. In other words, since 20 is larger than 5, it can be shown that there is less fluctuation from experiment to experiment in $\hat{\eta}$ than in the corresponding y-array mean, ignoring regression analysis. Further, the regression equation may be used to predict mean crop yield for other amounts of fertilizer. For example, with an application of 125 pounds of fertilizer, the estimated mean yield is 1,655 pounds. However, the regression equation should not be used for less than 50 or more than 200 pounds of fertilizer unless the subject matter specialist "knows" the fitted linear relationship extends to such values.

An unbiased estimate of the common variance of the y arrays is

$$s^2 = MSres = 73.94$$

It has $n - 2 = 18$ degrees of freedom since two estimators, a and b, were used in finding the residuals. In Section 11.3 we discuss the distribution properties of a and b and show how $MSres$ is useful in making inferences. For example, if the four assumptions of Section 11.1 hold, we show that b has a normal distribution with mean β and variance σ^2/SSx, or that $(b - \beta)\Big/\sqrt{\dfrac{MSres}{SSx}}$ has a Student t distribution with $n - 2$ degrees of freedom. Thus, for part (c) the 95 percent confidence limits of β are

$$b \pm t_{.025}\sqrt{\dfrac{MSres}{SSx}} = 0.224 \pm 2.101\sqrt{\dfrac{73.94}{62{,}500}}$$

$$= 0.224 \pm 2.101(0.03440)$$

$$= 0.224 \pm 0.0723$$

So the required confidence interval for β is

$$0.152 < \beta < 0.296$$

11.3. DISTRIBUTION OF SAMPLE REGRESSION COEFFICIENT b

In Example 11.3 of Section 11.2 we introduced the sampling distribution of b in a confidence interval problem. Now we take a closer look at b and its properties. It is certainly desirable that the reader know when each of the four assumptions at the end of Section 11.1 is required and why it is introduced.

First, we recall that each pair of values (x_i, y_i) in a sample satisfies the *model equation*

(11.24) $\qquad y_i = \mu_y + \beta(x_i - \bar{x}) + \epsilon_i \qquad i = 1, 2, \cdots, n$

as well as the *estimating equation*

(11.25) $\qquad y_i = a + b(x_i - \bar{x}) + e_i \qquad i = 1, 2, \cdots, n$

In Equation (11.24) μ_y and β are fixed unknown parameters, $\bar{x}$ is the sample mean of x values, and y_i and ϵ_i are random variables. Note that $\bar{x}$ in Equation (11.24) replaces μ_x in Equation (11.5); a known value $\bar{x}$ replaces an unknown value μ_x. For each x_i, the random variable y_i is distributed with mean $E(y_i|x_i) = \mu_y + \beta(x_i - \bar{x})$ and variance $V(y_i|x_i) = \sigma^2$ when assumptions 2 and 3 hold. Further, for each x_i, ϵ_i is distributed with mean $E(\epsilon_i) = 0$ and variance $V(\epsilon_i) = \sigma^2$. [The property $E(\epsilon_i) = 0$ comes from the assumption that the mean of the y array at x_i is η_i and that y values larger than η_i balance out with y values smaller than η_i. That is, the mean deviation of y values around η_i is zero. Also, the variance of the deviations ϵ_i at x_i should be the same as the variance of the y values.] However, in Equation (11.25), a, b, and e_i are all random variables. We call a and b least squares estimators of μ_y and β, respectively, and e_i is the residual of the ith pair (x_i, y_i).

The estimator b has been expressed in several ways already, but there is still another very important expression for b. Leading to this expression, we first observe that

$$SP = \sum_i (x_i - \bar{x})(y_i - \bar{y})$$
$$= \sum_i (x_i - \bar{x})y_i - \sum_i (x_i - \bar{x})\bar{y}$$

or

(11.26) $$SP = \sum_i (x_i - \bar{x})y_i$$

since

$$\sum_i (x_i - \bar{x})\bar{y} = \bar{y} \sum_i (x_i - \bar{x}) = \bar{y}(0) = 0$$

Thus, we may write the following equations for b:

$$b = \frac{SP}{SSx} = \frac{\sum_i (x_i - \bar{x})y_i}{SSx}$$

or

(11.27) $$b = \left(\frac{x_1 - \bar{x}}{SSx}\right)y_1 + \cdots + \left(\frac{x_n - \bar{x}}{SSx}\right)y_n$$

We assume $x_1, x_2, \ldots, x_n$ are fixed values, so $\bar{x}$ and SSx are also fixed. Thus, the multipliers

$$\frac{x_1 - \bar{x}}{SSX} \qquad \frac{x_2 - \bar{x}}{SSx} \qquad \cdots \qquad \frac{x_n - \bar{x}}{SSx}$$

are constants and b is a linear combination of random variables $y_1, y_2, \ldots, y_n$. In much the same way that we showed that $E(\bar{y}) = E\left(\frac{1}{n}y_1 + \cdots + \frac{1}{n}y_n\right) = \mu$, it can be shown that

(11.28) $$E(b) = \beta$$

that is, b is an *unbiased estimator of* β provided assumption 1 (x fixed and y random) and assumption 2 (regression is linear) hold.

Also, following the method that led to the property $V(\bar{y}) = \sigma_{\bar{y}}^2 = \sigma^2/n$, it can be shown that

(11.29) $$V(b) = \sigma_b^2 = \frac{\sigma^2}{SSx}$$

provided assumption 3 (equal variance) as well as assumptions 1 and 2 hold. Finally, if assumption 4 (normality) along with the other three assumptions hold, then the statistic

(11.30) $$z = \frac{b - \beta}{\sqrt{\frac{\sigma^2}{SSx}}}$$

has the standard normal distribution and the statistic

(11.31) $$t = \frac{b - \beta}{\sqrt{\frac{MSres}{SSx}}}$$

has the Student t distribution with $n - 2$ degrees of freedom.

For a fixed value of σ^2, we see from Equation (11.29) that the variance of b (or an estimator of the variance of b) can be made smaller by increasing SSx. The physical meaning of this fact can be seen by looking at a graph of a fitted regression line (see Figures 11.3 and 11.5 for examples). Since b is the slope of an estimated regression line, any variation in b results in a rotation of the line about the point $(\bar{x}, \bar{y})$ as the center. Increasing the sample size n does not necessarily stabilize the line, i.e., make b vary less from the true value β. For example, additional values of x at $x = \bar{x}$ would not affect b. However, additional points toward the ends of the line where $x_i - \bar{x}$ has greater value would be effective in cutting down variation in b. Thus, the accuracy of a sample regression coefficient b as an estimate of β does not depend entirely on the sample size. The combination of sample size and the spread of the x values as reflected in the value of SSx determines the accuracy of b. The importance of this point can be seen in Example 11.4 as we illustrate test of hypothesis procedure for β.

EXAMPLE 11.4. Use the data of Example 11.3 to determine if the estimated slope is significantly larger than $\frac{1}{7} \doteq 0.143$.

From past experience it might be known that for common varieties of the particular crop in the analysis that at least 10 pounds in yield is gained for each additional 7 pounds of fertilizer applied. In this particular experiment we want to know if a new variety gives larger increase in yield. Following the general test procedure we have:

1. $H_o: \beta \leq \frac{1}{7} \doteq 0.143$

   $H_a: \beta > 0.143$

2. The four assumptions of Section 11.1 hold. That is, 20 y values are randomly drawn from 4 normally distributed y arrays with means which fall on a straight line and variances which are equal. Let $\alpha = 0.05$.

3. Use

$$t = \frac{b - \beta}{\sqrt{\frac{MSres}{SSx}}} = \frac{b - 0.143}{\sqrt{\frac{MSres}{62,500}}}$$

which has the Student t distribution with $n - 2 = 20$ degrees of freedom.

4. The critical region is

$$CR = \{t \mid t > 1.734\}$$

5. The value of t obtained from the sample of 20 pairs is

$$t_c = \frac{0.224 - 0.143}{\sqrt{\frac{73.94}{62,500}}} = \frac{0.081}{0.0344} = 2.4$$

6. Since $t_c = 2.4$ falls in the critical region, we reject H_o and conclude that the sample regression coefficient $b = 0.224$ is significantly larger than 0.143. That is, on the basis of the experiment, we expect the new variety of sugar beets to produce at least 10 pounds more for each additional 7 pounds increase in fertilizer.

Note: Suppose the x values in Example 11.4 had been 75, 100, 125, 150, and the corresponding y values changed so that b and $MSres$ remained the same. Then the new sum of squares for x becomes

$$SSx = 15,625$$

and

$$t_c = \frac{0.224 - 0.143}{\sqrt{\frac{73.94}{15,625}}} = \frac{0.081}{0.0688} \doteq 1.2$$

Since $t_c = 1.2$ falls in the noncritical region, we fail to reject the null hypothesis. Since this conclusion is different from that of Example 11.4, it illustrates how important the spacing of the x values can be. In fact, the choice in spacing x values is as important for making inferences about β as the sample size is for making inferences about μ.

The test of $H_o: \beta = 0$ is often used. This test is applied to decide whether there is a relationship between the two variables x and y, and thus the test indicates whether we should use values of x to predict means of y's. A test on the value of μ_y, described in Section 11.4, is not so vital a test as one concerning β.

11.4. DISTRIBUTION OF ADJUSTED MEAN $\hat{\eta}$

In Example 11.3 we found point estimates of η, but we need to know the distribution of $\hat{\eta}$ in order to make inferences about η. The estimate $\hat{\eta}$ is given as a linear combination of two random variables $\bar{y}$ and b and can be written as

$$\hat{\eta} = \bar{y} + b(x - \bar{x})$$

or as

(11.32) $$\hat{\eta} = \bar{y} - b(\bar{x} - x)$$

The estimated mean of an array is called the **adjusted mean**, since its value depends on the particular y array. By contrast, the estimated mean $\bar{y}$ of the entire population is called the **unadjusted mean**. The distribution of the adjusted mean is presented in this section. However, it is desirable that we review the properties of $\bar{y}$ first.

Since

$$a = \bar{y} = \left(\frac{1}{n}\right)y_1 + \left(\frac{1}{n}\right)y_2 + \cdots + \left(\frac{1}{n}\right)y_n$$

is a linear combination of the random variables $y_1, y_2, \ldots, y_n$ it follows that

(11.33) $$E(a) = \mu_y$$

and

(11.34) $$V(a) = \sigma_a^2 = \frac{\sigma^2}{n}$$

Thus, we see that $a = \bar{y}$ has the usual properties of $\bar{y}$. The unbiased property is based on assumptions 1 and 2 and the variance property of Equation (11.34) is based on assumptions 1, 2, and 3. Finally, when all four assumptions of Section 11.1 hold, the statistic

(11.35) $$z = \frac{a - \mu_y}{\sqrt{\frac{\sigma^2}{n}}}$$

has the standard normal distribution and the statistic

(11.36) $$t = \frac{a - \mu_y}{\sqrt{\frac{MSres}{n}}}$$

has the Student t distribution with $n - 2$ degrees of freedom.

When the four assumptions of Section 11.1 hold, the adjusted mean $\hat{\eta}$ is a linear combination of two *independent* random variables $\bar{y}$ and b whose distributions are known. Thus, it follows immediately that

(11.37) $$E(\hat{\eta}) = \mu_y + \beta(x - \bar{x}) = \eta = E(y|x)$$

and

$$V(\hat{\eta}) = \sigma_{\hat{\eta}}^2 = \frac{\sigma^2}{n} + \frac{(x - \bar{x})^2 \sigma^2}{SSx}$$

or

(11.38) $$V(\hat{\eta}) = \sigma^2 \left(\frac{1}{n} + \frac{(x - \bar{x})^2}{SSx}\right)$$

Distribution of Adjusted Mean $\hat{\eta}$

where x can have any value between c and d. From Equation (11.37) we observe that $\hat{\eta}$ is an *unbiased estimator* of η where η is expressed in terms of the sample mean $\bar{x}$. Again, the property of unbiasedness is based on assumptions 1 and 2. The variance property of Equation (11.38) requires that assumptions 1, 2, 3, and 4 hold. The reader should realize that the variance $V(\hat{\eta})$ is a function of x and as such is often expressed in the more cumbersome notation as

$$V(\hat{\eta}) = V(\hat{\eta}_{y \cdot x})$$

Further, it should be observed that $V(\hat{\eta})$ is smallest when $x = \bar{x}$, and it increases as x increases its distance from $\bar{x}$. In terms of the crop yield experiment of Examples 11.3 and 11.4 this means that the estimate of mean crop yield is less accurate as we move further away from $\bar{x} = 125$ pounds of fertilizer.

Finally, if all four assumptions of Section 11.1 hold, the statistic

(11.39) $$z = \frac{\hat{\eta} - \mu_y - \beta(x - \bar{x})}{\sqrt{\sigma^2 \left(\frac{1}{n} + \frac{(x - \bar{x})^2}{SSx} \right)}}$$

has the standard normal distribution and the statistic

(11.40) $$t = \frac{\hat{\eta} - \mu_y - \beta(x - \bar{x})}{\sqrt{MSres \left(\frac{1}{n} + \frac{(x - \bar{x})^2}{SSx} \right)}}$$

has the Student t distribution with $n - 2$ degrees of freedom. The t statistic is useful in testing the hypothesis that the mean of an array is equal to a given value and in finding a confidence interval for the mean of an array.

EXAMPLE 11.5. Use the data of Example 11.3 to find and compare 95 percent confidence intervals for η corresponding to $x = 125$ and $x = 50$.

From past experience we know the symmetric 95 percent confidence limits of $\eta = \mu_y + \beta(x - 125)$, obtained from Equation (11.40), are given by

(11.41) $$\hat{\eta} \pm t_{\alpha/2} \sqrt{MSres \left(\frac{1}{n} + \frac{(x - \bar{x})^2}{SSx} \right)}$$

For our example, substituting in Expression (11.41), the limits are

(11.42) $$165.5 + 0.224(x - 125) \pm 2.101 \sqrt{73.94 \left(\frac{1}{20} + \frac{(x - 125)^2}{62{,}500} \right)}$$

When $x = 125$, we find confidence limits of

$$165.5 + 0 \pm 2.101 \sqrt{73.94 \left(\frac{1}{20} + \frac{0}{62{,}500} \right)}$$
$$= 165.5 \pm 2.101(1.923)$$
$$= 165.5 \pm 4.04$$

Thus, for $x = 125$ the required 95 percent confidence interval is

$$161.5 < \eta < 169.5$$

and it has length 8.0. Similarly, when $x = 50$ we find by substituting in Expression (11.42) that the required 95 percent confidence interval is

$$141.9 < \eta < 155.5$$

and it has length 13.6. Clearly, the length of the 95 percent confidence interval of η at $x = 50$ is longer than it is at $x = \bar{x} = 125$. It is easy to see that the confidence interval of η at $x = 200$ has the same length as at $x = 50$ since $|x - 125| = 75$ in both cases. Indeed, *for values of x symmetric about $\bar{x}$ the 0.95 confidence intervals of η are of equal length.* In the crop yield problem we know that for any x in the interval from 50 to 200 the length of the 0.95 confidence interval is between 8.0 and 13.6.

We have shown how to compute a confidence interval for the mean η of a y array corresponding to a specific x. Sometimes, we want a confidence interval for a particular observation y' corresponding to a specific x'. For example, in the crop yield data we might want to know what yield we could expect for a particular plot with 90 pounds of fertilizer. In such a case the standard error of y would obviously be larger than the standard error of $\hat{\eta}$. It can be shown that for any particular observation

$$y' = a + b(x' - \bar{x}) + e'$$

that the variance is

$$V(y') = \sigma^2 \left(\frac{1}{n} + \frac{(x' - \bar{x})^2}{SSx} + 1 \right)$$

Hence an unbiased estimate of y' is given by

(11.43) $$s_{y'}^2 = MSres\left(1 + \frac{1}{n} + \frac{(x' - \bar{x})^2}{SSx}\right)$$

Thus, when the four assumptions of Section 11.1 are satisfied the statistic

(11.44) $$t = \frac{y' - \mu_y - \beta(x' - \bar{x})}{\sqrt{MSres\left(1 + \frac{1}{n} + \frac{(x' - \bar{x})^2}{SSx}\right)}}$$

has the Student t distribution with $n - 2$ degrees of freedom.

EXAMPLE 11.6. Use the crop yield data of Example 11.3 to find a 95 percent confidence interval for the yield corresponding to an application of 90 pounds of fertilizer to a particular plot.

In this problem a subject matter specialist may want to use data gathered in the controlled experiment to predict the yield Mr. Jones can expect when he applies 90 pounds of fertilizer to a plot of the same size as an experimental plot. The point estimate is given by Equation (11.23) of Example 11.3. That is, for $x = 90$ the predicted yield is

$$y' = \hat{\eta} = 165.5 + 0.224(90 - 125) = 157.7$$

Thus, an unbiased point estimate of the predicted yield is $y' = 1,577$ pounds per plot. However, this single value does not give any indication of how much variation from 1,577 pounds Mr. Jones can expect. For this we use Equation (11.43) to find the estimate of standard error of y' to be

$$s_{y'} = \sqrt{73.94\left(1 + \frac{1}{20} + \frac{(90 - 125)^2}{62,500}\right)} = 8.893$$

Thus, using Equation (11.44) and t with 18 degrees of freedom we find the 95 percent confidence interval for y' to be

$$157.7 - 2.101\,(8.893) < y' < 157.7 + 2.101\,(8.893)$$

or

$$139.0 < y' < 176.4$$

That is, the subject matter specialist can with 95 percent confidence assure Mr. Jones that the yield will be between 1,390 and 1,764 pounds per plot, if he applies 90 pounds of fertilizer.

Only an introduction to regression analysis has been presented. Some other topics of interest might be a test of linearity of regression, multiple linear regression, polynomial regression, and two or more simple regressions. Information on such topics may be found in Draper and Smith [7] and Ezekiel and Fox [9].

11.5. EXERCISES

11.1. A bivariate population consists of the following values:

x	1	1	1	2	2	2	4	4	7
y	1	2	3	2	2	4	3	5	6

(a) Plot the scatter diagram of the bivariate population.
(b) Draw the regression line which passes through the means of all y arrays.
(c) Find the slope of the regression line of part (b), and write the regression equation in the form of Equation (11.4).

11.2. Find the linear regression equation of y on x for the following population of pairs of values:

Temperature in Degrees Centigrade x	-40	-30	-30	-10	-10	20
Tensil Strength in Pounds y	5	6	9	12	13	20

11.3. It is known that $\mu_x = 30$ and $\mu_y = 20$ for a specific bivariate distribution. Find the regression line of y on x for this distribution if the line contains the point $(x, y) = (37, 6)$. What is the value of η when $x = 5$?

11.4. Suppose a population of adult males has mean height and weight of 69 inches and 160 pounds, respectively. Further, suppose that for each increase of one inch in height, the mean weight increase is 6.5 pounds. For this population, find the regression of weights on heights and determine the mean weight of all males who are 6 feet 3 inches tall.

11.5. The regression line of y on x is

$$\eta = 3 - 2x \qquad \text{for } -3 \leq x \leq 7$$

and $\mu_x = -1$. Find μ_y and write the equation in the form of Equation (11.4).

11.6. Suppose three y's with corresponding specified x values are randomly drawn from the population in Exercise 11.1.
(a) Which of the assumptions for linear regression of y on x (see page 361) are satisfied? Explain.
(b) If the three pairs drawn are $(1, 3)$, $(2, 4)$, and $(4, 3)$, find the fitted regression of y on x.

11.7. (a) Plot the scatter diagram for the following sample data

x	2	4	4	5	7	7	9
y	7	6	5	4	4	1	1

(b) Use the data in part (a) to find the fitted regression line of y on x.
(c) Use the line in part (b) to estimate the mean of y at $x = 8$.

C11.8. For a sample of 25 pairs (x, y) it was found that $\sum x = 1{,}946$, $\sum y = 11{,}302$, $\sum x^2 = 157{,}196$, $\sum y^2 = 5{,}654{,}974$ and $\sum xy = 927{,}352$. Find the fitted regression line of y on x and the residual mean square. Estimate the mean y at $x = 55$; at $x = 115$.

11.9. (a) At $x = 2$ it is known that the random variable y has a normal distribution with mean 3 and variance 16 and at $x = 8$ the random variable y has a normal distribution with mean 6 and variance 16. Find the equation for the true linear regression of y on x if $\mu_x = 5$. Making the usual assumptions of linear regression, what can you say about the distribution of the y array at $x = 3$?

(b) The following sample data were obtained from the regression in part (a)

x	2	5	5	8	9	9
y	4	5	4	5	6	5

Find the estimated linear regression of y on x. At $x = 9$, find the true and estimated regression means, the array mean, and then compare the true and estimated error effects for both observed values.

(c) Find b by Equation (11.15); by Equation (11.27).
(d) Find a 0.90 confidence interval for the true slope. Does the true slope fall in this interval?

11.10. In Exercise 11.8, x denotes the stand height of upland oak in feet and y denotes the volume per plot in cubic feet. If $\hat{\eta} = -195.86 + 8.323x$ for $50 \leq x \leq 125$, $\bar{y} = 452.08$, $SP = 47,604.32$, and $MSres = 6,492.94$
(a) write the fitted regression equation in the form $\hat{\eta} = a + b(x - \bar{x})$;
(b) test $H_o: \beta \leq 6$ against $H_a: \beta > 6$ at the 0.05 level.

E11.11. The fitted regression line of y on x for the four points (1, 1), (2, 3), (4, 5), and (5, 7) is $\hat{\eta} = 4 + 1.4(x - 3)$ or $\hat{\eta} = -0.2 + 1.4x$
(a) Show that $SSres = 0.40$, the smallest possible value of $\sum e_i^2$ for these four points.
(b) Find $\sum e_i^2$ for $\hat{\eta} = 4 + 1.5(x - 3)$; for $\hat{\eta} = 4 + 1.3(x - 3)$.
(c) Find $\sum e_i^2$ for each of two lines different from the three above.

C11.12. Let x denote age in years (ranging from 30 to 60) and y denote blood pressure (systolic) in millimeters. Data for a sample of adult males are as follows:

$n = 20$ $\sum x = 870$ $\sum y = 2,426$
$\sum x^2 = 41,670$ $\sum y^2 = 299,800$ $\sum xy = 108,532$

(a) Find the linear regression of y on x.
(b) Making the assumptions of Section 11.1, find a 0.95 confidence interval of mean blood pressure when $x = 45$ years.

11.13. Suppose the admissions director of a college uses the high school grades and college freshman merit point ratios of 300 current college sophomores to find the following linear regression of y on x:

$$\hat{\eta} = -4.20 + .080x \qquad \text{for } 75 \leq x \leq 98$$

where x denotes mean high school grade and y denotes college merit point ratio for the freshman year. Also,

$$SSx = 4,768 \qquad SSy = 32.452 \qquad SSres = 1.937$$

(a) Find the 0.95 confidence interval for β.

(b) Assuming merit point ratios of incoming freshmen can be predicted by the above regression equation, find the 0.95 confidence interval for the merit point ratio of a particular student with a high school mean grade of 88 if $\bar{x} = 85$.

11.14. For a random sample of 25 families in City A it was found that the fitted regression line of "percentage of annual income spent for medical care" (y) on "the family income in thousands of dollars" (x) was

$$\hat{\eta} = 7.0 - 0.21x \qquad \text{for } 5 \leq x \leq 20$$

Also,

$$\bar{x} = 13, \ SS_x = 156.25 \text{ and } SSres = 2.222$$

(a) Test $H_o: \beta \geq -0.15$ at the 0.05 level.
(b) Suppose a family with an annual income of $15,000 spent $1,000 on medical care. Using statistical inference, decide whether this should be considered an unusual expenditure.

^E**11.15.** Starting with Equation (11.19), derive Formulas (11.22a) and (11.22b). *Hint:* First show that $\hat{\eta}_i - \bar{y} = b(x_i - \bar{x})$ for each i.

^C**11.16.** The modulus of rupture (y) and the specific gravity (x) of 10 similar fiberboards are as follows:

Board	x	y	Board	x	y
1	0.923	517	6	0.976	535
2	0.980	562	7	1.035	617
3	1.025	595	8	0.921	499
4	0.963	548	9	1.008	590
5	0.991	568	10	0.905	421

(a) Find the fitted regression line of the modulus of rupture on the specific gravity of fiberboards.
(b) Test the hypothesis that the mean modulus of rupture is not affected by the specific gravity of fiberboard.
(c) Fiberboard 11 has specific gravity of 0.950. Find 0.95 confidence limits for the modulus of rupture for this fiberboard.

^E**11.17.** (a) Use the data of Exercise 11.7 to find the fitted linear regression of x on y.
(b) On the same graph draw the line for part (a) and the line of Exercise 11.7(b). Explain why these lines are different.

12

CORRELATION ANALYSIS

12.1. THE PEARSON COEFFICIENT OF CORRELATION

In studies of relationships between two variables, x and y, we may wish to know the *strength* (or the *degree*) *of association between these variables* as well as (or in place of) an equation which allows us to predict a value of one variable from the value of the other. Obviously, if the strength of the association is very weak, there is no point in predicting y from x or x from y. In some investigations, particularly in new fields, one of the first problems is to decide which pairs of variables from among a large number are most closely related. Then, once a pair (or several pairs) is identified as having a strong degree of association, a regression equation can be found for predictive purposes.

Of the many measures of association which have been defined we first introduce the *Pearson product-moment coefficient of correlation r*, or briefly, the *coefficient of correlation*, because of its connection with linear regression and because it is the measure often applied. The numerical coefficient r measures the closeness of points to the fitted least squares equation of Chapter 11.

We can be specific concerning the connection between a scatter diagram and the corresponding value of r by using the definition of r or one of its equivalent forms.

The **coefficient of correlation** between x and y for a sample of n pairs (x_i, y_i) is defined by

(12.1)
$$r = \frac{\sum (x_i - \bar{x})(y_i - \bar{y})}{\sqrt{\sum (x_i - \bar{x})^2 \sum (y_i - \bar{y})^2}}$$

We may express Equation (12.1) in a more familiar computational form by using the symbolism of Chapter 11. That is,

(12.2)
$$r = \frac{SP}{\sqrt{SSxSSy}}$$

where

$$SP = \frac{n \sum x_i y_i - (\sum x_i)(\sum y_i)}{n}$$

$$SSx = \frac{n \sum x_i^2 - (\sum x_i)^2}{n} \qquad SSy = \frac{n \sum y_i^2 - (\sum y_i)^2}{n}$$

From Equation (12.2) we may write

(12.3) $$r^2 = \frac{SP^2}{SSx SSy}$$

or

(12.4) $$r^2 = \frac{SSreg}{SSy}$$

where

$$SSreg = \frac{SP^2}{SSx}$$

From Equation (12.4) we know that r^2, called the **coefficient of determination**, must be a number between 0 and 1, since SSy is equal to the regression sum of squares plus the residual sum of squares. Thus, r has values between -1 and 1. Further, the sign of r is the same as the sign of b, since SP is in the numerator of both and their denominators are both positive. That is, the sign of r indicates whether the corresponding estimated line of regression has a positive or negative slope.

The scatter diagrams of Figure 12.1 illustrate how r changes with different patterns of points. When points lie *close to a straight line* with positive slope, as in Figure 12.1(a), there is high positive correction and r is close to 1. That is, as x increases y increases and the points do not deviate far from the straight line. When points do not show a linear trend, as in Figures 12.1(b) and (c), the correlation coefficient r is near zero. Note that a correlation coefficient r may have a value of zero (or a value near zero) and still the two variables may be closely associated, as in Figure 12.1(c), but they will not be linearly related. When points lie on a straight line with negative slope, shown in Figure 12.1(d), the coefficient of correlation is -1, regardless how steep the line.

When $r = 1$ or -1, all points fall on a line. That is, with the aid of the regression line we can use one number to compute (or predict) the other number. When $r = 0$, the points are so scattered and the fit of a regression line so poor that knowledge of x, say, does not aid in the least in predicting the corresponding value of y. It is much more difficult to interpret the meaning of other values of r. Certainly, the reader should be cautioned, a correlation coefficient of $r = 0.60$, say, is *not* "twice as strong" as $r = 0.30$. Neither is the value $r = 0.90$ "three times as strong" as $r = 0.30$.

There is a sense in which we can compare values of r^2. Referring to Equation (12.4):

The Pearson Coefficient of Correlation

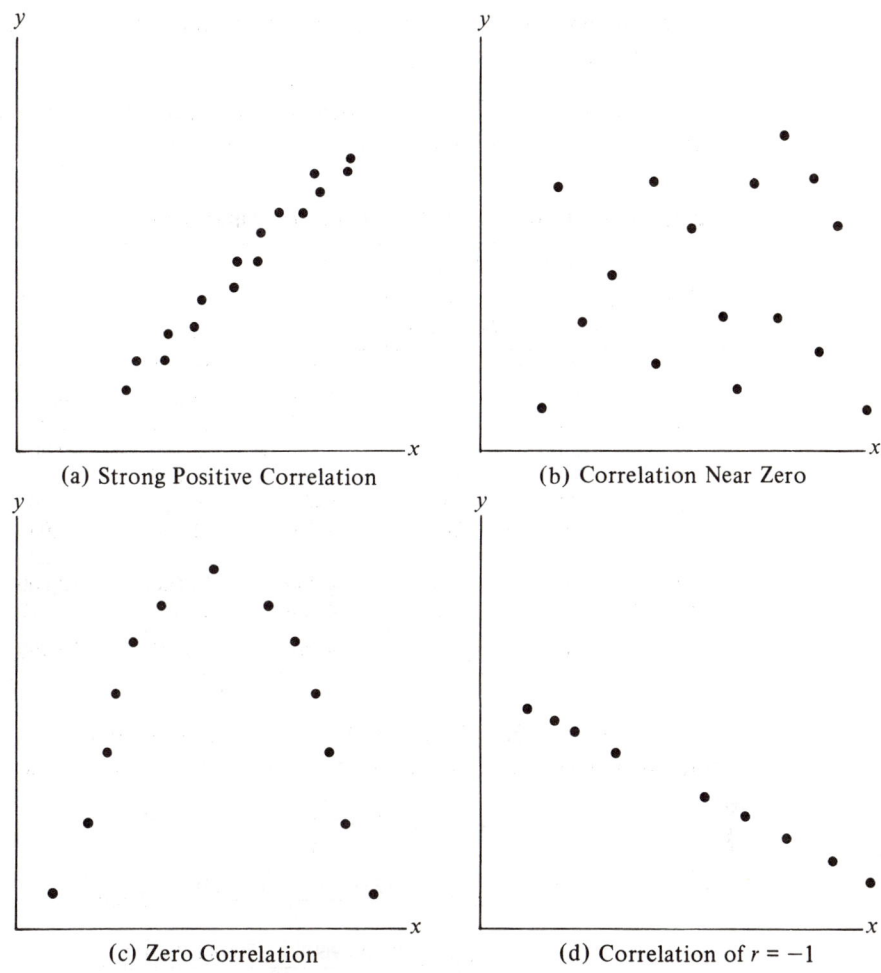

Figure 12.1. Scatter Diagrams Showing Different Strengths and Directions of Association Between x and y

(12.5) We interpret $100r^2$ as the percent of the variation in the sample y values that can be attributed to the line of regression of y on x, i.e., by the relationship of y to x.

For example, if all points fall on a straight line, then $SSres = 0$ and $SSreg = SSy$. Thus $r^2 = 1$, and we say 100 percent of the variation in the y values is due to the linear relationship with x. Thus, when $r = 0.60$, only 36 percent of the variation in the y values is attributed to the line of regression of y on x and when $r = 0.30$, only 9 percent of the variation in the y values is accounted for by linear regression of y on x. So, in terms of the "percentage of variation due to linear regression" we can say for two samples of the same size that $r = 0.60$ is approximately *four times as strong* as $r = 0.30$ and that $r = 0.90$ is approximately *nine times as strong* as $r = 0.30$.

EXAMPLE 12.1. Find the coefficient of correlation for the heights and weights of the 10 adult males shown in Table 12.1.

Table 12.1. Heights and Weights of Ten Adult Males

Male	Height in Inches (x)	Weight in Pounds (y)	x^2	y^2	xy
1	72	185	5,184	34,225	13,320
2	68	160	4,624	25,600	10,880
3	70	170	4,900	28,900	11,900
4	66	150	4,356	22,500	9,900
5	70	165	4,900	27,225	11,550
6	74	190	5,476	36,100	14,060
7	76	210	5,776	44,100	15,960
8	68	165	4,624	27,225	11,220
9	70	180	4,900	32,400	12,600
10	66	145	4,356	21,025	9,570
Total	700	1,720	49,096	299,300	120,960

The first three columns of Table 12.1 show the data, and the remaining columns are for the computation of r. Using the five totals of Table 12.1 we find

$$SSx = \frac{10(49,096) - (700)^2}{10} = 96$$

$$SSy = \frac{10(299,300) - (1,720)^2}{10} = 3,460$$

$$SP = \frac{10(120,960) - (700)(1,720)}{10} = 560$$

$$r = \frac{560}{\sqrt{(96)(3,460)}} = \frac{560}{576.3} \doteq 0.972$$

A coefficient of $r = 0.972$ indicates a *strong positive association* between height and weight. That is, on the average, the taller the person the more he weighs. But in the absence of a linear regression equation we cannot predict a weight from a height nor a height from a weight. Since $100r^2 = 94.4784$, we can say that approximately 94.5 percent of the variation in the weights can be explained by the fact that the men have different heights.

By dividing both numerator and denominator of r [see Equation (12.2)] by $n - 1$ we obtain

(12.6) $$r = \frac{cov(x, y)}{\sqrt{s_x^2 s_y^2}} = \frac{cov(x, y)}{s_x \times s_y}$$

where

(12.7) $\quad s_x^2 = \dfrac{SSx}{n-1}, \quad s_y^2 = \dfrac{SSy}{n-1}, \quad \text{and} \quad cov(x, y) = \dfrac{SP}{n-1}$

The covariance between x and y, denoted by $cov(x,y)$, actually measures the strength of association between x and y in terms of the units of x and y. (In Example 12.1, $cov(x,y) = 560/(10 - 1) = 62.2$ is in terms of inches-pounds which would be difficult to compare with other measures of strength of association between two variables.) In order to have a measure of association which is free of the units of measurement of the two variables, we divide by the product of the standard deviations of the two variables to obtain r. So r is expressed in terms of an *abstract* or *pure* or *dimensionless number*.

Sometimes it is desirable that the data be coded before r is calculated. This is especially true of large collections of data which have been grouped into classes (see Exercises 12.4 and 12.7). To be specific, it can be shown that

When $u = \dfrac{x - x'}{c}$ and $v = \dfrac{y - y'}{k}$ where x', $c \neq 0$, y', and $k \neq 0$ are constants, the correlation coefficient of u and v, r_{uv}, is equal to the correlation coefficient of x and y, r_{xy}. That is,

(12.8) $\quad\quad\quad\quad\quad\quad\quad\quad\quad r_{uv} = r_{xy}$

EXAMPLE 12.2. For the data of Table 12.1, let

$$u = \dfrac{x - 66}{2} \quad \text{and} \quad v = \dfrac{y - 170}{5}$$

and compute the correlation coefficient r_{uv}.

Using the coding formulas we obtain the totals shown in Table 12.2.

Table 12.2. Computation of the Correlation Coefficient of Example 12.1 Using Coded Heights and Weights

Male	u	v	u²	v²	uv
1	3	3	9	9	9
2	1	−2	1	4	−2
3	2	0	4	0	0
4	0	−4	0	16	0
5	2	−1	4	1	−2
6	4	4	16	16	16
7	5	8	25	64	40
8	1	−1	1	1	−1
9	2	2	4	4	4
10	0	−5	0	25	0
Total	20	4	64	140	64

Then we find

$$SSu = \frac{10(64) - (20)^2}{10} = 24 \qquad SSv = \frac{10(140) - (4)^2}{10} = 138.4$$

$$SP = \frac{10(64) - 20(4)}{10} = 56 \qquad r_{uv} = \frac{56}{\sqrt{24(138.4)}} \doteq 0.972$$

This is the same value found in Example 12.1.

If all the points lie along either a vertical or a horizontal line, the correlation coefficient does not exist. For example, for the three points (1, 2), (1, 3), (1, 4) we find

$$SSx = \sum (x_i - \bar{x})^2 = \sum (x_i - 1)^2 = 0$$

and

$$SP = \sum (x_i - \bar{x})(y_i - \bar{y}) = \sum (x_i - 1)(y_i - 3)$$
$$= (0)(-1) + (0)(0) + (0)(1) = 0$$

Thus

$$r = \frac{0}{\sqrt{0 SSy}} = \frac{0}{0}$$

which is an indeterminant form.

Caution: A large numerical value of r does not imply a cause-effect relationship between the x and y variables. For example, a high correlation between ministers' salaries and the consumption of liquor over the years does not mean that an increase in ministers' salaries *causes* an increase in consumption of liquor—the high correlation very likely results from the fact that both variables are affected by an overall standard of living. This example is not intended to encourage the reader to take lightly a high correlation between two variables when a cause-effect relationship may be in doubt. The fact that an "increase in smoking" is highly correlated to "decrease in length of life due to lung cancer" (i.e., early death due to lung cancer) is very sobering regardless whether smoking *causes* early death or not.

12.2. INFERENCES ABOUT THE CORRELATION COEFFICIENT

In both the linear regression and correlation analyses we sample from a bivariate population of pairs (x, y). For linear regression of y on x when we say a sample of n pairs

(12.9) $\qquad (x_1, y_1), (x_2, y_2), \cdots, (x_n, y_n)$

is random we mean the x values are fixed or controlled and the y values are random. However, in correlation analysis when we say the n pairs of Expression (12.9) are random we mean that both members of the pair are random variables. In a regression analysis of weights (y) on heights (x) in Example 12.1 we would specify the number of people at each height, then we would randomly select the required number of people at each height to obtain a random sample of weights. In correlation analysis we would randomly select from all people in the population and then determine the height and weight of each person.

In the regression analysis we use the sample line of regression

$$\hat{\eta} = a + b(x - \bar{x})$$

to make statements or inferences about the population line of regression

$$\eta = \mu_y + \beta(x - \bar{x})$$

In correlation analysis we use the sample correlation coefficient r to make a statement or inference about the correlation coefficient ρ of the bivariate population from which the sample was drawn. If a random sample of n pairs is drawn from a bivariate population with correlation coefficient ρ, it can be shown that r is usually a biased estimator of ρ. Still, for reasons beyond the scope of this books, r is usually considered the best point estimator of ρ. Thus, in Example 12.1, $r = 0.972$ is a point estimate of the correlation coefficient ρ of the population from which the 10 males were randomly selected. However, we need to study the distribution properties of the random variable r before we can say how close to the true value $r = 0.972$ is likely to be.

Before we present a discussion of the complicated distribution of r, it is desirable that we indicate specific relations between the slope of a regression line and the correlation coefficient. If the pair of random variables (x, y) has a *bivariate normal distribution* with correlation coefficient ρ, then it can be shown that the y arrays for given x have means

$$\eta = \mu_y + \rho \frac{\sigma_y}{\sigma_x}(x - \mu_x)$$

Where μ_x and μ_y denote the means and σ_x and σ_y, the standard deviations of the random variables x and y, respectively. Thus, it follows that

(12.10) $$\beta = \rho \frac{\sigma_y}{\sigma_x}$$

where β is the coefficient of linear regression of y on x. Similarly, if β' denotes the coefficient of linear regression of x on y, we have

(12.11) $$\beta' = \rho \frac{\sigma_x}{\sigma_y}$$

Thus, we find that

(12.12) $$\rho^2 = \beta\beta' \quad \text{or} \quad \rho = \pm\sqrt{\beta\beta'}$$

Also, from Equation (12.3) we write

(12.13) $$r^2 = \frac{SP}{SSx} \frac{SP}{SSy} = bb' \quad \text{or} \quad r = \pm\sqrt{bb'}$$

where b and b' are least squares estimators of β and β', respectively. So, for both the population and a sample, the correlation coefficient is the geometric mean of the slopes of the two regression lines.

We start our discussion of the distribution of r with the special important case where $\rho = 0$, because of connections with regression analysis. From Equation (12.10) we observe that $\rho = 0$ if and only if $\beta = 0$ provided the bivariate distribution is not degenerate, that is, provided $\sigma_x \neq 0$ and $\sigma_y \neq 0$. Thus, we may test the null hypothesis that $\rho = 0$ by the statistic used to test the null hypothesis that $\beta = 0$. It can be shown (see Exercise 12.8) that the t statistic $(b - \beta)/\sqrt{MSres/SSx}$ reduces to $r/\sqrt{(1 - r^2)/(n - 2)}$ when $\beta = 0$. We summarize this in the following statement:

Theorem 12.1. If a random sample of n pairs

$$(x_1, y_1), (x_2, y_2), \ldots, (x_n, y_n)$$

is drawn from a bivariate normal population with correlation coefficient $\rho = 0$, then the statistic

(12.14) $$t = \frac{r}{\sqrt{\frac{1 - r^2}{n - 2}}}$$

has the Student t distribution with $(n - 2)$ degrees of freedom.

Since Theorem 12.1 applies only when $\rho = 0$, we cannot use it to establish confidence limits for ρ. The theorem may be used only to decide whether a particular sample correlation coefficient r is significantly different from zero.

EXAMPLE 12.3. A sample of 18 pairs of values (x, y) has a correlation coefficient of $r = 0.370$. Is the population correlation coefficient positive?
By the general test procedure we write:

1. $H_o: \rho \leq 0$ and $H_a: \rho > 0$
2. The 18 pairs of values are randomly drawn from a bivariate normal population. (Under the temporary assumption of the null hypothesis that ρ is actually zero the y arrays are independent of x.) Let the significance level be 0.05.

3. Use the statistic

$$t = \frac{r}{\sqrt{\frac{1-r^2}{n-2}}} = \frac{4r}{\sqrt{1-r^2}}$$

which has the Student t distribution with 16 degrees of freedom.

4. We would want to reject H_o only when r is a large positive value. Thus

$$CR = \{t \,|\, t > 1.746\}$$

5. For $r = 0.370$, we find that

$$t_c = \frac{4(0.370)}{\sqrt{1-(0.370)^2}} \doteq 1.59$$

6. Since $t_c = 1.59$ falls in the noncritical region, we fail to reject H_o. Thus, $r = 0.370$ is not significantly larger than $\rho = 0$ at the 0.05 level.

Note: If the sample size had been 27, the critical region would have been $CR = \{t \,|\, t > 1.708\}$. For $r = 0.370$ we would find

$$t_c = \frac{5(0.370)}{\sqrt{1-(0.370)^2}} \doteq 1.99$$

Thus, in this case a sample correlation coefficient of the same size as in Example 12.3 is declared significantly greater than $\rho = 0$. This indicates that r is less variable as the sample size increases. Put another way, we need stronger evidence to conclude significant correlation with a small sample than we do with a large sample.

When $\rho = 0$, r has a symmetric distribution for values of r between -1 and 1, and, as we have already observed,

$$t = \frac{r\sqrt{n-2}}{\sqrt{1-r^2}}$$

has the Student t distribution for values of t between $-\infty$ and ∞ provided $n > 2$. (For $n = 2$ the sample correlation coefficient is either -1 or 1 unless the two sample points coincide and then r is indeterminant. In either case the result is of no use in making inferences about ρ.)

When ρ differs from zero, the sampling distribution of r is not symmetric. In fact, as ρ approaches either -1 or 1 the asymmetry increases, as is shown in Figure 12.2. The distribution of r depends on the sample size n as well as ρ. For small samples r is quite variable particularly when ρ is in the neighborhood of zero.

Since the sampling distributions of r are complicated, the percentage points are difficult to compute directly. Fortunately, due to the work of David

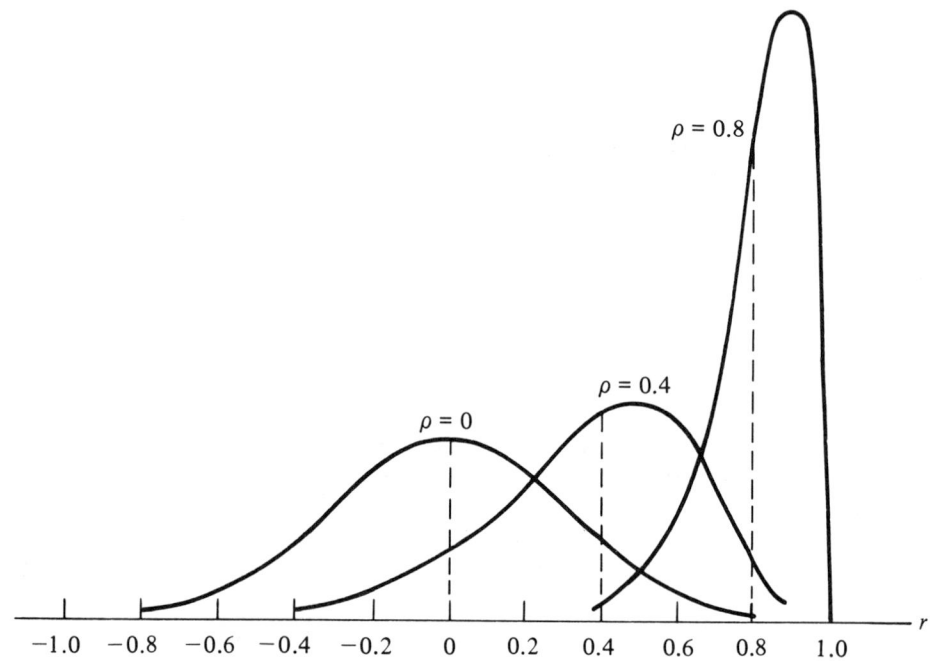

Figure 12.2. Distributions of r for $\rho = 0$, 0.4, and 0.8 when $n = 10$

[5] and Fisher [10], we can use charts and simple transformations to determine confidence intervals and to test hypotheses.

For selected values of n smaller than 401, David [5] computed the distributions of r. From these extensive calculations David prepared charts, reproduced in Table VI in the Appendix, which are useful in finding 95 and 99 percent confidence intervals and in testing hypotheses about ρ.

EXAMPLE 12.4. A random sample of size 15 has a correlation coefficient of 0.7. Find the 95 percent confidence interval for the population correlation coefficient ρ.

Using Table VI, construct a vertical line through $r = 0.7$ and note where it crosses the two curves corresponding to $n = 15$. At these points of intersection construct two horizontal lines which cross the vertical axis ρ. The two values of ρ on the vertical axis are 0.28 and 0.88. Thus, the required 95 percent confidence interval for ρ based on $n = 15$ and $r = 0.7$ is

$$0.28 < \rho < 0.88$$

The two limits are definitely not symmetric about $r = 0.7$, indicating that the distribution of r tails off to the left (i.e., is skewed to the left).

We may use Table VI to test $H_o: \rho = \rho_o$ against $H_a: \rho \neq \rho_o$ at the 5 or 1 percent level or to test H_o against $H_a: \rho > \rho_o$ (or $H_a: \rho < \rho_o$) at the 2.5 or 0.5 percent level, where ρ_o is some specified constant in the interval from -1 to 1. A point on the appropriate curve on the chart is a critical point for a test.

The charts of David are not always satisfactory. For example, an experimenter may require different percentage points or more accuracy than can be obtained from the charts. In such cases the transformation

(12.15) $$z' = \frac{1}{2} \log_e \left(\frac{1+r}{1-r} \right) \quad \text{for } -1 < r < 1$$

due to Fisher [10], may usually be applied. See Freund [12] for illustrations.

12.3. RANK CORRELATION

In Chapter 9 we gave some distribution-free methods for dealing with characteristics of one or more univariate distributions. Now, we introduce a distribution-free method, called rank correlation, for measuring the strength of association between two variables whose bivariate distribution is unknown. In this situation all that is required is that we be able to *order* the observations for each of two variables. As in Chapter 9, there may be a loss in information when ranks are used, but the computations are simplified and fewer assumptions are required in tests of significance.

Suppose we make two observations on each of n objects. Denote the n pairs of observations by

$$(x_1, y_1), (x_2, y_2), \cdots, (x_n, y_n)$$

If the x values are not already in terms of rank, arrange them in order of magnitude, assigning the rank of 1 to the largest or best value, the rank of 2 to the next largest or next best value, and so on. Do the same for the y values. For the ith pair, expressed in terms of rank, let d_i denote the difference of the rank assigned to y_i minus the rank assigned to x_i. By substituting the ranks for the x and y values in the Pearson product moment correlation coefficient formula, Equation (12.1), it can be shown that the coefficient of the correlation of ranks, r_s, becomes

(12.16) $$r_s = 1 - \frac{6 \sum d_i^2}{n(n^2 - 1)}$$

The rank correlation coefficient r_s is also known as Spearman's rank-correlation coefficient, in honor of the man who introduced it. In case of ties, assign to each of the tied values the mean of the ranks these values would have had if they had not been tied. (In this case there is usually a small but negligible difference in the values of r and r_s; otherwise, r and r_s are identical when ranks are used in both formulas.)

EXAMPLE 12.5. In identical races of two track meets, twelve boys had the relative positions shown in Table 12.3. Find the coefficient of rank correlation.

Table 12.3. Order of Finish of 12 Boys in Two Races and Calculations for Rank Correlation

Boy	Order of Finish in Race One(x)	Two(y)	d	d²
A	5	6	1	1
B	11	10	−1	1
C	6	4	−2	4
D	2	1	−1	1
E	4	5	1	1
F	9	11	2	4
G	10	9	−1	1
H	1	2	1	1
I	7	8	1	1
J	12	12	0	0
K	3	3	0	0
L	8	7	−1	1
			$\Sigma d^2 =$	16

$$r_s = 1 - \frac{6(16)}{12(143)} = 1 - \frac{8}{143} = \frac{135}{143} = 0.944$$

Since the x and y values are already in terms of ranks, we compute the d's directly from the original data, showing the computations in Table 12.3. Otherwise, we would first replace the x and y values by ranks. The correlation coefficient of $r_s = 0.944$ shows a strong positive correlation between the races.

It will be left to the reader (see Exercise 12.13) to show that when the ranks are substituted directly in Equation (12.1) the value of the product moment correlation coefficient is $r = 0.944$. However, in any problem where the x and y values must first be transformed to ranks, the product moment correlation coefficient of the original values is almost certainly different from the product moment correlation coefficient of the ranks (see Exercise 12.14).

The statistic r_s may be used to test the hypothesis that the two variables x and y are independent or that $\rho = 0$. It can be shown that, when x and y are independently distributed the standard error of r_s is

(12.17) $$\sigma_{r_s} = \frac{1}{\sqrt{n-1}}$$

as the sampling distribution of r_s approaches the normal distribution as n increases. That is,

(12.18) $$z = \frac{r_s - 0}{1/\sqrt{n-1}} = r_s\sqrt{n-1}$$

is distributed approximately as a standard normal distribution for large n, say larger than 10. In Example 12.5

$$r_s \sqrt{n-1} = 0.944 \sqrt{11} \doteq 3.13$$

is large enough to be significant at the 0.002 level since $z_{.001} = 3.09$.

For $n \leq 10$, the exact sampling distribution in terms of n and $\sum d^2$ should be applied. Such a table can be found in Dixon and Massey [6]. (See Exercise 12.16 for a method of computing the sampling distribution of r_s and Exercise 12.17 for an application.)

12.4. OTHER MEASURES OF CORRELATION

Often it is desirable that we have a measure of association between a continuous variable and a dichotomous variable. Such a correlation is defined by the point biserial correlation coefficient, or briefly **biserial correlation**, and denoted by r_p. See reference [24] for applications.

In many problems a measure of the strength of association between two dichotomous variables would be desirable. For example, a test of independence in a 2×2 contingency table may lead to the conclusion that the two variables are dependent, but it does not give a measure of the strength of dependence. The **phi** and the **tetrachoric** coefficients are two useful measures of correlation between two dichotomous variables. See Exercises 12.18 for an application.

The **tau** coefficient, introduced by Kendall, may be used in place of the Spearman rank correlation coefficient. It is generally more difficult to compute but has the great advantage that its sampling distribution is known. References may be found in [24].

Whenever the association between two variables, at least one being continuous, is not linear the **correlation ratio** may be applied. This correlation coefficient can be developed by generalizing Equation (12.4). First, since $SSreg = SSy - SSres$, Equation 12.4 may be expressed as

$$r^2 = 1 - \frac{SSres}{SSy}$$

or

$$r^2 = 1 - \frac{\sum (y_i - \hat{\eta}_i)^2}{\sum (y_i - \bar{y})^2}$$

If $\hat{\eta}$ denotes the fitted mean of y on x whether the regression is linear or not, we have the sample correlation ratio r' defined by

$$r' = \sqrt{1 - \frac{\sum (y_i - \hat{\eta}_i)^2}{\sum (y_i - \bar{y})^2}}$$

Note that r' is always between 0 and 1, inclusive.

The degree of association among more than two variables is measured by partial and multiple correlation coefficients. References to these and other topics of correlation are found in [9].

12.5. EXERCISES

12.1. In Example 11.3, twenty pairs of values were used to find $SSx = 62{,}500$, $SSy = 4{,}467$, $SP = 14{,}000$, and the linear regression of yield on amount of fertilizer.
(a) Find the coefficient of correlation between amount of fertilizer applied and yield of crop.
(b) Find the coefficient of determination and indicate what percentage of the variation in yield can be explained by the regression of yield on amount of fertilizer.

12.2. (a) If $SSres = 15$ and $SSreg = 60$ for simple linear regression, what is the coefficient of determination; the coefficient of correlation? Also, what percentage of the variation in the y values can be explained by deviation of observations *from* linear regression of y on x?
(b) If the percentage of variation explained by linear regression of y on x were only half as great as in part (a), what would be the value of the correlation coefficient?

12.3. For each of the following bivariate distributions find the coefficient of correlation and plot the scatter diagram:

(a) x	y	(b) x	y	(c) x	y
2	1	2	1	2	5
3	3	3	4	3	4
4	3	4	5	4	3
5	4	5	4	5	3
6	5	6	1	6	1

12.4. (a) For the data of Table 12.1 let

$$u = \frac{x - 68}{2} \quad \text{and} \quad v = \frac{y - 160}{5}$$

and compute the correlation coefficient r_{uv}.
(b) Find the slope of the fitted regression of weight on height; of height on weight.

12.5. (a) Suppose a population consists of the five pairs of values in Exercise 12.3(a). Find the correlation coefficient of each of the five samples of size four which can be drawn without replacement. Is r an unbiased estimator of ρ in this case?
(b) Do part (a) using the population of Exercise 12.3(b).

12.6. (a) A sample of 11 pairs of values (x, y) has a correlation coefficient of $r = -0.50$. Is the population correlation coefficient negative? Use a 0.05 level test.
(b) Use the sample information of part (a) to find a 0.95 confidence interval of ρ.

12.7. (a) The final grade in a course is determined from four tests and an examination. Compute the correlation coefficient r_{uv} for the following data by first letting $u = x - 75$ and $v = (y - 75)/3$.

Test 1	(x)	71	85	85	59	82	57	59	70	88	84
Examination	(y)	84	93	90	87	90	57	60	72	84	75

(b) Test, at the 0.05 level, the hypothesis that the population correlation coefficient between test 1 and the examination is equal to zero.
(c) Find a 0.95 confidence interval for ρ.

E12.8. Show that the t statistic $(b - \beta)/\sqrt{MSres/SSx}$ reduces to $r/\sqrt{(1 - r^2)/(n - 2)}$ when $\beta = 0$.

12.9. Find a 0.99 confidence interval for ρ if $n = 25$ and $r = 0.30$.

12.10. An official of an insurance company states that the correlation between annual family income and amount of life insurance is in excess of 0.40. For a random sample of 50 policies in the company files it was found that $r = 0.63$. Is the sample value of r consistent with the official's statement.

12.11. Over a number of years it has been found that the correlation between scores on a mathematics placement test and final grades in the first college mathematics course at Beta College does not deviate much from 0.75. However, this year the correlation coefficient, based on a sample of 100 students, is 0.67. Can it be claimed that ρ is no longer at least 0.75?

12.12. Suppose 11 evening television programs are ranked by two different rating systems as follows:

Program	a	b	c	d	e	f	g	h	i	j	k
System I	10	9	2	8	1	4	3	7	11	5	6
System II	7	11	1	6	5	4	2	10	9	3	8

Compute r_s and, using a 0.05 level test, decide whether the computed value is significantly different from zero.

12.13. Use Equation (12.1) to show that the product moment correlation coefficient of the data of Example 12.5 is $r = 0.944$.

12.14. Find r_s for the data of Exercise 12.7(a) and compare its value with the value of r found in Exercise 12.7(a).

12.15. Suppose a class of 18 students has the 10 pairs of grades listed in Exercise 12.7(a) plus the 8 pairs of grades which follow:

Test 1 (x)	95	75	78	76	68	65	91	87
Examination (y)	89	82	85	75	85	70	95	83

Compute r_s as a measure of the consistency of the two grades and decide whether the value of r_s is significantly larger than zero.

E12.16. Letting $w = \sum_i d_i^2$, we write $r_s = 1 - 6w/[n(n^2-1)]$. When $n = 3$, $r_s = 1 - (w/4)$ and every sample of three pairs of numbers may be expressed as

$$(1, y_1) \qquad (2, y_2) \qquad (3, y_3)$$

where y_1, y_2, and y_3 have the values 1, 2, and 3 in some order. Since there are $3! = 6$ ways to arrange 1, 2, 3, there are 6 ways to find a sample of the type indicated above. For the sample (1, 1), (2, 3), (3, 2) we find $w = 0^2 + 1^2 + (-1)^2 = 2$ and $r_s = 0.50$. In a similar way we may find w and r_s for each sample. Assuming the six samples to be equally likely, we obtain Table 12.4.

Table 12.4. Probability Sampling Distribution of w (or of r_s) for Samples of Size Three

w	r_s	$f(w)$ or $f(r_s)$	Cum. freq.
0	1.00	$\frac{1}{6} = .167$	.167
2	.50	$\frac{2}{6} = .333$	.500
6	−.50	$\frac{2}{6} = .333$	.833
8	−1.00	$\frac{1}{6} = .167$	1.000

When $n = 4$, show that the probability sampling distribution of w is

w	0	2	4	6	8	10	12	14	16	18	20
$f(w)$	$\frac{1}{24}$	$\frac{3}{24}$	$\frac{1}{24}$	$\frac{4}{24}$	$\frac{2}{24}$	$\frac{2}{24}$	$\frac{2}{24}$	$\frac{4}{24}$	$\frac{1}{24}$	$\frac{3}{24}$	$\frac{1}{24}$

E12.17. Two taste testers rank four products as shown in the following table.

Product	a	b	c	d
Tester I	1	2	3	4
Tester II	1	2	4	3

(a) Compute w and r_s (see Exercise 12.16).
(b) Use the table of Exercise 12.16 to find the probability of r_s being equal to or larger than the value found in part (a).

E12.18. In Section 8.8.3 we discussed tests of independence of the two variables in a contingency table. Now, we wish to introduce a measure of dependence in the case where each of the variables is dichotomized as shown in Table 12.5. We could assign any two values to success and failure, but the computations are minimized by using 1 and 0, respectively. By using the formula for grouped data it can be shown that the product moment correlation coefficient becomes

Table 12.5.

Variable y	Variable x		Total
	Success (1)	Failure (0)	
Success (1)	a	b	a + b
Failure (0)	c	d	c + d
Total	a + c	b + d	n

$$\phi = \frac{ad - bc}{\sqrt{(a+b)(c+d)(a+c)(b+d)}}$$

and is called the **phi correlation coefficient**. Referring to Exercise 8.36, we observe that

$$\chi'^2 = n\phi^2$$

where χ'^2 is distributed approximately as chi-square with one degree of freedom (or $\sqrt{n}\,\phi$ is distributed approximately as a standard normal). When n is small Yates' correction may be applied [see Exercise 8.36(c)].

(a) Omit Manufacturer A from Table 8.9 and compute the phi coefficient. Decide whether ϕ is significantly different from zero at the 0.05 level.

(b) Omitting the "more than 40" column from Table 8.11, compute the phi coefficient for the remaining 165 students and decide whether ϕ is significantly different from zero.

(c) In a public opinion survey answers are tabulated to the following questions:

1. Did you vote in the last presidential election?
2. Do you plan to vote in this presidential election?

Use the following table to compute the value of ϕ and decide whether the responses to the questions are independent at the 0.01 level.

	Question 1	
Question 2	Yes	No
Yes	60	10
No	5	25

ANNOTATED REFERENCES AND BIBLIOGRAPHY

1. AUBLE, D., "Extended Tables for the Mann-Whitney Statistic," *Bulletin of Institution of Educational Research*, Indiana University, Vol. 1, No. 2 (1953).
2. BROSS, I. D. J., *Design for Decisions*. New York: The Macmillan Company, 1953. May be useful as a supplement to Chapter 7.
3. *Careers in Statistics*, 2nd ed., Committee of Presidents of Statistical Societies. Washington, D.C.: American Statistical Association, 1966. Sections I, II, and V are quoted in Chapter 1.
4. *Careers in Statistics*, Committee of Presidents of Statistical Societies. Washington, D.C.: American Statistical Association, 1973. Quotations from pages 6–13 appear in Chapter 1.
5. DAVID, F. N., *Tables of the Ordinates and Probability Integral of the Distribution of the Correlation Coefficient in Small Samples*. London: Cambridge University Press, 1938.
6. DIXON, W. J., and F. J. MASSEY, *Introduction to Statistical Analysis*, 3rd ed. New York: McGraw-Hill, Inc., 1969.
7. DRAPER, N. R., and H. SMITH, *Applied Regression Analysis*. New York: John Wiley & Sons, Inc., 1966. A detailed treatment of problems of regression.
8. DWYER, P. S., *Linear Computations*. New York: John Wiley & Sons, Inc., 1951. Careful discussion of range numbers and other approximations.
9. EZEKIEL, M., and K. A. FOX, *Methods of Correlation and Regression Analysis*, 3rd ed. New York: John Wiley & Sons, Inc., 1959. A detailed treatment of problems in regression and correlation.
10. FISHER, R. A., "On the 'Probable Error' of a Coefficient of Correlation Deduced from a Small Sample," *Metron*, Vol. 1 (1921), pp. 3–32.
11. FREUND, J. E., *Introduction to Probability*. Encino, Calif.: Dickenson Publishing Company, Inc., 1973. Elementary treatment of rules of probability.

12. FREUND, J. E., *Modern Elementary Statistics*, 4th ed. Englewood Cliffs, N.J.: Prentice-Hall, Inc., 1973.
13. LIEBERMAN, G. J., and D. B. OWEN, *Tables of the Hypergeometric Probability Distribution*. Stanford, Calif.: Stanford University Press, 1960.
14. LINDLEY, D. V., *Introduction to Probability and Statistics from a Bayesian Viewpoint. Part 1, Probability*. London: Cambridge University Press, 1965. Subjective probabilities are discussed.
15. MILLER, L. H., "Table of Percentage Points of Kolmogorov Statistics," *Journal of American Statistical Association*, Vol. 51 (1956), pp. 113–121. Table 9.5 is abridged from this publication.
16. MILLS, F. C., *Introduction to Statistics*. New York: Holt, Rinehart and Winston, Inc., 1956. Sheppard's correction for the grouping error introduced when calculating the variance.
17. MOSTELLER, F., R. E. K. ROURKE, and G. B. THOMAS, *Probability with Statistical Applications*, 2nd ed. Reading, Mass.: Addison-Wesley Publishing Company, Inc., 1970. Elementary treatment of probability.
18. SANDLER, J., "A Test of the Significance of the Difference Between the Means of Correlated Measures, Based on a Simplification of Student's t," *British Journal of Psychology*, Vol. 46 (1955), pp. 225–226. See Exercises 10.8 and 10.9 for the A test.
19. SAVAGE, I. R., *Bibliography of Nonparametric Statistics*. Cambridge, Mass.: Harvard University Press, 1962.
20. SIEGEL, A. E., "Film-mediated Fantasy Aggression and Strength of Aggressive Drive," *Child Development*, Vol. 27 (1956), pp. 365–378. See Exercise 9.45.
21. SIEGEL, S., *Nonparametric Statistics for the Behavioral Sciences*. New York: McGraw-Hill, Inc., 1956. Elementary treatment of many nonparametric test procedures.
22. SWED, F. S., and C. EISENHART, "Tables for Testing Randomness of Grouping in a Sequence of Alternatives," *Annals of Mathematical Statistics*, Vol. 14 (1943), pp. 66–87. Tables 9.6 and 9.8 are abridged from this publication.
23. TANUR, J. M., and others (eds.), *Statistics: A Guide to the Unknown*. San Francisco: Holden-Day, Inc., 1972. Short, nontechnical essays on a wide range of subjects.
24. WALKER, H. M., and J. LEV, *Statistical Inference*. New York: Holt, Rinehart and Winston, Inc., 1953. Simple applications of several correlation coefficients, including rank, biserial, phi, tetrachoric, and tau.
25. WALSH, J. E., *Handbook of Nonparametric Statistics*, Vols. I (1962), II (1965), and III (1968). Princeton, N.J.: D. Van Nostrand Company, Inc.

26. WHITE, C., "The Use of Ranks in a Test of Significance for Comparing Two Treatments," *Biometrics*, Vol. 8 (1952), pp. 33–41. Table 9.9 is taken from this publication.

27. WILCOXON, F., and R. A. WILCOX, *Some Rapid Approximate Statistical Procedures*. Pearl River, N.Y: Lederle Laboratories, revised 1964. Table 9.3 is taken from this publication.

STATISTICAL TABLES

Table I. Poisson Probabilities*

[Entries in the table are values of $p(y; \mu) = e^{-\mu}\mu^y/y!$ for indicated values of y and μ. When $y = 0$, $p(0; \mu) = e^{-\mu}$]

y \ μ	0.1	0.2	0.3	0.4	0.5	0.6	0.7	0.8	0.9	1.0
0	.9048	.8187	.7408	.6703	.6065	.5488	.4966	.4493	.4066	.3679
1	.0905	.1637	.2222	.2681	.3033	.3293	.3476	.3595	.3659	.3679
2	.0045	.0164	.0333	.0536	.0758	.0988	.1217	.1438	.1647	.1839
3	.0002	.0011	.0033	.0072	.0126	.0198	.0284	.0383	.0494	.0613
4	.0000	.0001	.0003	.0007	.0016	.0030	.0050	.0077	.0111	.0153
5	.0000	.0000	.0000	.0001	.0002	.0004	.0007	.0012	.0020	.0031
6	.0000	.0000	.0000	.0000	.0000	.0000	.0001	.0002	.0003	.0005
7	.0000	.0000	.0000	.0000	.0000	.0000	.0000	.0000	.0000	.0001

y \ μ	1.1	1.2	1.3	1.4	1.5	1.6	1.7	1.8	1.9	2.0
0	.3329	.3012	.2725	.2466	.2231	.2019	.1827	.1653	.1496	.1353
1	.3662	.3614	.3543	.3452	.3347	.3230	.3106	.2975	.2842	.2707
2	.2014	.2169	.2303	.2417	.2510	.2584	.2640	.2678	.2700	.2707
3	.0738	.0867	.0998	.1128	.1255	.1378	.1496	.1607	.1710	.1804
4	.0203	.0260	.0324	.0395	.0471	.0551	.0636	.0723	.0812	.0902
5	.0045	.0062	.0084	.0111	.0141	.0176	.0216	.0260	.0309	.0361
6	.0008	.0012	.0018	.0026	.0035	.0047	.0061	.0078	.0098	.0120
7	.0001	.0002	.0003	.0005	.0008	.0011	.0015	.0020	.0027	.0034
8	.0000	.0000	.0001	.0001	.0001	.0002	.0003	.0005	.0006	.0009
9	.0000	.0000	.0000	.0000	.0000	.0000	.0001	.0001	.0001	.0002

y \ μ	2.1	2.2	2.3	2.4	2.5	2.6	2.7	2.8	2.9	3.0
0	.1225	.1108	.1003	.0907	.0821	.0743	.0672	.0608	.0550	.0498
1	.2572	.2438	.2306	.2177	.2052	.1931	.1815	.1703	.1596	.1494
2	.2700	.2681	.2652	.2613	.2565	.2510	.2450	.2384	.2314	.2240
3	.1890	.1966	.2033	.2090	.2138	.2176	.2205	.2225	.2237	.2240
4	.0992	.1082	.1169	.1254	.1336	.1414	.1488	.1557	.1622	.1680
5	.0417	.0476	.0538	.0602	.0668	.0735	.0804	.0872	.0940	.1008
6	.0146	.0174	.0206	.0241	.0278	.0319	.0362	.0407	.0455	.0504
7	.0044	.0055	.0068	.0083	.0099	.0118	.0139	.0163	.0188	.0216
8	.0011	.0015	.0019	.0025	.0031	.0038	.0047	.0057	.0068	.0081
9	.0003	.0004	.0005	.0007	.0009	.0011	.0014	.0018	.0022	.0027
10	.0001	.0001	.0001	.0002	.0002	.0003	.0004	.0005	.0006	.0008
11	.0000	.0000	.0000	.0000	.0000	.0001	.0001	.0001	.0002	.0002
12	.0000	.0000	.0000	.0000	.0000	.0000	.0000	.0000	.0000	.0001

*This table is abridged from Table III.3 of *Handbook of Tables for Probability and Statistics*, 2nd ed., published by the Chemical Rubber Company, with permission of the publishers.

Table I. Poisson Probabilities (continued)

y \ μ	3.1	3.2	3.3	3.4	3.5	3.6	3.7	3.8	3.9	4.0
0	.0450	.0408	.0369	.0334	.0302	.0273	.0247	.0224	.0202	.0183
1	.1397	.1304	.1217	.1135	.1057	.0984	.0915	.0850	.0789	.0733
2	.2165	.2087	.2008	.1929	.1850	.1771	.1692	.1615	.1539	.1465
3	.2237	.2226	.2209	.2186	.2158	.2125	.2087	.2046	.2001	.1954
4	.1734	.1781	.1823	.1858	.1888	.1912	.1931	.1944	.1951	.1954
5	.1075	.1140	.1203	.1264	.1322	.1377	.1429	.1477	.1522	.1563
6	.0555	.0608	.0662	.0716	.0771	.0826	.0881	.0936	.0989	.1042
7	.0246	.0278	.0312	.0348	.0385	.0425	.0466	.0508	.0551	.0595
8	.0095	.0111	.0129	.0148	.0169	.0191	.0215	.0241	.0269	.0298
9	.0033	.0040	.0047	.0056	.0066	.0076	.0089	.0102	.0116	.0132
10	.0010	.0013	.0016	.0019	.0023	.0028	.0033	.0039	.0045	.0053
11	.0003	.0004	.0005	.0006	.0007	.0009	.0011	.0013	.0016	.0019
12	.0001	.0001	.0001	.0002	.0002	.0003	.0003	.0004	.0005	.0006
13	.0000	.0000	.0000	.0000	.0001	.0001	.0001	.0001	.0002	.0002
14	.0000	.0000	.0000	.0000	.0000	.0000	.0000	.0000	.0000	.0001

y \ μ	4.1	4.2	4.3	4.4	4.5	4.6	4.7	4.8	4.9	5.0
0	.0166	.0150	.0136	.0123	.0111	.0101	.0091	.0082	.0074	.0067
1	.0679	.0630	.0583	.0540	.0500	.0462	.0427	.0395	.0365	.0337
2	.1393	.1323	.1254	.1188	.1125	.1063	.1005	.0948	.0894	.0842
3	.1904	.1852	.1798	.1743	.1687	.1631	.1574	.1517	.1460	.1404
4	.1951	.1944	.1933	.1917	.1898	.1875	.1849	.1820	.1789	.1755
5	.1600	.1633	.1662	.1687	.1708	.1725	.1738	.1747	.1753	.1755
6	.1093	.1143	.1191	.1237	.1281	.1323	.1362	.1398	.1432	.1462
7	.0640	.0686	.0732	.0778	.0824	.0869	.0914	.0959	.1002	.1044
8	.0328	.0360	.0393	.0428	.0463	.0500	.0537	.0575	.0614	.0653
9	.0150	.0168	.0188	.0209	.0232	.0255	.0280	.0307	.0334	.0363
10	.0061	.0071	.0081	.0092	.0104	.0118	.0132	.0147	.0164	.0181
11	.0023	.0027	.0032	.0037	.0043	.0049	.0056	.0064	.0073	.0082
12	.0008	.0009	.0011	.0014	.0016	.0019	.0022	.0026	.0030	.0034
13	.0002	.0003	.0004	.0005	.0006	.0007	.0008	.0009	.0011	.0013
14	.0001	.0001	.0001	.0001	.0002	.0002	.0003	.0003	.0004	.0005
15	.0000	.0000	.0000	.0000	.0001	.0001	.0001	.0001	.0001	.0002

Table II. Cumulative Standard Normal Distribution

Probabilities for Selected Values of z

z'	$P(z < z')$	z'	$P(z < z')$	z'	$P(z < z')$	z'	$P(z < z')$
−4.0	.0000	−2.0	.0228	0	.5000	2.0	.9772
−3.9	.0000	−1.9	.0287	.1	.5398	2.1	.9821
−3.8	.0001	−1.8	.0359	.2	.5793	2.2	.9861
−3.7	.0001	−1.7	.0446	.3	.6179	2.3	.9893
−3.6	.0002	−1.6	.0548	.4	.6554	2.4	.9918
−3.5	.0002	−1.5	.0668	.5	.6915	2.5	.9938
−3.4	.0003	−1.4	.0808	.6	.7257	2.6	.9953
−3.3	.0005	−1.3	.0968	.7	.7580	2.7	.9965
−3.2	.0007	−1.2	.1151	.8	.7881	2.8	.9974
−3.1	.0010	−1.1	.1357	.9	.8159	2.9	.9981
−3.0	.0013	−1.0	.1587	1.0	.8413	3.0	.9987
−2.9	.0019	− .9	.1841	1.1	.8643	3.1	.9990
−2.8	.0026	− .8	.2119	1.2	.8849	3.2	.9993
−2.7	.0035	− .7	.2420	1.3	.9032	3.3	.9995
−2.6	.0047	− .6	.2743	1.4	.9192	3.4	.9997
−2.5	.0062	− .5	.3085	1.5	.9332	3.5	.9998
−2.4	.0082	− .4	.3446	1.6	.9452	3.6	.9998
−2.3	.0107	− .3	.3821	1.7	.9554	3.7	.9999
−2.2	.0139	− .2	.4207	1.8	.9641	3.8	.9999
−2.1	.0179	− .1	.4602	1.9	.9713	3.9	1.0000

Values of z for Selected Probabilities

z'	$P(z < z')$	z'	$P(z < z')$
−3.090	.001	0.6745	.750
−2.576	.005	1.282	.900
−2.326	.010	1.645	.950
−1.960	.025	1.960	.975
−1.645	.050	2.326	.990
−1.282	.100	2.576	.995
−0.6745	.250	3.090	.999

Table III. Random Digits*

Line	1–5	6–10	11–15	16–20	21–25	26–30	31–35	36–40	41–45	46–50
1	10480	15011	01536	02011	81647	91646	69179	14194	62590	36207
2	22368	46573	25595	85393	30995	89198	27982	53402	93965	34095
3	24130	48360	22527	97265	76393	64809	15179	24830	49340	32081
4	42167	93093	06243	61680	07856	16376	39440	53537	71341	57004
5	37570	39975	81837	16656	06121	91782	60468	81305	49684	60672
6	77921	06907	11008	42751	27756	53498	18602	70659	90655	15053
7	99562	72905	56420	69994	98872	31016	71194	18738	44013	48840
8	96301	91977	05463	07972	18876	20922	94595	56869	69014	60045
9	89579	14342	63661	10281	17453	18103	57740	84378	25331	12566
10	85475	36857	43342	53988	53060	59533	38867	62300	08158	17983
11	28918	69578	88231	33276	70997	79936	56865	05859	90106	31595
12	63553	40961	48235	03427	49626	69445	18663	72695	52180	20847
13	09429	93969	52636	92737	88974	33488	36320	17617	30015	08272
14	10365	61129	87529	85689	48237	52267	67689	93394	01511	26358
15	07119	97336	71048	08178	77233	13916	47564	81056	97735	85977
16	51085	12765	51821	51259	77452	16308	60756	92144	49442	53900
17	02368	21382	52404	60268	89368	19885	55322	44819	01188	65255
18	01011	54092	33362	94904	31273	04146	18594	29852	71585	85030
19	52162	53916	46369	58586	23216	14513	83149	98736	23495	64350
20	07056	97628	33787	09998	42698	06691	76988	13602	51851	46104
21	48663	91245	85828	14346	09172	30168	90229	04734	59193	22178
22	54164	58492	22421	74103	47070	25306	76468	26384	58151	06646
23	32639	32363	05597	24200	13363	38005	94342	28728	35806	06912
24	29334	27001	87637	87308	58731	00256	45834	15398	46557	41135
25	02488	33062	28834	07351	19731	92420	60952	61280	50001	67658
26	81525	72295	04839	96423	24878	82651	66566	14778	76797	14780
27	29676	20591	68086	26432	46901	20849	89768	81536	86645	12659
28	00742	57392	39064	66432	84673	40027	32832	61362	98947	96067
29	05366	04213	25669	26422	44407	44048	37937	63904	45766	66134
30	91921	26418	64117	94305	26766	25940	39972	22209	71500	64568
31	00582	04711	87917	77341	42206	35126	74087	99547	81817	42607
32	00725	69884	62797	56170	86324	88072	76222	36086	84637	93161
33	69011	65797	95876	55293	18988	27354	26575	08625	40801	59920
34	25976	57948	29888	88604	67917	48708	18912	82271	65424	69774
35	09763	83473	73577	12908	30883	18317	28290	35797	05998	41688
36	91567	42595	27958	30134	04024	86385	29880	99730	55536	84855
37	17955	56349	90999	49127	20044	59931	06115	20542	18059	02008
38	46503	18584	18845	49618	02304	51038	20655	58727	28168	15475
39	92157	89634	94824	78171	84610	82834	09922	25417	44137	48413
40	14577	62765	35605	81263	39667	47358	56873	56307	61607	49518
41	98427	07523	33362	64270	01638	92477	66969	98420	04880	45585
42	34914	63976	88720	82765	34476	17032	87589	40836	32427	70002
43	70060	28277	39475	46473	23219	53416	94970	25832	69975	94884
44	53976	54914	06990	67245	68350	82948	11398	42878	80287	88267
45	76072	29515	40980	07391	58745	25774	22987	80059	39911	96189
46	90725	52210	83974	29992	65831	38857	50490	83765	55657	14361
47	64364	67412	33339	31926	14883	24413	59744	92351	97473	89286
48	08962	00358	31662	25388	61642	34072	81249	35648	56891	69352
49	95012	68379	93526	70765	10593	04542	76463	54328	02349	17247
50	15664	10493	20492	38391	91132	21999	59516	81652	27195	48223

*This table is abridged from "A Table of 14,000 Random Units" (Table XII.4), *Handbook of Tables for Probability and Statistics*, 2nd ed., published by the Chemical Rubber Company, with permission of the publishers.

Table III. Random Digits (Continued)

Line	1–5	6–10	11–15	16–20	21–25	26–30	31–35	36–40	41–45	46–50
51	16408	81899	04153	53381	79401	21438	83035	92350	36693	31238
52	18629	81953	05520	91962	04739	13092	97662	24822	94730	06496
53	73115	35101	47498	87637	99016	71060	88824	71013	18735	20286
54	57491	16703	23167	49323	45021	33132	12544	41035	80780	45393
55	30405	83946	23792	14422	15059	45799	22716	19792	09983	74353
56	16631	35006	85900	98275	32388	52390	16815	69298	82732	38480
57	96773	20206	42559	78985	05300	22164	24369	54224	35083	19687
58	38935	64202	14349	82674	66523	44133	00697	35552	35970	19124
59	31624	76384	17403	53363	44167	64486	64758	75366	76554	31601
60	78919	19474	23632	27889	47914	02584	37680	20801	72152	39339
61	03931	33309	57047	74211	63445	17361	62825	39908	05607	91284
62	74426	33278	43972	10119	89917	15665	52872	73823	73144	88662
63	09066	00903	20795	95452	92648	45454	09552	88815	16553	51125
64	42238	12426	87025	14267	20979	04508	64535	31355	86064	29472
65	16153	08002	26504	41744	81959	65642	74240	56302	00033	67107
66	21457	40742	29820	96783	29400	21840	15035	34537	33310	06116
67	21581	57802	02050	89728	17937	37621	47075	42080	97403	48626
68	55612	78095	83197	33732	05810	24813	86902	60397	16489	03264
69	44657	66999	99324	51281	84463	60563	79312	93454	68876	25471
70	91340	84979	46949	81973	37949	61023	43997	15263	80644	43942
71	91227	21199	31935	27022	84067	05462	35216	14486	29891	68607
72	50001	38140	66321	19924	72163	09538	12151	06878	91903	18749
73	65390	05224	72958	28609	81406	39147	25549	48542	42627	45233
74	27504	96131	83944	41575	10573	08619	64482	73923	36152	05184
75	37169	94851	39117	89632	00959	16487	65536	49071	39782	17095
76	11508	70225	51111	38351	19444	66499	71945	05422	13442	78675
77	37449	30362	06694	54690	04052	53115	62757	95348	78662	11163
78	46515	70331	85922	38329	57015	15765	97161	17869	45349	61796
79	30986	81223	42416	58353	21532	30502	32305	86482	05174	07901
80	63798	64995	46583	09765	44160	78128	83991	42865	92520	83531
81	82486	84846	99254	67632	43218	50076	21361	64816	51202	88124
82	21885	32906	92431	09060	64297	51674	64126	62570	26123	05155
83	60336	98782	07408	53458	13564	59089	26445	29789	85205	41001
84	43937	46891	24010	25560	86355	33941	25786	54990	71899	15475
85	97656	63175	89303	16275	07100	92063	21942	18611	47348	20203
86	03299	01221	05418	38982	55758	92237	26759	86367	21216	98442
87	79626	06486	03574	17668	07785	76020	79924	25651	83325	88428
88	85636	68335	47539	03129	65651	11977	02510	26113	99447	68645
89	18039	14367	61337	06177	12143	46609	32989	74014	64708	00533
90	08362	15656	60627	36478	65648	16764	53412	09013	07832	41574
91	79556	29068	04142	16268	15387	12856	66227	38358	22478	73373
92	92608	82674	27072	32534	17075	27698	98204	63863	11951	34648
93	23982	25835	40055	67006	12293	02753	14827	22235	35071	99704
94	09915	96306	05908	97901	28395	14186	00821	80703	70426	75647
95	50937	33300	26695	62247	69927	76123	50842	43834	86654	70959
96	42488	78077	69882	61657	34136	79180	97526	43092	04098	73571
97	46764	86273	63003	93017	31204	36692	40202	35275	57306	55543
98	03237	45430	55417	63282	90816	17349	88298	90183	36600	78406
99	86591	81482	52667	61583	14972	90053	89534	76036	49199	43716
100	38534	01715	94964	87288	65680	43772	39560	12918	86537	62738

Table III. Random Digits (Continued)

Line	1–5	6–10	11–15	16–20	21–25	26–30	31–35	36–40	41–45	46–50
101	13284	16834	74151	92027	24670	36665	00770	22878	02179	51602
102	21224	00370	30420	03883	96648	89428	41583	17564	27395	63904
103	99052	47887	81085	64933	66279	80432	65793	83287	34142	13241
104	00199	50993	98603	38452	87890	94624	69721	57484	67501	77638
105	60578	06483	28733	37867	07936	98710	98539	27186	31237	80612
106	91240	18312	17441	01929	18163	69201	31211	54288	39296	37318
107	97458	14229	12063	59611	32249	90466	33216	19358	02591	45263
108	35249	38646	34475	72417	60514	69257	12489	51924	86871	92446
109	38980	46600	11759	11900	46743	27860	77940	39298	97838	95145
110	10750	52745	38749	87365	58959	53731	89295	59062	39404	13198
111	36247	27850	73958	20673	37800	63835	71051	84724	52492	22342
112	70994	66986	99744	72438	01174	42159	11392	20724	54322	36923
113	99638	94702	11463	18148	81386	80431	90628	52506	02016	85151
114	72055	15774	43857	99805	10419	76939	25993	03544	21560	83471
115	24038	65541	85788	55835	38835	59399	13790	35112	01324	39520
116	74976	14631	35908	28221	39470	91548	12854	30166	09073	75887
117	35553	71628	70189	26436	63407	91178	90348	55359	80392	41012
118	35676	12797	51434	82976	42010	26344	92920	92155	58807	54644
119	74815	67523	72985	23183	02446	63594	98924	20633	58842	85961
120	45246	88048	65173	50989	91060	89894	36063	32819	68559	99221
121	76509	47069	86378	41797	11910	49672	88575	97966	32466	10083
122	19689	90332	04315	21358	97248	11188	39062	63312	52496	07349
123	42751	35318	97513	61537	54955	08159	00337	80778	27507	95478
124	11946	22681	45045	13964	57517	59419	58045	44067	58716	58840
125	96518	48688	20996	11090	48396	57177	83867	86464	14342	21545
126	35726	58643	76869	84622	39098	36083	72505	92265	23107	60278
127	39737	42750	48968	70536	84864	64952	38404	94317	65402	13589
128	97025	66492	56177	04049	80312	48028	26408	43591	75528	65341
129	62814	08075	09788	56350	76787	51591	54509	49295	85830	59860
130	25578	22950	15227	83291	41737	79599	96191	71845	86899	70694
131	68763	69576	88991	49662	46704	63362	56625	00481	73323	91427
132	17900	00813	64361	60725	88974	61005	99709	30666	26451	11528
133	71944	60227	63551	71109	05624	43836	58254	26160	32116	63403
134	54684	93691	85132	64399	29182	44324	14491	55226	78793	34107
134	25946	27623	11258	65204	52832	50880	22273	05554	99521	73791
136	01353	39318	44961	44972	91766	90262	56073	06606	51826	18893
137	99083	88191	27662	99113	57174	35571	99884	13951	71057	53961
138	52021	45406	37945	75234	24327	86978	22644	87779	23753	99926
139	78755	47744	43776	83098	03225	14281	83637	55984	13300	52212
140	25282	69106	59180	16257	22810	43609	12224	25643	89884	31149
141	11959	94202	02743	86847	79725	51811	12998	76844	05320	54236
142	11644	13792	68190	01424	30078	28197	55583	05197	47714	68440
143	06307	97912	68110	59812	95448	43244	31262	88880	13040	16458
144	76285	75714	89585	99296	52640	46518	55486	90754	88932	19937
145	55322	07589	39600	60866	63007	20007	66819	84164	61131	81429
146	78017	90928	90220	92503	83375	26986	74399	30885	88567	29169
147	44768	43342	20696	26331	43140	69744	82928	24988	94237	46138
148	25100	19336	14605	86603	51680	97678	24261	02464	86563	74812
149	83612	46623	62876	85197	07824	91392	58317	37726	84628	42221
150	41347	81666	82961	60413	71020	83658	02415	33322	66036	98712

Table IV. Percentage Points of the Chi-square Distributions*

ν \ α	.995	.990	.975	.950	.900	.750
1	$.0^4393$	$.0^3157$	$.0^3982$	$.0^2393$	.0158	.102
2	.0100	.0201	.0506	.103	.211	.575
3	.0717	.115	.216	.352	.584	1.21
4	.207	.297	.484	.711	1.06	1.92
5	.412	.554	.831	1.15	1.61	2.67
6	.676	.872	1.24	1.64	2.20	3.45
7	.989	1.24	1.69	2.17	2.83	4.25
8	1.34	1.65	2.18	2.73	3.49	5.07
9	1.73	2.09	2.70	3.33	4.17	5.90
10	2.16	2.56	3.25	3.94	4.87	6.74
11	2.60	3.05	3.82	4.57	5.58	7.58
12	3.07	3.57	4.40	5.23	6.30	8.44
13	3.57	4.11	5.01	5.89	7.04	9.30
14	4.07	4.66	5.63	6.57	7.79	10.2
15	4.60	5.23	6.26	7.26	8.55	11.0
16	5.14	5.81	6.91	7.96	9.31	11.9
17	5.70	6.41	7.56	8.67	10.1	12.8
18	6.26	7.01	8.23	9.39	10.9	13.7
19	6.84	7.63	8.91	10.1	11.7	14.6
20	7.43	8.26	9.59	10.9	12.4	15.5
21	8.03	8.90	10.3	11.6	13.2	16.3
22	8.64	9.54	11.0	12.3	14.0	17.2
23	9.26	10.2	11.7	13.1	14.8	18.1
24	9.89	10.9	12.4	13.8	15.7	19.0
25	10.5	11.5	13.1	14.6	16.5	19.9
26	11.2	12.2	13.8	15.4	17.3	20.8
27	11.8	12.9	14.6	16.2	18.1	21.7
28	12.5	13.6	15.3	16.9	18.9	22.7
29	13.1	14.3	16.0	17.7	19.8	23.6
30	13.8	15.0	16.8	18.5	20.6	24.5
40	20.7	22.2	24.4	26.5	29.1	33.7
50	28.0	29.7	32.4	34.8	37.7	42.9
60	35.5	37.5	40.5	43.2	46.5	52.3

*This table is abridged from Table 8 of *Biometrika Tables for Statisticians, Vol. I*, published at the Cambridge University Press, with permission of the Biometrika Trustees.

Table IV. Percentage Points of the Chi-square Distributions (cont.)

α	.500	.250	.100	.050	.025	.010	.005	ν
	.455	1.32	2.71	3.84	5.02	6.63	7.88	1
	1.39	2.77	4.61	5.99	7.38	9.21	10.6	2
	2.37	4.11	6.25	7.81	9.35	11.3	12.8	3
	3.36	5.39	7.78	9.49	11.1	13.3	14.9	4
	4.35	6.63	9.24	11.1	12.8	15.1	16.7	5
	5.35	7.84	10.6	12.6	14.4	16.8	18.5	6
	6.35	9.04	12.0	14.1	16.0	18.5	20.3	7
	7.34	10.2	13.4	15.5	17.5	20.1	22.0	8
	8.34	11.4	14.7	16.9	19.0	21.7	23.6	9
	9.34	12.5	16.0	18.3	20.5	23.2	25.2	10
	10.3	13.7	17.3	19.7	21.9	24.7	26.8	11
	11.3	14.8	18.5	21.0	23.3	26.2	28.3	12
	12.3	16.0	19.8	22.4	24.7	27.7	29.8	13
	13.3	17.1	21.1	23.7	26.1	29.1	31.3	14
	14.3	18.2	22.3	25.0	27.5	30.6	32.8	15
	15.3	19.4	23.5	26.3	28.8	32.0	34.3	16
	16.3	20.5	24.8	27.6	30.2	33.4	35.7	17
	17.3	21.6	26.0	28.9	31.5	34.8	37.2	18
	18.3	22.7	27.2	30.1	32.9	36.2	38.6	19
	19.3	23.8	28.4	31.4	34.2	37.6	40.0	20
	20.3	24.9	29.6	32.7	35.5	38.9	41.4	21
	21.3	26.0	30.8	33.9	36.8	40.3	42.8	22
	22.3	27.1	32.0	35.2	38.1	41.6	44.2	23
	23.3	28.2	33.2	36.4	39.4	43.0	45.6	24
	24.3	29.3	34.4	37.7	40.6	44.3	46.9	25
	25.3	30.4	35.6	38.9	41.9	45.6	48.3	26
	26.3	31.5	36.7	40.1	43.2	47.0	49.6	27
	27.3	32.6	37.9	41.3	44.5	48.3	51.0	28
	28.3	33.7	39.1	42.6	45.7	49.6	52.3	29
	29.3	34.8	40.3	43.8	47.0	50.9	53.7	30
	39.3	45.6	51.8	55.8	59.3	63.7	66.8	40
	49.3	56.3	63.2	67.5	71.4	76.2	79.5	50
	59.3	67.0	74.4	79.1	83.3	88.4	92.0	60

Table V. Percentage Points of the Student t Distributions*

ν \ α	.20	.10	.05	.025	.01	.005
1	1.376	3.078	6.314	12.706	31.821	63.657
2	1.061	1.886	2.920	4.303	6.965	9.925
3	.978	1.638	2.353	3.182	4.541	5.841
4	.941	1.533	2.132	2.776	3.747	4.604
5	.920	1.476	2.015	2.571	3.365	4.032
6	.906	1.440	1.943	2.447	3.143	3.707
7	.896	1.415	1.895	2.365	2.998	3.499
8	.889	1.397	1.860	2.306	2.896	3.355
9	.883	1.383	1.833	2.262	2.821	3.250
10	.879	1.372	1.812	2.228	1.764	3.169
11	.876	1.363	1.796	2.201	2.718	3.106
12	.873	1.356	1.782	2.179	2.681	3.055
13	.870	1.350	1.771	2.160	2.650	3.012
14	.868	1.345	1.761	2.145	2.624	2.977
15	.866	1.341	1.753	2.131	2.602	2.947
16	.865	1.337	1.746	2.120	2.583	2.921
17	.863	1.333	1.740	2.110	2.567	2.898
18	.862	1.330	1.734	2.101	2.552	2.878
19	.861	1.328	1.729	2.093	2.539	2.861
20	.860	1.325	1.725	2.086	2.528	2.845
21	.859	1.323	1.721	2.080	2.518	2.831
22	.858	1.321	1.717	2.074	2.508	2.819
23	.858	1.319	1.714	2.069	2.500	2.807
24	.857	1.318	1.711	2.064	2.492	2.797
25	.856	1.316	1.708	2.060	2.485	2.787
26	.856	1.315	1.706	2.056	2.479	2.779
27	.855	1.314	1.703	2.052	2.473	2.771
28	.855	1.313	1.701	2.048	2.467	2.763
29	.854	1.311	1.699	2.045	2.462	2.756
30	.854	1.310	1.697	2.042	2.457	2.750
40	.851	1.303	1.684	2.021	2.423	2.704
60	.848	1.296	1.671	2.000	2.390	2.660
120	.845	1.289	1.658	1.980	2.358	2.617
∞	.842	1.282	1.645	1.960	2.326	2.576

*Table V is taken from Table III of Fisher and Yates, *Statistical Tables for Biological, Agricultural and Medical Research*, published by Longman Group, Ltd., London (previously published by Oliver & Boyd, Edinburgh), and by permission of the authors and publishers.

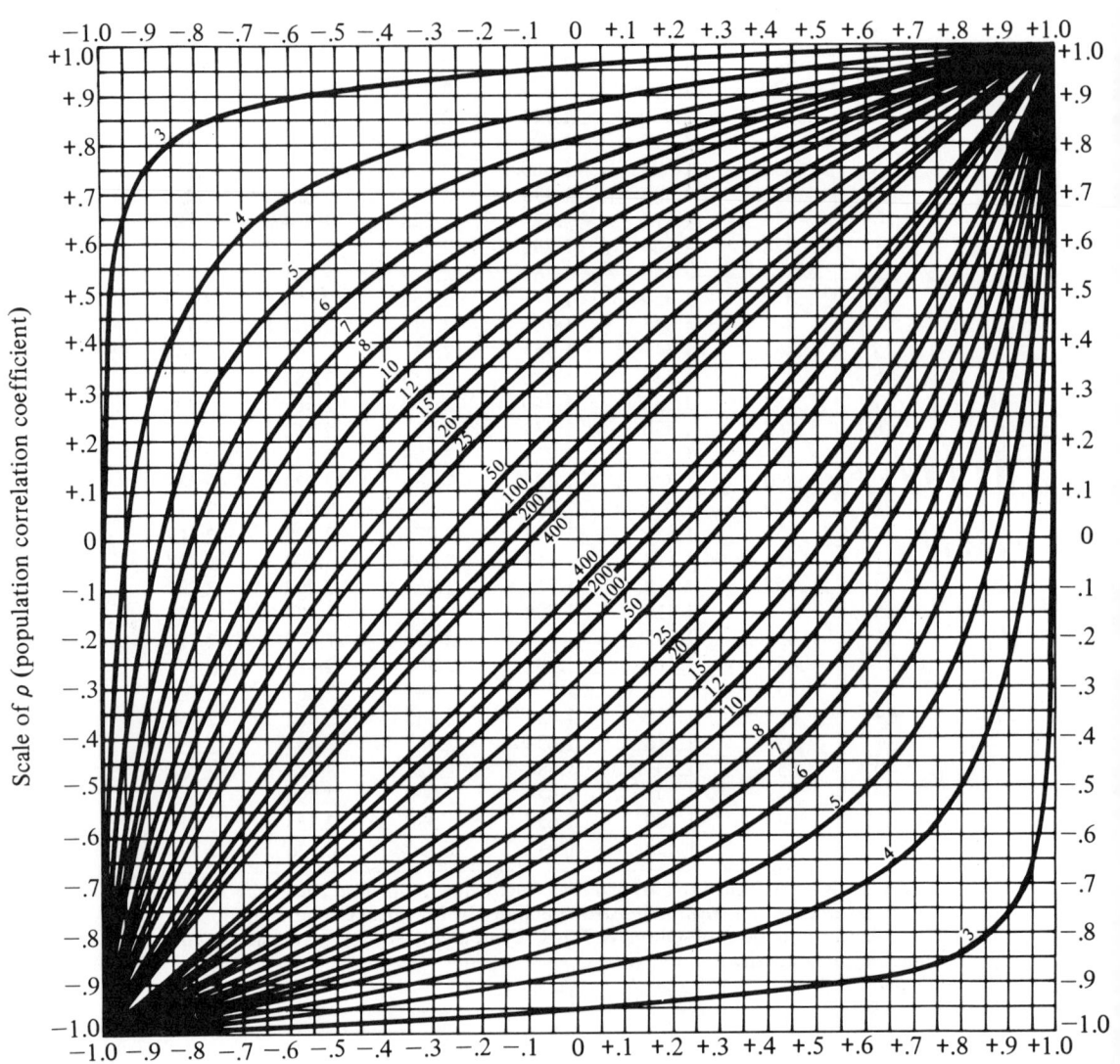

Table VI. Confidence Belts for the Correlation Coefficient ρ*
Confidence Coefficient 0.95

Scale of ρ (population correlation coefficient)

Scale of r (sample correlation coefficient)

The numbers on the curves indicates sample size. The chart can also be used to determine upper and lower 2·5% significance points for r, given ρ.

*This chart is reproduced from Table 15 of *Biometrika Tables for Statisticians*, Vol. I, published at the Cambridge University Press, with permission of the Biometrika Trustees.

Table VI. Confidence Belts for the Correlation Coefficient ρ
Confidence Coefficient 0.99

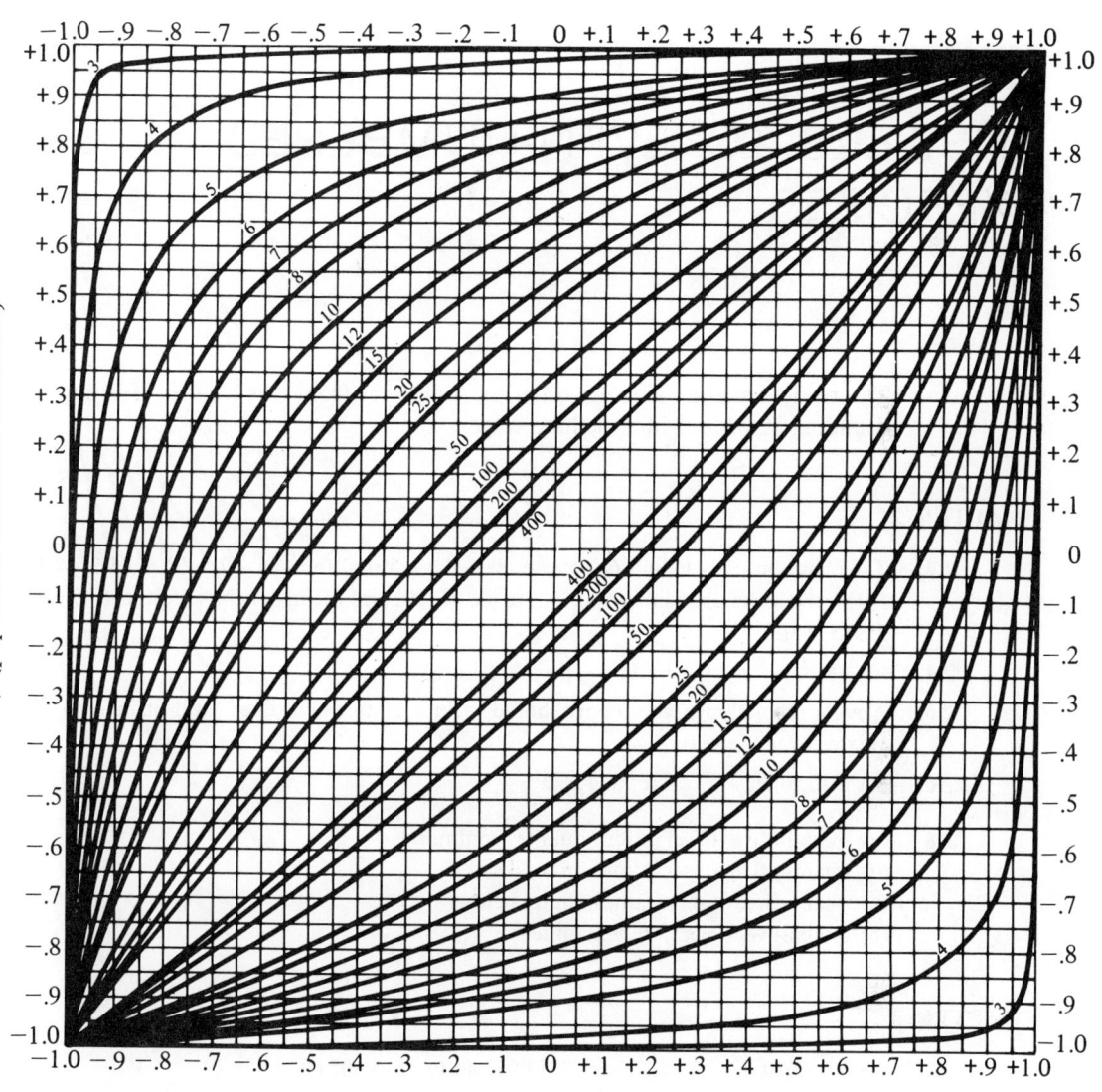

The numbers on the curves indicates sample size. The chart can also be used to determine upper and lower 0.5% significance points for r, given ρ.

ANSWERS TO EVEN-NUMBERED EXERCISES

The reader may get answers differing somewhat from those given here due to rounding at various intermediate stages.

CHAPTER 2

2.2. The common class length is 13; the class boundaries are 36.5, 49.5, 62.5, 75.5, 88.5, 101.5; the lower class limits are 37, 50, 63, 76, 89; and the upper class limits are 49, 62, 75, 88, 101.

2.4. (a) 16   (b) 11   (c) 35   (d) 16

2.6. Among other things, the report should include a statement like "level of income increases over the years."

2.10. (d) 0.10; 0.76   (e) 6.4; 5.6

2.18. (a) quantitative, observation, infinite, continuous
(c) quantitative, observation, finite, discrete
(e) qualitative, observation, infinite, neither
(h) qualitative, object, finite, neither

CHAPTER 3

3.2. (a) $315,000   (b) from $313,977.40 to $315,305.00
(c) $157,986.76 or more

3.4. (a) 0, 0   (b) 0, −0.5   (c) 0, 1.5   (d) −86.6, 0
(e) 86.6, 0   (f) 0, 0

3.8. (a) 6.4   (b) 0.0857   (c) 48   (d) 43 cents, yes
(e) 330 miles per hour

3.10. 5.4 which is less than both the mean and the median

3.12. (b) 22.50, 22.11, 21.86

3.14. (a) 64.8   (b) 64.9
3.16. (a) −0.672, 68.64   (b) 8.64, 68.64
3.18. (a) 71.2   (b) 71.5
3.20. (a) 92   (b) not possible
3.22. 41
3.24. 4.73

3.28. (a) $\sum_{i=2}^{9} y_i^2$ (b) $\sum_{i=1}^{N} (y_i - 4) f_i$
       (c) $\left[\sum_{i=1}^{3} (y_i - 4) \right]^2$ (d) $\sum_{i=1}^{N} (y_i - \mu)^2 f_i / N$
       (e) $\sum_{i=1}^{2} \sum_{j=1}^{4} y_{ij}$ (f) $\sum_{i=1}^{3} \sum_{j=1}^{2} (y_{ij} - \mu_i)^2$

3.30. (a) 100   (b) 2.45   (c) 6.8   (d) 26.48
3.42. (a) 2   (b) 6   (c) 2.14
3.44. As indicated by each of the three measures of dispersion, the distribution of Exercise 3.42 is more disperse than the distribution of Exercise 3.43.
3.46. (a) $\frac{13}{7}$ (b) $\frac{13}{7}$
3.48. Weighted means in standard units are 1.875, 0.790, 1.475.
3.50. (a) The weighted means in Z-scores are 68.75, 57.90, 64.75.
       (b) The weighted means in Z-scores are 65.6, 61.2, 63.5, 70.8.
3.52. The variation in weights of rats ($V = 0.049$) is less than for cows ($V = 0.084$).
3.54. 26.67   6.67   60.00
3.58. For data grouped into classes 1–2, 3–4, $\cdots$, 11–12 the resulting variance of 4.529, when adjusted by Sheppard's correction, becomes 4.196. Note that 4.196 is only slightly greater than 4.111, the correct value.
3.60. (a) 1.30   (b) 0.67, 0.94, 1.00
3.62. (a) $\begin{bmatrix} 7.35 \\ 7.05 \end{bmatrix}$ (b) $\begin{bmatrix} 0.75 \\ 0.45 \end{bmatrix}$ (c) $\begin{bmatrix} 7.3275 \\ 6.8375 \end{bmatrix}$
       (d) $\begin{bmatrix} 1.89655 \\ 1.70968 \end{bmatrix}$ (e) $\begin{bmatrix} 13.86725 \\ 9.29411 \end{bmatrix}$ (f) $\begin{bmatrix} 0.84729 \\ 0.71490 \end{bmatrix}$
3.64. (a) 30(1.5)   (b) 30(1.5)   (c) 2.10(.92)   (d) 2.115(.495)
       (e) 8.535(.085)

CHAPTER 4

4.4. The seminar is cancelled when only 3 or 2 or 1 or 0 members appear. This can happen in $35 + 21 + 7 + 1 = 64$ ways.
4.6. $S = \{AM_1, AM_2, AF_1, AF_2, BM_3, BF_3, BF_4, BF_5\}$ where A, B, M_i, and F_j denote room A, room B, boy i, and girl j, respectively.

Answers to Even-Numbered Exercises

4.8. (a) $A, B, C; C, D$ (b) A, C (c) none (d) B, D
(e) none (f) 8 like $\{HHH\}, \{HHT\}, \cdots, \{TTT\}$
(g) 28 like $\{HHH, HHT\}, \cdots, \{HHH, TTT\}, \{HHT, HTH\}, \cdots, \{TTH, TTT\}$.

4.10. (a) $S = \{(1,1), (1,2), (1,3), (2,1), (2,2), (2,3), (3,1), (3,2), (3,3)\}$
$A = \{(1,2), (2,1), (2,3), (3,2)\}$
$B = \{(1,1), (1,2), (1,3), (3,1), (3,2), (3,3)\}$
$C = \{(1,2), (2,2), (3,2)\}$
(b) $\{(1,2), (3,2)\}, \{(2,1), (2,3)\}, \{(2,2)\}$
(c) D means "number on white die is even," E means "total on both dice is three," and F means "number on each die is odd."
(d) $\bar{B}$, not possible, $\overline{A \cup C}$

4.12. (a) $\{(1,1), (1,2), (1,3), (2,1), (2,2), (2,3), (3,1), (3,2), (4,1)\}; \{(1,3), (2,2), (3,1)\};$
$\{(1,2), (1,3), (2,3)\}; \{(3,1), (3,2), (4,1)\}; \{1,1), (2,2)\}$
(b) $\emptyset; \{(1,1), (1,2), (2,1), (2,3), (3,2), (4,1)\}; \{(1,1), (2,1), (2,2), (3,1), (3,2), (4,1)\};$
$\{(1,1), (1,2), (1,3), (2,1), (2,2), (2,3)\}; \{(1,2), (1,3), (2,1), (2,3), (3,1), (3,2), (4,1)\}$
(c) $\{(1,2), (2,3)\}; \{(2,2), (3,1)\}$
(d) $\{(1,2), (2,2), (2,3), (3,1)\}$
(e) $\{(1,1), (2,1), (3,2), (4,1)\}$
(f) $\{(1,3), (3,1)\}$ (g) $\{(2,1)\}$

4.18. (a) $A \cap B \cap C = \{1\}$, $A \cap B \cap \bar{C} = \{7\}$, $A \cap \bar{B} \cap C = \{5, 9\}$, $\bar{A} \cap B \cap C = \emptyset$,
$A \cap \bar{B} \cap \bar{C} = \{3\}$, $\bar{A} \cap B \cap \bar{C} = \{4\}$, $\bar{A} \cap \bar{B} \cap C = \emptyset$, $\bar{A} \cap \bar{B} \cap \bar{C} = \{2, 6, 8\}$
(b) $\{1, 7\}; \{1, 5, 9\}; \{2, 6, 8\}; \{2, 6, 8\}; \{1, 2, 4, 6, 7, 8\}; \{4\}$
(c) $(A \cap \bar{B} \cap \bar{C}) \cup (A \cap \bar{B} \cap C), (A \cap \bar{B} \cap \bar{C}) \cup (\bar{A} \cap B \cap \bar{C}),$
$(A \cap B \cap \bar{C}) \cup (A \cap \bar{B} \cap C), (A \cap B \cap C) \cup (\bar{A} \cap \bar{B} \cap \bar{C})$

4.20. 110

4.22. 0.68

4.24. (a) 5 to 4 (b) 4 to 13

4.26. (a) The probability that a car traveled over 60-miles per hour; (b) the probability of a car not traveling on an interstate highway; (c) the probability of a car traveling over 60-miles per hour or of traveling on an interstate highway; (d) the probability of a car traveling over 60-miles per hour on an interstate highway; (e) the probability of a car traveling 60-miles per hour or less on an interstate highway; (f) the probability of not traveling over 60-miles per hour or not traveling on an interstate highway (also, the probability of a car not traveling over 60-miles per hour on an interstate highway).

4.28. $\frac{4}{13}$

4.30. $\frac{1}{24}, \frac{1}{24}$

4.32. between $\frac{2}{7}$ and $\frac{3}{13}$

4.34. (a) $\frac{17}{18}$ (b) $\frac{1}{9}$ (c) $\frac{1}{9}$ (d) $\frac{5}{9}$
(e) $\frac{1}{9}$ (f) $\frac{4}{9}$

Answers to Even-Numbered Exercises

4.36. (a) no (b) $B = S$

4.38. (a) $A, D; C, D$ (b) $A, B; A, C; B, C$
(c) $P(A \cup B) = \frac{1}{2}; P(C \cup D) = \frac{1}{3}$ (d) $\frac{1}{2}$

4.40. (a) none (b) all

4.42. (a) yes (b) yes (c) yes

4.46. (a) The probability that a high school graduate will major in science given he has gone to college; (b) the probability that a person who does not go to college will get married within one year; (c) the probability that a person who goes to college will get married within one year and major in science; (d) the probability that a person who goes to college and does not get married will major in science; (e) the probability that a person who does not get married within one year will go to college and major in science; (f) $P(\bar{C}|M)$; (g) $P(\overline{C \cup O}|\bar{M})$; (h) $P(C|\bar{M} \cup \bar{O})$

4.48. 0.20, 0.67, 0.50

4.50. (a) $\frac{3}{4}$ (b) 1 (c) 0 (d) $\frac{1}{4}$ (e) $\frac{1}{2}$
(f) $\frac{1}{4}$ (g) $\frac{3}{4}$ (h) $\frac{1}{2}$ (i) $\frac{1}{2}$

4.52. (a) $\frac{1}{4}$ (b) $\frac{1}{4}$ (c) $\frac{1}{8}$ (d) $\frac{1}{8}$ (e) $\frac{1}{8}$ (f) 0

4.54. (a) $\frac{6}{203}$ (b) $\frac{95}{609}$ (c) $\frac{45}{203}$ (d) $\frac{11}{4,060}$

4.56. (a) $P(A) \times P(B|A) \times P(C|A \cap B); P(A) \times P(C|A) \times P(B|A \cap C);$
$P(B) \times P(A|B) \times P(C|A \cap B); P(B) \times P(C|B) \times P(A|B \cap C);$
$P(C) \times P(A|C) \times P(B|A \cap C); P(C) \times P(B|C) \times P(A|B \cap C)$
(b) $\frac{9}{20} \cdot \frac{3}{9} \cdot \frac{1}{3} = \frac{1}{20}$, and so on

4.62. $\frac{18}{49}$

4.64. (a) 0.931 (b) 0.068

4.72. 0.14; 0.39; 0.52; 0.53; 0.48; 0.25

4.74. $P(\{b\}) = 0.30; P(\{c\}) = 0.05; P(\{d\}) = 0.40$

4.76. I. acceptable II. acceptable III. not acceptable
IV. not acceptable V. acceptable VI. acceptable
VII. not acceptable VIII. not acceptable

4.78. (a) $\frac{39}{53}$ (b) 1 (c) 0 (d) $\frac{13}{48}$ (e) $\frac{1}{2}$
(f) $\frac{13}{48}$ (g) $\frac{35}{48}$ (h) $\frac{13}{25}$ (i) $\frac{1}{2}$

4.80. 0.117

4.82. (c) 0.125 (d) 0.10

CHAPTER 5

5.4. $c = 1/m$

5.6. $c = 0.2$

5.8. (a) Letting y denote the number of heads, the probability function of y resulting from 16 equally likely outcomes of the experiment is

y	0	1	2	3	4
$f(y)$	$\frac{1}{16}$	$\frac{4}{16}$	$\frac{6}{16}$	$\frac{4}{16}$	$\frac{1}{16}$

(d) $\frac{7}{8}$

5.10. 53,130

5.12. $7^{20} = 79{,}792{,}266{,}297{,}612{,}001$

5.14. 720

5.16

y	1	2	3	4	5	6
$f(y)$	0.05	0.40	0.10	0.25	0.05	0.15

5.18. (a) 1,260   (b) 2,142

5.22. 113,379,295,270

5.24.

Number of Correct Matches	0	1	2	3	4
Probability	$\frac{9}{24}$	$\frac{8}{24}$	$\frac{6}{24}$	$\frac{0}{24}$	$\frac{1}{24}$

5.26. (a) 4   (b) 10

5.30. $(1-\pi)^n$; $n(1-\pi)^{n-1}\pi$; π^n

5.32. (a) 0.1172   (b) 0.1719   (c) 0.1719

5.36. (a) 12   (b) 0.65

5.38. 0.0746

5.40. (a) 0.6488   (b) 0.10; 0.9744   (c) 10   (d) 0.9995

5.42. (a) 0.9486   (b) 0.0015

5.50. (a) $\frac{7}{2}$; $\frac{35}{12}$; 1.708   (b) 0.57; 1.00

5.52. (a) 0.30   (b) 0.50

5.54. 0.837; 0.0000911

5.56. (a) 0.688   (b) 0.224   (c) 0.047

5.58. $h(y+1) = \dfrac{A-y}{1+y} \times \dfrac{n-y}{B-n+1+y} h(y)$

5.60. (a) $\frac{1}{64}$ (b) 0.059   (c) 10   (d) 7

5.62. 0.15625

5.64. (a) 0.6268   (b) 0.1912   (c) 0.6991   (d) 0.0057

5.68. (a) $m/2$; $m^2/12$ (b) 0.577; 0.866; 1   (c) $\frac{1}{2}$; $\frac{1}{3}$; $\frac{1}{3}$; $\frac{1}{6}$

5.70. (b) 0.8647   (c) 0.9502   (d) 0.3012

Answers to Even-Numbered Exercises

5.72. 0.6826; 0.9544; 0.9974

5.74. (a) 0.2119 (b) 0.7213 (c) 0.8621 (d) 0.1379
(e) 43 (f) 73.26

5.76. 54

5.78. (a) 0.93 (b) 274

5.80. 69 and above; 66, 67, 68; 59, 60, $\cdots$, 65; 56, 57, 58

5.82. 65, using interpolation

5.84. (a) 30.8 (b) 78.8; 67.6; 55.7; 46.3, by interpolation

5.86. (a) 38.2 (b) 61.8

5.88. (a) 0.9755; 0.0013 (b) 0.0111; 0.00002

5.90. (a) 0.022, by interpolation (b) 0.8746, by interpolation

5.92. 43

5.94. (a) 0.426 (b) 0.375

CHAPTER 6

6.2. (a) 8; $3\sqrt{2}$

(b)
$\bar{y}$	5	6	7	8	9	10	11
$f(\bar{y})$	.1	.1	.2	.2	.2	.1	.1

(c) 8.0; $\sqrt{3.0}$

6.4. (a) 5.5; $3\sqrt{5}/2$

(b)
$\bar{y}$	1	2	3	4	5	6	7	8	9	10
$f(\bar{y})$	$\frac{1}{64}$	$\frac{3}{64}$	$\frac{6}{64}$	$\frac{10}{64}$	$\frac{12}{64}$	$\frac{12}{64}$	$\frac{10}{64}$	$\frac{6}{64}$	$\frac{3}{64}$	$\frac{1}{64}$

(c) 5.5; $\sqrt{15}/2$ (d) $\frac{1}{64}$ (e) .0875; 0.957

6.6. (a) $\frac{10}{3}$; $2\sqrt{5}/3$

(b)
$\bar{y}$	2	3	4	5
$f(\bar{y})$	$\frac{3}{15}$	$\frac{6}{15}$	$\frac{4}{15}$	$\frac{2}{15}$

(c) $\frac{10}{3}$; $2\sqrt{2}/3$ (d) 0.723

(e)
$\bar{y}$	2	3	4	5	6
$f(\bar{y})$	$\frac{9}{36}$	$\frac{12}{36}$	$\frac{10}{36}$	$\frac{4}{36}$	$\frac{1}{36}$

(f) $\frac{10}{3}$; $\sqrt{10}/3$

6.8. (a)

m	1	4	7	10
$f(m)$	$\frac{5}{32}$	$\frac{11}{32}$	$\frac{11}{32}$	$\frac{5}{32}$

(b) 5.5; 2.81   (c) $1.25\sigma/\sqrt{n} = 2.42$

6.10. 0.84

6.12. 43

6.14. (b) 0.6250; 0.6874; 0.7948

6.16. (a) 0.0228   (b) 0.8400

6.18. 25

6.20. (a) Sample size must be four times larger; (b) sample size must be nine times larger for smaller standard error

6.22. $2.03

6.24. (a)

v_1	1	5	9
$f(v_1)$	.25	.50	.25

(b)

v_2	0	1	4	9
$f(v_2)$	.250	.375	.250	.125

6.26. (a)

s^2	0	3	9	12	21	27
f	4	18	12	12	12	6

(b)

v_2	0	2	6	8	14	18
f	4	18	12	12	12	6

(c) s^2 is an unbiased estimator of σ^2, but v_2 is not

(d)

$\frac{2s^2}{\sigma^2}$	0	$\frac{8}{15}$	$\frac{24}{15}$	$\frac{32}{15}$	$\frac{56}{15}$	$\frac{72}{15}$
Prob.	.06250	.28125	.18750	.18750	.18750	.09375

(e) 2

6.28. (a) Since $E(\bar{d}) = -2$ and $\sigma_{\bar{d}} = \sqrt{4.5} \doteq 2.12$, $u = (\bar{d} + 2)/2.12$ and

u	−2.83	−2.36	−1.89	−1.42	−0.94	−0.47	0	0.47	0.94	1.42	1.89
f	1	5	15	32	53	71	78	71	53	32	15

u	2.36	2.83
f	5	1

(b) 0; 1.00   (c) 0.046

6.30. 0.97

6.32. 14

6.34. $n_1 = 7$ and $n_2 = 13$

6.48. (a)

$\bar{y}$	64	65	66	67	68	69	70	71	72	73
f	1	1	2	2	2	2	2	1	1	1

(b)

$\bar{y}$	66	67	68	69	70	71
f	1	2	2	2	1	1

(c)

$\tilde{y}$	64	65	66	70	72	73
$f(\tilde{y})$	$\frac{2}{9}$	$\frac{2}{9}$	$\frac{2}{9}$	$\frac{1}{9}$	$\frac{1}{9}$	$\frac{1}{9}$

(d) 68.33; 68.33; 67.22

CHAPTER 7

7.10. (a) Since $E(t_1') = 4$, $E(t_2') = \frac{53}{12} \doteq 4.42$, and $E(t_3') = 5.25$, only t_1' is an unbiased estimator of $\mu = 4$.
(b) $M_1^2 = \frac{5}{2}$; $M_2^2 = \frac{25}{9}$; $M_3^2 = 5$; $M_1^2 < M_2^2 < M_3^2$
(c) The efficiency of t_1' relative to t_2' is $\frac{10}{9} \doteq 1.11$, of t_1' relative to t_3' is 2.00, and of t_2' relative to t_3' is 1.80.

7.12. (a) Only $\bar{y}$ is an unbiased estimator of μ; none of the three statistics is an unbiased estimator of either the population median or population mid-range.
(b) The sample mid-range is most efficient; the median least efficient.
(c) None of the statistics is an unbiased prime estimator of the population mean; only the median is an unbiased prime estimator of the population median; only the mid-range is an unbiased prime estimator of the population mid-range.
(d) None of the statistics is an unbiased double prime estimator of either the population mean or the population median; all are unbiased double prime estimators of the population mid-range.
(e) The efficiency of the mid-range is greatest; the efficiency of the median is least.
(f) The efficiency of the mid-range is greatest; the efficiency of the median is least.

7.14. (b)

n	2	3	4	5	6	7	8	9	10
Statistic	$51\bar{y}$	$34.3\bar{y}$	$26\bar{y}$	$21\bar{y}$	$17.7\bar{y}$	$15.3\bar{y}$	$13.5\bar{y}$	$12.1\bar{y}$	$11\bar{y}$

7.16. 34.0
7.18. 0.9544
7.20. 0.9447
7.22. 27
7.24. 0.6572
7.26. 1,068 or less

7.28. The symmetric 0.95 confidence interval is $52.13 < \mu < 54.87$ and has length 2.74; since $z_{.01} = 2.33$ and $z_{.04} = 1.75$ we find another 0.95 confidence interval to be $51.87 < \mu < 54.72$ with length 2.85. The shortest 0.95 confidence interval has length 2.74, being a symmetric confidence interval.

7.30. $-33.1 < \mu_A - \mu_B < 5.1$

7.32. $\{p | p > 0.43\}$

7.34. $\{\bar{y} | \bar{y} < 37.23 \text{ or } \bar{y} > 42.77\}$

7.36. $\{z | z < -1.96 \text{ or } z > 1.96\}$; $z_o = -1.8$; fail to reject $H_o: \mu = 70$

7.38. $z_o = -2.16$; conclude that $\mu \neq 70$

7.40. $z_o = -1.73$; conclude that $\pi < \frac{1}{2}$, i.e., the coin is biased in favor of tails

7.42. 0.7881

7.44. $\alpha = 0.125$; $\beta(.3) = 0.657$

7.46. (a)

μ	29.4	29.5	29.6	29.7	29.8	29.9
$OC(\mu)$	.01	.05	.16	.37	.63	.84

(b)

μ	29.70	29.75	29.80	29.85	29.90	29.95
$OC(\mu)$	.01	.04	.14	.35	.61	.82

(c) As n changes from 16 to 64 the largest bacteria count changes from 29.56 to 29.78 indicating that when the two types of errors are fixed the critical distance between the sample mean and the population mean becomes less with increased sample size.

7.48. $z_o = 1.8$; conclude that $\mu_1 > \mu_2$ when $\alpha = 0.05$

7.50. $z_o = -1.89$; conclude that $\mu_A < \mu_B$

CHAPTER 8

8.2. (a) 25.64 (b) 16

8.4. (a)

χ^2	2	4	6	8	9.49	12
s^2	$0.5\sigma^2$	$1\sigma^2$	$1.5\sigma^2$	$2\sigma^2$	$2.37\sigma^2$	$3\sigma^2$

(b) $0.178\sigma^2$

8.8. (a) $2.91 < \sigma < 4.47$ (b) $2.97 < \sigma < 4.44$

8.10. $CR = \{\chi^2 | \chi^2 < 4.40 \text{ or } \chi^2 > 23.3\}$; $\chi_o^2 = 23.0$; $s^2 = 191.7$ is not significantly different from $\sigma^2 = 100$.

8.12. (a) $0.16 < \sigma < 0.39$
(b) $CR = \{\chi^2 | \chi^2 > 19.7\}$; $\chi^2 = 17.57$; $s = 0.23$ is not significantly larger than $\sigma = 0.18$.

8.14. 4.75
8.16. (a) 0.4514 (b) 0.1936 (c) 0.95 (d) 1.64 (e) 3.17
8.18. (a) $\chi_o'^2 = 6.00$; decide $\pi \neq 0.5$
(b) $\chi_o'^2 = 3.00$; fail to reject $\pi = 0.5$
8.22. Since $\chi_o'^2 = 7.00$, we cannot deny the claim that the programs are about equally popular.
8.24. (a) Since $\chi_o'^2 = 4.36$, the data seems consistent with the model.
(b) Since $\chi_o'^2 = 3.6$, we fail to reject $H_o: \pi_1 = \frac{2}{3}, \pi_1 = \frac{1}{3}$.
8.26. Since $\chi_o'^2 = 0.50$, decide the observed frequencies do not deviate significantly from expected frequencies.
8.28. $\chi_o'^2 = 1.275$; the experimental sampling distribution is consistent with the theoretical sampling distribution.
8.30. (a) Since $z_o = 1.06$, the proportion of men favoring the proposition is not significantly different from the corresponding proportion for women; (b) 1.125
8.34. Since $z_o = 1.14$, the data does not confirm the claim of Manufacturer A.
8.36. (a) 16   (c) 15.21, which is smaller by 0.79   (d) 0.781
8.38. $\chi_o'^2 = 3.98$; cannot say the three defective rates are significantly different.
8.40. (a) $\chi_o'^2 = 9.04$; decide male and female students have different study habits in terms of hours per week.
8.42. (a) $CR = \{\chi^2 | \chi^2 > 15.5\}$; $\chi_o'^2 = 12.2$; decide that "grade" is independent of "school."
8.44. (a) 46.51   (b) 27.3, by sex; 16.4, by school

CHAPTER 9

9.2. Decide that neither athelete is superior to the other in the 100 yard dash.
9.6. (a) Since $T = 2$, decide that there is greater gain with Diet II.
(b) Since $T = 13$, reach same conclusion as in part (a).
9.8. (a) $T = 8$; two methods of analysis are not different
(b) Same conclusion as in part (a)
9.10. The sign test fails to reject $H_o: \mu_1 - \mu_2 = -3$; the signed rank test rejects H_o.
9.12. (b) When $n = 30$,

α	.025	.010	.005
T	137	120	108

When $n = 35$,

α	.025	.010	.005
T	195	174	159

424 Answers to Even-Numbered Exercises

9.14. (a) $z_o = -2.46$; decide running time is less than 30 seconds
(b) $z_o = -1.80$; decide running time is not significantly different from 30 seconds

9.16. (a) 6.125; decide stimulus does effectively change blood pressure of students
(b) 7.111; decide there is a change in examination grade

9.18. $D = 0.217$; decide sample could have been drawn from the normal distribution

9.20. $D = 0.333$; decide die is not well-balanced

9.22. (a) $D^* = 0.2804$; fail to reject H_o: y has a normal distribution with mean 50

9.24. Computed values are uniformly larger than tabled values.

9.26. $r = 22$; decide sample could be nonrandom

9.28. (a) $\frac{1}{63}$; $\frac{1}{14}$; $\frac{1}{14}$

9.30. (a) $z'_o = 0.868$; even and odd digits seem to appear in random order
(b) $z'_o = 1.06$; the sixes seem to be in random order

9.32. $r = 6$; conclude the sequence is nonrandom

9.34. (a) $d' = 0.50$; $z' = 2.22$ (b) 3.18; -1.74 (c) 1.78; 0.33

9.36. $z'_o = -1.16$; Typist II does not make significantly more errors than Typist I

9.38. Cannot decide that median for County A is smaller.

9.40. 7; the samples are not significantly different

9.42. (a) $z'_o = 1.95$; cannot distinguish between two distributions
(b) $z'_o = -3.70$; conclude that the two distributions are different

9.44. 24; cannot say median for A is smaller than median for B

9.46. (a)

T	3	4	5	6	7	8	9
$f(T)$	.1	.1	.2	.2	.2	.1	.1

c9.48. Since $\chi'^2_o = 29.29$, the 4 distributions are not all the same

9.50. $U = 38$; $z'_o = -2.38$; decide the location parameter of distribution 2 is smaller than the corresponding location parameter of distribution 3

E9.52. (a)

U	0	1	2	3	4	5	6	7	8	9
$P_{3,3}(U)$	.05	.05	.10	.15	.15	.15	.15	.10	.05	.05

U	0	1	2	3	4	5	6	7	8	9	10	11	12
$P_{3,4}(U)$	$\frac{1}{35}$	$\frac{1}{35}$	$\frac{2}{35}$	$\frac{3}{35}$	$\frac{4}{35}$	$\frac{4}{35}$	$\frac{5}{35}$	$\frac{4}{35}$	$\frac{4}{35}$	$\frac{3}{35}$	$\frac{2}{35}$	$\frac{1}{35}$	$\frac{1}{35}$

(b)

U	0	1	2	3	4	5	6	7	8	9	10	$\cdots$	15
$P_{3,5}(U)$	$\frac{1}{56}$	$\frac{1}{56}$	$\frac{2}{56}$	$\frac{3}{56}$	$\frac{4}{56}$	$\frac{5}{56}$	$\frac{5}{56}$	$\frac{6}{56}$	$\frac{6}{56}$	$\frac{6}{56}$	$\frac{5}{56}$	$\cdots$	$\frac{1}{56}$

CHAPTER 10

10.2. (a) -1.950 does belong to the set $\{t \mid t < -1.833\}$
(b) 50 does fall in the interval $46.1 < \mu < 50.3$

10.4. Since $t_o = 2.247$, conclude that 70.6 is not significantly different from 68.

10.6. $t_o = 2.308$; the salesman's claim is correct

10.10. $CR = \{A \mid A < 0.276\}$; $A_o = 0.278$; fail to reject $H_o: \mu - 68 = 0$

10.12. $t_o = -1.96$; decide $\mu_1 < \mu_2$

10.14. $t_o = -1.24$; decide the mean maximum temperatures of Counties A and B are not different

10.16. (a) $t'_o = 2.20$; decide the population means are different
(b) $t_o = 1.96$; decide the sample means are not significantly different

10.18. (a)

n_1	25	20	1
$V(\bar{y}_1 - \bar{y}_2)$	$.080\sigma^2$	$.083\sigma^2$	$1.020\sigma^2$

(b)

n_2	10	100	1,000	10,000
$V(\bar{y}_1 - \bar{y}_2)$	$.15\sigma^2$	$.06\sigma^2$	$.051\sigma^2$	$.0501\sigma^2$

The solution is $n_1 = 40$ and $n_2 = 41$ (or $n_1 = 41$ and $n_2 = 40$).

10.20. $611.3 < \mu_1 + 3\mu_2 < 735.9$; $152.8 < (\mu_1 + 3\mu_2)/4 < 184.0$

10.22. $t_o = 0.57$; decide that Test II was not more difficult than Test I

10.24. $t_o = -3.33$; decide the stimulus increases the mean blood pressure more than 5 units

CHAPTER 11

11.2. $\eta = \frac{65}{6} + \frac{1}{4}\left(x + \frac{50}{3}\right)$ for $x = -40, -30, -10, 20$

11.4. $\eta = 160 + 6.5(x - 69)$; 199 pounds

11.6. (a) 1 and 2
(b) $\hat{\eta} = \frac{10}{3} - \frac{1}{14}\left(x - \frac{7}{3}\right)$

11.8. $\hat{\eta} = 452.08 + 8.3234(x - 77.84)$; 6,492.94; 262.0; 761.4

11.10. (a) $\hat{\eta} = 452.08 + 8.323(x - 77.85)$ for $50 \leq x \leq 125$
(b) $t_o = 2.18$; decide the slope is greater than 6

11.12. (a) $\hat{\eta} = 121.3 + 0.7846(x - 43.50)$ for $30 \leq x \leq 60$
(b) $116.2 < \eta < 128.8$

11.14. (a) $t_o = -2.41$; conclude that $\beta < -0.15$
(b) Should be considered an unusual expenditure

11.16. (a) $\hat{\eta} = 545.2 + 1180(x - 0.9727)$ for $0.905 \le x \le 1.035$
(b) $t_o = 7.34$; decide $\beta > 0$ or that the mean modulus of rupture is affected by the specific gravity of fiberboard.
(c) $465.5 < y' < 571.3$

CHAPTER 12

12.2. (a) 0.80; ± 0.89; 20% (b) ± 0.63

12.4. (a) 0.972 (b) 5.833; 0.1618

12.6. (a) $t_o = -1.732$; cannot conclude that ρ is negative

12.10. $r = 0.63$ is consistent with the official's statement that $\rho > 0.40$

12.12. $r_s = 0.745$; since $z_o = 2.367$, decide that $\rho > 0$

12.14. $r_s = 0.639$; r_s is approximately 4.5% smaller than $r = 0.669$

12.18. (a) $\phi = 0.0745$; $\chi_o'^2 = 3.33$; decide 0.0745 is not significantly different from zero
(b) $\phi = -0.2228$; $\chi_o'^2 = 8.19$; decide -0.2228 is significantly less than zero. With correction, $\phi = -0.209$ is significantly less than 0.
(c) $\phi = 0.641$; $\chi_o'^2 = 41.03$; decide 0.641 is highly significant

INDEX

Absolute value, 58
Abstract number, 60
Actuarial science, 9
Alpha, α, significance level of test, 239
Arithmetic mean. *See* Mean
Array, 358
Association, strength of, 383
A-test, 345
Auble, D., 332
Average, 36

Bar graph, 25
Bayes' formula, 124
Behrens–Fisher test, 348
Best estimator, 218
Best unbiased estimator, 219
Beta, β, probability of type II error, 244
Beta, β, regression coefficient, 359
Binomial distribution:
    assumptions, 147
    coefficients, 133, 139
    formula, 138
    histograms, 169
    mean, 144
    normal approximation, 168–72
    parameters, 138
    probabilities (tables), 140, 141
    recursion formula, 138
    variance, 145
Binomial index of dispersion, 295
Bivariate normal distribution, 389
Blood groups, 279

Cartesian product, 77
Categorical distribution (table), 14
Categories, 13
Causal relationship, 290
Central limit theorem, 184–85
Chebyshev's theorem, 60, 147, 184

Chi-square distributions, 257–59
    curves, 258
    degrees of freedom, 258
    table, 409–10
    variance, 258
Chi-square test:
    contingency table, 280–90
    goodness-of-fit, 272–78
    of independence, 284–90
    proportions, 277, 281–82
    variances, 260–61
Class, 15–18
    boundary, 15, 16
    frequency, 45
    length, 15
    limit, 15
    mark, 17, 45
    number of, 16
    open, 15
Classes, number of, 16
Cluster sampling, 208
Coding, 47, 67
Coefficient of correlation. *See* Correlation coefficient
Coefficient of determination, 384
Coefficient of regression, 360
Coefficient of variation, 64
Combinations, 130
Comparison of two t-tests, 352
Confidence interval, 215, 231–35
    adjusted mean, 377–78
    array mean, 378
    correlation coefficient, 392
    degree of confidence, 231
    difference in means, 348, 352
    difference in proportions, 284
    length of, 231
    limits, 231, 233
    mean, 232–35
    regression coefficient, 372

Confidence interval (*cont.*)
  standard deviation, 262–64
  symmetric, 232
  variance, 262–63
Consistency, 223–24
Contingency table, 280–90
Continuous distribution, 154–64
Continuous variable, 29
Control charts, 313, 314
Correlation coefficient:
  biserial, 395
  Kendall tau, 395
  multiple and partial, 396
  Pearson's product-moment, 383
  phi and tetrachoric, 395, 399
  ratio, 395
  Spearman's rank, 393
Correlated samples, 297
Countably infinite, 116
Covariance, 386–87
Critical point, 242
Critical region, 237, 239, 241
Critical values (tables):
  chi-square, 409–10
  Kolmogorov-Smirnov, 312
  number of runs, 316, 326
  rank sum, 328–29
  sign, 299
  standard normal, 405
  student t, 411
  Wilcoxon's signed rank, 302
Cumulative distribution, 18, 310
Cumulative frequency polygon, 25

Data:
  grouped, 16–28, 43–49, 65–68
  quantitative and qualitative, 13
  raw, 15
  transformed, 46
David, F. N., 393
Decile, 48–49
Degrees of freedom:
  chi-square distribution, 258
  meaning, 266–67
  student t distribution, 340
  variance, 266
De Moivre, A., 158, 168
Density function, 155
Dependent events, 100
Descriptive statistics, 13–68
Design of experiments, 4
Deviation from mean, 57–58
Deviation from median, 57, 62

Difference in means, 194–97
Difference in proportions, 281–84
Discrete variable, 29
Dispersion, measure of, 57
Distribution. *See also* Frequency; Sampling; Theoretical distributions
  bell-shaped, 59
  binomial, 136–43
  chi-square, 257–59
  class of, 125
  continuous, 154–64
  cumulative, 310
  exponential, 165
  frequency, 13
  geometric, 147, 152
  hypergeometric, 147, 151
  negative binomial, 147, 153
  normal, 160
  Poisson, 147, 153
  probability, 130
  runs, 313
  student t, 339–42
  uniform, 134, 147, 150, 157
Distribution-free tests, 296–338
Draper, N. R., 379
Dwyer, P. S., 72

Efficiency, relative, 218–19
Energy crisis, 6
Equally likely, 93–111
Equation:
  estimating, 372
  linear, 360
  model, 360
Errors, type I and type II, 243
Errors in calculations, 68–73
Estimate:
  interval, 231
  point, 215
Estimation, 212–20
  errors and risk in, 226
  principle of, 216
Estimator:
  best, 218–19
  consistent, 223–24
  efficient, 218–20
  statistic, 191, 218–19
  sufficient, 223–24
  unbiased, 191
Event, 81–87
  certain to occur, 95
  complement of, 84
  compound, 83

null, 83
occurs, 82
probability of, 93
simple, 83
space, 83
Events, 81–87
   combination of, 84
   independent, 100–101
   intersection of, 84, 85
   mutually exclusive, 85, 97
   union of, 85, 86
Expected value, 143, 190
Experiment, 76–79, 137–38
Experimental outcome, 76–79
Experiments:
   dichotomous, 137–38
   selection of units, 353
   simulated, 199–206
Ezekiel, M., 379

Factorial, 126
Fisher, R. A., 393
Fox, K. A., 379
Fractiles, 48–49
Frequency distribution:
   bell-shaped, 59
   categorical, 14, 285–88
   class, 15–16
   cumulative, 18
   graphs, 23–27
   grouped (table), 14, 16
   percentage, 13
   skewed, 30
   symmetric, 30
Frequency ogive, 25
Frequency polygon, 27
Freund, J. E., 393
Function:
   probability, 128
   set and number, 93

Gallup poll, 202
Galton, F., 159
Gauss, K. F., 159
Gaussian distribution. *See* Normal distribution
Geometric mean, 41
Geometric progression, 117
Goodness of fit, 272–78
Gosset, W. S., 339
Graphs:
   binomial distribution, 142, 169
   chi-square distribution, 258–59
   frequency distribution, 23–27
   normal distribution, 159–63
   student t distribution, 340

Harmonic mean, 41
Histogram, 22
Homogeneity of a sequence of successes, 288
Hypothesis:
   alternative, 237
   nonparametric, 272–78, 296–338
   null, 237
   one-sided, 237
   parametric, 237
   statistical, 236
   two-sided, 238
Hypothesis testing. *See* Tests of hypothesis

Identity:
   regression sum of squares, 368
   summation, 52–53
Independent events, 100–101
Independent samples, 195
Index of dispersion, binomial, 295
Interpolation, linear, 162
Interstate Commerce Commission, 200
Interval estimate, 231

Jurors, women, 7

Kolmogorov–Smirnov tests, 310–12, 337

Laplace, P. S., 159
Least squares, method of, 224, 362
Level of significance, 239
Linear combination of two means, 349
Linear interpolation, 162
Linear regression, 360
Linear trend, 358
Literary Digest poll, 200
Location, measure of, 35

McNemar test, 308
Mann-Whitney test, 331
Matched pairs, 297
Maximum likelihood, method of, 224
Mean:
   adjusted, 376
   arithmetic, 36
   array, 357
   binomial distribution, 144
   chi-square, 258
   combined data, 39
   confidence interval for, 343
   deviation, 58

Mean (*cont.*)
  deviation about the median, 62
  exponential distribution, 166
  geometric, 41
  harmonic, 41
  random variable, 143
  sample and population, 36, 143
  sampling distribution of the mean, 183
  second moment about, 221
  standard error of, 183
  uniform distribution, 158
  weighted, 39, 47, 61
Mean square successive difference, 320
Measurement:
  standard, 60
  true, 69
Median:
  class, 48
  definition, 37, 47
  efficiency of, 220
  grouped data, 47
  sampling distribution, 188, 217
  second moment about, 222
  standard error of, 189
Method of least squares, 224, 362
Method of maximum likelihood, 224
Mid-range, 62
Mills, F. C., 74
Model equation, 372
Monte Carlo methods, 203, 205
Mu, μ, population mean, 36
Multiple correlation, 396
Mutually exclusive events, 85
Mutually independent events, 182

Nonparametric test, 272–78, 296–338
Normal distribution:
  applications of Table II, 161–64
  curves, 159–63
  density function, 159–60
  parameters, 159
  standard form, 160
  table, 405
Normal approximation to:
  binomial distribution, 168–72
  distribution of s, 260
  distribution of U, 332
  number of runs, 317, 326
  rank correlation, 394
  rank-sum distribution, 330
Null hypothesis, 237
Number:
  abstract, 60

  approximate, 70
  approximate-error form, 74
  of combinations, 130–33
  components of the range, 70
  of permutations, 130–33
  probability of, 125
  random, 199
  random (table), 406–408
  range, 70
  of runs, 313
  significant digits, 69

Objects, 30
Observations, 28
Odds, 94
Ogive, 25
One-sided hypothesis, 237
One-tailed test, 240
Open class, 15
Operating characteristic curve, 246

Parameter, 138, 191
Partial correlation, 396
Partition of set, 96, 118
Pascal's triangle, 148
Pearson coefficient of correlation, 383
Percentile, 49
Permutation, 130
Pie chart, 31
Point estimation, properties, 215
Polio vaccine, 6
Population:
  array of, 358
  definition, 11, 28
  mean, 36, 143
  objects, 30
  parent, 177
  sample, and, 11, 28–29
  size of, 29
  standard deviation, 58, 144
  univariate, 175
Power curve, 247
Power of a test, 247, 343
Probability:
  area under curve, and, 155
  conditional, 106–11
  cumulative distribution, 310
  definition:
    classical, 93
    mathematical, 115
    relative frequency, 112
  density, 155
  distribution, 130

estimates of, 112
function, 126, 128
number, 125
properties, 108
of r runs, 319
sampling plans, 209
of U, 337
Probability for 2 × 2 contingency table, 292
Properties of point estimation, 215-20
   consistent, 223-24
   efficient, 218-20
   sufficient, 223-24
   unbiased, 191
Proportion(s):
   confidence interval for, 284
   standard error of a, 229
   standard error of difference between, 281
   successes in a sample, 229

Quantiles, 48
Quartile, 48
Quetelet, A., 159

Rand Corporation, the, 200
Random interval, 232
Randomness test, 313-18
Random number, 199
Random number (table), 406-408
Random sample, 177, 181
Random variable, 126, 128
Range, 57
Rank correlation, 393
Rank-sum test, 327-31
Recursion formula, binomial, 138
Reduced sample space, 106
Regression:
   assumptions, 361
   coefficient, β, 360
   confidence interval for β, 372
   equation, 358
   fitted line, 362-72
   slope of fitted line of, 363
   slope of line of, 359
   sum of squares, 368
   sum of squares identity, 368
   unbiased estimator of β, 373
   unbiased estimator of η, 377
Relative efficiency, 218
Relative frequencies, 18
Residual mean square, 367
Rule of elimination, 124
Runs:
   number and length, 313

probability of, 319
test for randomness, 313

Sample:
   definition, 11, 29
   drawn without replacement, 100, 175
   drawn with replacement, 99, 175
   plans, 206-209
   points, 79
   population, and, 11, 28-29
   random, 177, 181
   size, 174, 231
   space, 76-81, 106
   variance, 191
Sampling:
   area or cluster, 208
   experimentally, 175
   probability, 209
   stratified random, 207
   systematic, 208
   theoretically, 175
Sampling distribution:
   adjusted mean, $\hat{\eta}$, 375-79
   difference in means, 194-97, 346
   difference in proportions, 281
   mean, 175, 177
   median, 188, 217
   regression coefficient, b, 373-74
   standard deviation, 260
   theoretical, 176
   variance, 191-94
Savage, I. R., 334
Scatter diagram, 358-59
Seatbelts, 8
Series expansion, 168
Set of ordered pairs, 126
Sheppard's correction, 74
Siegel, S., 334
Sigma, Σ, summation notation, 52
Significance level of test, 239
Significant digits, 69
Significantly larger, 243
Sign test, 298
Simulated experiments, 199-206
Spearman rank correlation, 393
Standard deviation:
   binomial distribution, 145
   confidence interval for, 262-64
   definition, 58, 144
   random variable, 144
Standard error:
   adjusted mean $\hat{\eta}$, 377
   difference in means, 195-96

Standard error (cont.)
  difference in proportions, 281
  mean, 183
  median, 189
  regression coefficient, b, 373
  statistic, 190
  variance, 190
Standard units, 60
Statistical inference, 212
Student t distributions, 339–42
  curves, 340
  table, 411
  variance, 341
Student t-test:
  correlation coefficient, 390–91
  difference in two means, 346–47
  linear combination of two means, 349–50
  mean, 342–43
  means for correlated samples, 350–54
  power, 343
  slope of regression line, β, 374–75
Sufficiency, 223–24
Summation, 52, 54
Summation identities, 52–55
Supersonic transport, 7

Tanur, J., 12
Test of hypothesis:
  adjusted mean, η, 377
  Behrens-Fisher, 348
  correlation coefficient, 390, 392
  difference in two means, 252, 346, 350
  difference in two proportions, 281–82
  equality of p proportions, 285
  homogeneity of a sequence of successes, 288
  independence, 284–90
  independent t, 345–50
  Kolmogorov–Smirnov, 310–12, 337
  McNemar, 308
  Mann-Whitney, 331
  matched (or paired) t, 350–54
  mean, 339–43
  median, 321–24
  nonparametric, 296–97
  one- and two-tailed, 240
  outline for, 250
  paired-comparison, 298
  proportion, 237–39, 269–70
  randomness, 313
  rank-sum, 327–31
  rule for, 237, 243
  sign, 298
  slope of regression line, β, 374–75
  variance, 260

  Wald-Wolfowitz runs, 324–27
  Walsh's nonparametric, 304, 334
  Wilcoxon's signed rank, 301–304
Theoretical distribution. See also Probability
    distribution
  binomial, 136–43
  chi-square, 257–59
  exponential, 165
  geometric, 147, 152
  hypergeometric, 147, 151
  normal, 158–64
  probability, 128
  Poisson, 147, 153
  student t, 339–42
  uniform, 134, 147, 150, 157–58
  z^2, 267–69
Ties, 327, 330
Time series, 12
Tree diagram, 80
Triad, 136
Type II error, probability of, 244

Unbiased estimator, 191
Uniform distribution, continuous, 157–58
  density, 157
  mean and variance, 158
Uniform distribution, discrete, 134, 150
Unit cost, 208
U-test, 331

Variable:
  continuous, 29
  dependent and independent, 357
  discrete, 29
  distribution of, 28
  value, 15, 28
Variance:
  binomial distribution, 145
  computing form, 65–68, 149
  confidence interval for, 262–63
  definition, 58, 65, 144
  sample and population, 58, 144, 191
Venn diagram, 87–90

Wald-Wolfowitz runs test, 324–27
Walsh's nonparametric test, 304, 334
Weather, 5
Weighted mean, 39, 47, 61
Whales, vanishing, 8
Wilcoxon's signed rank test, 301–304

Yates' correction for continuity, 292

Z-score, 64

4407TH PA
08-CS-08 32397 MS 320

WITHDRAWN